Cancer Sensitizing Agents for Chemotherapy
NITRIC OXIDE (DONOR/INDUCED) IN CHEMOSENSITIZATION

VOLUME 1

Cancer Sensitizing Agents for Chemotherapy

NITRIC OXIDE (DONOR/INDUCED) IN CHEMOSENSITIZATION

VOLUME 1

Edited by

BENJAMIN BONAVIDA

Department of Microbiology, Immunology, and Molecular Genetics, David Geffen School of Medicine, University of California, Los Angeles, CA, 90095, USA

ELSEVIER

ACADEMIC PRESS

An imprint of Elsevier

Academic Press is an imprint of Elsevier
50 Hampshire Street, 5th Floor, Cambridge, MA 02139, United States
525 B Street, Suite 1800, San Diego, CA 92101-4495, United States
The Boulevard, Langford Lane, Kidlington, Oxford OX5 1GB, United Kingdom
125 London Wall, London EC2Y 5AS, United Kingdom

Library of Congress Cataloging-in-Publication Data
A catalog record for this book is available from the Library of Congress

British Library Cataloguing-in-Publication Data
A catalogue record for this book is available from the British Library

ISSN : 2468-3183
ISBN : 978-0-12-816471-6

For information on all Academic Press publications
visit our website at https://www.elsevier.com/books-and-journals

Publisher: Mica Haley
Acquisition Editor: Rafael Teixeira
Editorial Project Manager: Mariana Kuhl
Production Project Manager: Stalin Viswanathan
Cover Designer: Greg Harris

Typeset by SPi Global, India

Contents

Contributors

Nacer D. Abrouk Innovexe Inc., Mountain View, CA, United States

Ahlam A. Ali Queen's University Belfast, Belfast, United Kingdom

Ilham Alshiraihi Colorado State University, Fort Collins, CO, United States

Ali Bettaieb University of Burgundy-Franche Comté, LIIC EA7269, F-21000, Dijon; EPHE, PSL Research University, F-75014 Paris, France

Benjamin Bonavida University of California at Los Angeles, Los Angeles, CA, United States

Kathleen Brennan David Geffen School of Medicine at UCLA, Los Angeles, CA, United States

Mark A. Brown Colorado State University; Colorado School of Public Health, Fort Collins, CO, United States

Amanda Cañas Maimónides Institute of Biomedical Research of Córdoba (IMIBIC), Córdoba, Spain

Pedro Cabrales University of California at San Diego (UCSD), San Diego, CA, United States

Scott Caroen EpicentRx, Inc., San Diego, CA, United States

Corey A. Carter Walter Reed National Military Medical Center, Bethesda, MD, United States

Gautam Chaudhuri David Geffen School of Medicine at UCLA, Los Angeles, CA, United States

Regina M. Day Uniformed Services University, Bethesda, MD, United States

Melissa Edwards Colorado State University, Fort Collins, CO, United States

Mahnoush Esmaili City University of New York School of Medicine, New York, NY, United States

Pablo Garrido National University of Ireland Galway, Galway, Ireland

Sharon A. Glynn National University of Ireland Galway, Galway, Ireland

Sergio Huerta University of Texas Southwestern Medical Center; VA North Texas Health Care Center, Dallas, TX, United States

Jonathan Imran University of Texas Southwestern Medical Center, Dallas, TX, United States

Kirsten Jensen David Geffen School of Medicine at UCLA, Los Angeles, CA, United States

Saheli Jha EpicentRx, Inc., San Diego, CA; Uniformed Services University, Bethesda, MD, United States

Khosrow Kashfi City University of New York School of Medicine, New York, NY, United States

Michelle M. Kim University of Michigan, Ann Arbor, MI, United States

Min-Jung Lee National Cancer Institute, National Institutes of Health, Bethesda, MD, United States

Chiara Lenzi University of Pisa, Pisa, Italy

Mario Lo Bello University of Rome Tor Vergata, Rome, Italy

Laura M. López-Sánchez Maimónides Institute of Biomedical Research of Córdoba (IMIBIC), Córdoba, Spain

Sudjit Luanpitpong Mahidol University, Bangkok, Thailand

Valter Lubrano Fondazione G. Monasterio, CNR-Regione Toscana, Pisa, Italy

Amruta Manke RTI International, Durham, NC, United States

Helen O. McCarthy Queen's University Belfast, Belfast, United Kingdom

Cian M. McCrudden Queen's University Belfast, Belfast, United Kingdom

Arnold Oronsky InterWest Partners, Menlo Park, CA, United States

Bryan Oronsky EpicentRx, Inc., San Diego, CA, United States

Lorien J. Parker St. Vincent's Institute of Medical Research, Fitzroy; The University of Melbourne, Parkville, VIC, Australia

Michael W. Parker St. Vincent's Institute of Medical Research, Fitzroy; The University of Melbourne, Parkville, VIC, Australia

Catherine Paul University of Burgundy-Franche Comté, LIIC EA7269, F-21000, Dijon; EPHE, PSL Research University, F-75014 Paris, France

Jon Peñarando Maimónides Institute of Biomedical Research of Córdoba (IMIBIC), Córdoba, Spain

Stéphanie Plenchette University of Burgundy-Franche Comté, LIIC EA7269, F-21000, Dijon; EPHE, PSL Research University, F-75014 Paris, France

Sivakumar Ramadoss David Geffen School of Medicine at UCLA, Los Angeles, CA, United States

Tony R. Reid University of California at San Diego (UCSD), San Diego, CA, United States

Giorgio Ricci University of Rome Tor Vergata, Rome, Italy

Antonio Rodríguez-Ariza Maimónides Institute of Biomedical Research of Córdoba (IMIBIC); Oncology Department, Córdoba University Hospital, Córdoba, Spain

Yon Rojanasakul West Virginia University, Morgantown, WV, United States

Christian Schmidt Fraunhofer-Institute for Applied Polymer Research (IAP), Potsdam-Golm, Germany

Jan Scicinski EpicentRx, Inc., San Diego, CA, United States

Suvajit Sen David Geffen School of Medicine at UCLA, Los Angeles, CA, United States

Joachim Storsberg Fraunhofer-Institute for Applied Polymer Research (IAP), Potsdam-Golm, Germany

Thomas A. Summers Walter Reed National Military Medical Center, Bethesda, MD, United States

Jane B. Trepel National Cancer Institute, National Institutes of Health, Bethesda, MD, United States

Federica Vannini Institute of Life Sciences, Sant'Anna School of Advanced Studies; Institute of Clinical Physiology, CNR, Pisa, Italy

Elaine M. Walsh National University of Ireland Galway, Galway, Ireland

Liying Wang National Institute for Occupational Safety and Health, Morgantown, WV, United States

Claudia Weidensteiner Department of Radiology/Medical Physics, Medical Center - University of Freiburg, Faculty of Medicine, University of Freiburg, Freiburg, Germany

Aims and Scope for Series "Cancer Sensitizing Agents for Chemotherapy"

Current cancer management strategies fail to adequately treat malignancies with chemotherapy, with multivariable dose-restrictive factors such as systemic toxicity and multidrug resistance, hence limiting therapeutic benefits, quality of life, and complete long-term remission rates. The resistance of cancer cells to anticancer drugs is one of the major reasons for the failure of traditional cancer treatments. Cellular components and dysregulation of signaling pathways contribute to drug resistance. If modulated, such perturbations may restore the drug response and its efficacy. The recent understanding of the molecular mechanisms and targets that are implicated for cancer chemoresistance has paved the way to develop a large battery of small molecules (sensitizing agents) that can target resistant factors and reduce the threshold of resistance. Thus, allowing their combination with chemotherapeutic drugs to be effective and to reverse the chemoresistance, validated in preclinical and clinical settings.

The main objective of the proposed series *Cancer Sensitizing Agents for Chemotherapy* is to publish individualized and focused volumes whereby each volume is edited by an invited expert editor(s). Each volume will dwell on specific chemosensitizing agents with similar targeting activities. The combination treatment of the sensitizing agent with chemotherapy may result in a synergistic/additive activity and the reversal of tumor cells resistance to drugs.

The editors compile nonoverlapping review chapters on reported findings in various cancers, both experimentally and clinically, with particular emphases on underlying biochemical, genetic, and molecular mechanisms of the sensitizing agent(s) and the combination treatment.

The scope of the series is to provide scientists and clinicians with updated and clinical informations that will be valuable in their quest to investigate, develop, and apply novel combination therapies to reverse drug resistance and, thereby, prolong survival and even cure in cancer patients.

Benjamin Bonavida (Series Editor)

Aims and Scope for Volume 1 "Nitric Oxide (Donor/Induced) in Chemosensitization"

The physiological roles mediated by nitric oxide (NO) and derivatives have been the subject of many studies spanning embryogenesis, differentiation, morphogenesis, homeostasis, immune systems, etc. NO has also been used as a therapeutic in heart diseases and also in erectile dysfunction; however, the role of NO in cancer has been controversial. For many years, several reports indicated its protumorigenic activity and other reports of its antitumorigenic activity. This contradicting effect of NO was, in part, resolved by findings demonstrating that low levels of NO are protumorigenic while high levels are antitumorigenic on one hand and the type of tumor tissues being examined on the other hand. The potential therapeutic effect of NO in cancer began to be explored by several scientists.

The present volume titled *Nitric Oxide (Donor/Induced) in Chemosensitization* is the first volume that has been focused to address the biochemical, molecular, and genetic approaches in which NO and/or derivatives are used as chemosensitizing agents. Several contributors in this volume have reviewed different aspects related to the sensitizing properties mediated by NO in different settings and models.

Benjamin Bonavida (Editor)

About the Series Editor

Dr. Benjamin Bonavida, PhD (Series Editor) is currently *Distinguished Research Professor* at the University of California, Los Angeles (UCLA). His research career, thus far, has focused on basic immunochemistry and cancer immunobiology. His research investigations have ranged from the mechanisms of cell-mediated killing, sensitization of resistant tumor cells to chemo-/immunotherapy, characterization of resistant factors in cancer cells, cell-signaling pathways mediated by therapeutic anticancer antibodies, and characterization of a dysregulated NF-κB/Snail/YY1/RKIP/PTEN loop in many cancers that regulates cell survival, proliferation, invasion, metastasis, and resistance. He has also investigated the role of nitric oxide in cancer and its potential antitumor activity. Many of the above studies are centered on the clinical challenging features of cancer patients' failure to respond to both conventional and targeted therapies. The development and activity of various chemosensitizing agents and their modes of action to reverse chemo- and immunoresistance are highlighted in many refereed publications.

Preface

The physiological roles mediated by nitric oxide (NO) and derivatives have been the subject of many studies spanning embryogenesis, differentiation, morphogenesis, homeostasis, immune systems, etc. NO has also been used as a therapeutic in heart diseases and also in erectile dysfunction; however, the role of NO in cancer has been controversial. For many years, several reports indicated its protumorigenic activity and other reports of its antitumorigenic activity. This contradicting effect of NO was, in part, resolved by findings demonstrating that low levels of NO are protumorigenic while high levels are antitumorigenic on one hand and the type of tumor tissues being examined on the other hand. The potential therapeutic effect of NO in cancer began to be explored by several scientists, and its focus was initiated by the First International Conference on NO and Cancer in Paris, France, in 2007. Subsequently, the second through the Fourth International Workshops on NO and Cancer took place in Paris, France (2007); Kyoto, Japan (2010); Kingston, Canada (2013); and Seville, Spain (2015). Of interest, the Fifth International Workshop was held in Bologna, Italy, in Mar. 2017. In addition, two published books on NO and cancer have been edited by the editor of this volume [1,2].

The present volume titled *Nitric Oxide (Donor/Induced) in Chemosensitization* is the first volume that has been focused to address the biochemical, molecular, and genetic approaches in which NO and/or derivatives are used as chemosensitizing agents. Several contributors in this volume have reviewed different aspects related to the sensitizing properties mediated by NO in different settings and models. This volume consists of 14 review chapters that cover a wide spectrum of relevant topics spanning from studies in the laboratory and in animals and potential translational applications in clinical studies. Briefly, the highlights of each chapter are briefly presented below.

The first chapter by Edwards et al. titled "Nitric-Oxide-Mediated Chemosensitization: Gene Therapy Versus Exogenous Introduction of NO Donors" describes the potential application of NO as a hypersensitizing agent and its use in combination with chemotherapy to reverse drug resistance. The authors review the application of NO donors in comparison with gene therapy approaches. The authors caution of the use of gene therapy with NO for cancer applications due to several issues related to safety, reliability, specificity, and efficacy and also due to the microenvironment complexity and how gene expression will be manifested. Nevertheless, the authors are optimistic for the future of gene therapy with iNOS. They also believe that exogenous NO donor therapeutics will be more adaptable for chemosensitization in the clinic.

The chapter by Bonavida titled "Nitric-Oxide Donors Sensitize Resistant Cancer Cells to Apoptosis Induced by Chemotherapy: Molecular Mechanisms of Sensitization" discusses various studies performed by his group and by his collaborators and by others on the underlying biochemical and molecular mechanisms by which NO sensitizes drug-resistant tumor cells to chemotherapy-induced apoptosis. The discussion is centered on the inhibition of the constitutive expression of the cell survival/antiapoptotic NF-κB pathway in the majority of cancers. Treatment with NO donors inhibits NF-κB by S-nitrosylation of p50 and p65. In addition, a dysregulated NF-κB/Snail/YY1/RKIP/PTEN loop was reported in many cancers and whose role is to regulate cell survival, cell growth, and resistance to cytotoxic drugs. The introduction of NO donors modifies this resistant loop and results in chemosensitization of drug-resistant cancer cells. It is suggested of the potential clinical application of NO donors, under suitable conditions, in combination with low subtoxic doses of chemotherapy for the treatment of drug-resistant cancer cells.

The chapter by Scincinski et al. titled "RRx-001 Reset: Chemosensitization via NO-Mediated M1 Macrophage Repolarization" reviews the novel compound NO donor RRX-001 and its use in various in vitro model systems and also used in phase 1 and phase 2 clinical studies. They focus on the chemo- and radiosensitization properties of RRX-001 and demonstrate its chemosensitizing activity and also its role in arming tumor-associated macrophages. They also suggest that this new compound has the potential of its use in cancer patients who no longer have any treatment modality.

The chapter by Garrido et al. titled "Nitric Oxide and Nitric-Oxide Donors in Preclinical Studies of Breast and Prostate Cancer" reviewed the role of NO in different therapeutic approaches for breast and prostate cancers. While the authors are convinced of the merit of using NO in these cancers, they are careful in suggesting additional studies to better understand the role of NO in the different stages of the disease and its environment during the process of cancer progression. They suggest that NO be considered as an adjuvant in cancer therapy. They also suggest the prospective significance of iNOS overexpression in breast cancer and melanoma cancer and contradict the potential chemosensitizing activity of NO. However, they propose a need to investigate every cancer setting and microenvironment in order to determine whether to use NO donors or NO inhibitors.

The chapter by Ali et al. titled "Evaluation of the Impact of Nitric Oxide on Resistance to Platinum-Based Chemotherapeutics" reviewed the role of NO in tumor cells that are resistant to platinum-based chemotherapeutics. Platinum drugs are widely used in the treatments of various types of cancer and seldom used as monotherapy. Many patients are initially resistant or develop resistance to combination therapies. However, there are now agents to either sensitize or synergize with conventional chemotherapeutic drugs to reverse resistance. The authors discuss the underlying mechanism of the chemosensitizing effect mediated by NO. They also caution on the potential side effects of systemic NO administration and suggested that the combination of NO with nanoparticles may reduce some side effects. However, these NO nanosphere complexes must be designed and calibrated for each clinical administration.

The chapter by Plenchette et al. titled "Nitric Oxide and Platinum Derivative-Based Regimens for Cancer Treatment: From Preclinical Studies to Clinical Trials" reviews the literature on the therapeutic application of NO donors in combination with platinum compounds in both preclinical and clinical settings. They also reviewed the use of NO donors as sensitizing agents for other therapies including checkpoint inhibitor monoclonal antibodies. They also suggest that NTG as an NO donor is a good candidate for the use in combination therapy.

The chapter by Kashfi and Esmaili titled "NO-H_2S-Releasing Chimeras as a Multifaceted Approach to Cancer Therapy" reviews the roles of nitric oxide and hydrogen sulfide donors as chemosensitizing agents in experimental model systems, both in vitro and in mice. The authors describe various anticancer compounds that incorporate both NO and H_2S in complexes and focused on the recent H_2S-releasing pharmaceuticals in their anticancer activities. They also discuss the minimum toxicity mediated by the NO-H_2S complexes. Clearly, the authors suggest the potential clinical application of these compounds in the clinic.

The chapter by Vanini et al. titled "Nitric-Oxide-Based Anticancer Therapeutics: The New Technologies of the Nanoparticles" discusses the limitations of applying NO donors as chemosensitizing agents due to their short half-lives, the poor concentrations that reach the cancer cells, and also that they are highly nonspecific. They suggest that the application of nanoparticle-based delivery systems of NO and targeting them to the cancer cells will result in minimal side effects and overriding the above limitations by the NO donors. They also discuss and update the potential therapeutic applications of the combination of NO donors and chemotherapeutic drugs that are incorporated in nanoparticles.

Imran and Huerta's chapter titled, "Nitric Oxide in Rectal Cancer: From Mice to Patients" reviews the role of NO in rectal cancer in both preclinical and clinical settings. They discuss their findings in experimental rectal mouse models and their findings that the NO donor DETANOATE was a potent radiosensitizer. In addition, they report the findings in patients who were administered NO via NTG transdermal patches and report that this administration was safe when administered in combination with conventional treatments. Their findings have established the rationale for a phase 2 clinical trial.

Sen et al. chapter titled "Chemoprotective and Chemosensitizing Effects of Nitric Oxide and Other Biologically Active Gases in Breast Cancer Chemotherapy: Potential Implications" reviewed the literature on the various properties and effects of endogenous NO produced in breast cancer cells and breast tumors as well as the NO role in cancer homeostasis. They discuss how endogenous NO and H_2S may cross talk and, thus, may reveal potential novel interventions. They also discuss in the chapter the relationship between tumor-associated macrophages and NO.

The chapter by Manke et al. titled "Role of Nitric Oxide in Cancer Stem Cell Regulation and Metastasis" discusses the role of NO in the regulation of its effect on the cell physiology and its role in metastases. They discuss the oncogenic properties of NO in various types of cancers. They also discuss the molecular mechanisms by which cancer stem cells regulate tumor growth invasiveness and metastases. They also point out that NO regulates tumor metastases by affecting cancer stem cells regulated by proteins such as caveolin-1. Thus, their study suggests the potential use of NO inhibitors to prevent CSC-mediated tumor growth.

The chapter by Lo Bello et al. titled "Nitric Oxide Interacting with Glutathione Transferases" reviewed the interaction of NO with glutathione transferases in the presence of GSH and iron. The authors emphasize the importance of delineating the underlying mechanisms that are involved in the metabolism of NO present at high concentrations in the cells in order to initiate cellular signaling involved in its anticancer activity.

The chapter by Weidensteiner titled "Application of MRI to Study the Role of Nitric Oxide in Cancer Therapy" discussed the application of noninvasive imaging inhibiting its upregulation of antiapoptotic genes [39,40]. Likewise, S-nitrosylation of IκB blocks the phosphorylation of that protein allowing for cytoplasmic sequestration of NF-κB [41,42].

GENE THERAPEUTIC EXPRESSION OF NOS

To date, the findings related to gene therapeutic expression of inducible nitric oxide synthase (iNOS) (Fig. 2) as an adjuvant to chemotherapeutic interventions to treat malignancies has been quite promising. Coulter et al. [62] demonstrated that tumor-specific promoters, for human osteocalcin, and prostate-specific membrane antigen can effectively initiate an iNOS transgene and produce cytotoxicity in malignant prostate cells. They also reported significant improvements for cisplatin-induced apoptosis when administered in combination with an iNOS gene transfer application [63]. The iNOS gene transfer significantly increased expression of iNOS and induced a corresponding increased accumulation of NO in radiation-induced murine sarcoma cells (RIF-1) [64], human prostate cancer cells (DU145 and PC3) [65], and human colorectal adenocarcinoma cells (HT29) [66]. The treatment was cytotoxic in the absence of cisplatin and, when used in combination with cisplatin, enhanced cell death relative to the administration of cisplatin, alone [63]. These findings correspond with previous reports that iNOS gene transfer is effective as an adjuvant to radiotherapy by way of enhancing radiosensitization [67–69]. Worthington et al. also introduced and validated a method for the restricted expression of iNOS in the tumor using a radiation-inducible promoter to regulate activation of the transgene [70]. Using a p21$^{(WAF1)}$ promoter, this group improved upon

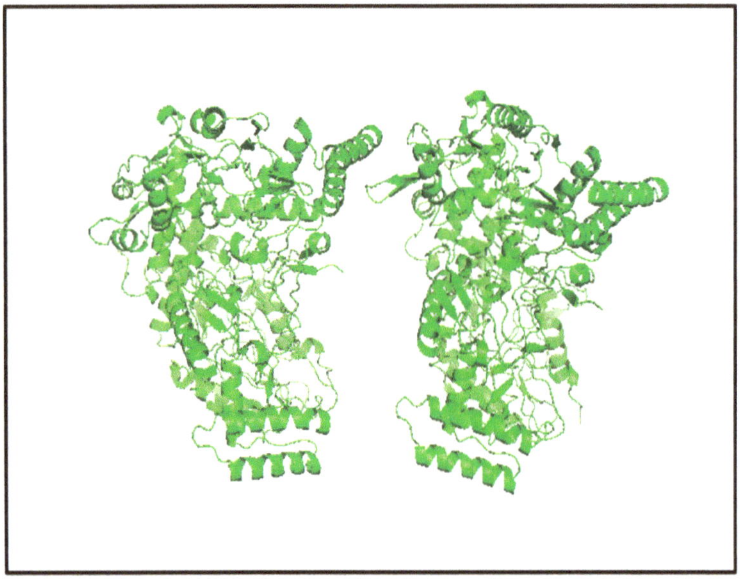

FIG. 2 Structural representation of iNOS: gene therapeutic expression of iNOS as an adjuvant to chemotherapeutics has proved to hypersensitize otherwise chemoresistant malignant cells [62–72].

their original iNOS gene transfer protocols in a way that eliminated expression of the transgene in healthy cells while achieving constitutive expression in tumor cells [71]. The iNOS transgene application has since been further improved upon by the introduction of a pE9 promoter system [72].

Gene therapeutic applications of INOS isoforms have been extensively investigated for noncancer applications including intimal hyperplasia, vascular tone, ischemia-reperfusion damage, and restenosis [73–103]. Many of these were in vivo studies, and several were clinical studies, representing a significantly advanced stage, relative to the current state of research related to NOS gene therapeutics for cancer applications. It is also noteworthy that most of the cardiovascular applications employ adenoviral vectors that are far more likely to represent any future clinical applications for cancer than the more limited gene transfer methods currently being tested for NO therapeutics with cancer indications. Indeed, cardiovascular applications of NOS gene therapeutics have informed and paved the way for corresponding oncology applications. Pitfalls have been identified; methodologies have been refined; in vivo and even clinical applications have been successfully employed.

INTRODUCTION OF NO DONORS

A wide range of nitric oxide donors have been studied for their potential as therapeutics for oncology indications [10,104,105]. Those that have exhibited the greatest capacity for achieving clinically relevant concentrations include metal-nitrosyl complexes (sodium nitroprusside, SNP), organic nitrites (glyceryl trinitrite, GTN), *S*-nitrosoglutathione (GSNO), *S*-nitrosothiols (*S*-nitroso-*N*-acetylpenicillamine (SNAP)), diazeniumdiolates, and sydnonimines (3-morpholinosydnonimine, SIN-1) [10,104,105]. The impacts of each of these on malignant cells provide evidence that any of them could serve as potential adjuvants for more traditional chemo- and/or radiotherapeutics.

GTN initiates apoptosis in malignant colon cells via enhanced expression of Fas and lowered expression of key inhibitors of apoptosis [106]. Similarly, minimal concentrations of NO-mimetic GTN inhibit the hypoxia-mediated metastatic capacity of murine melanoma cells [107] and intensify the chemosensitivity in xenografts of human prostate cancer [108]. SNP release of NO slows proliferation of human prostate and bladder cancer cells by inhibiting hypoxia-inducible factor 1 (HIF-1) and compromising mitochondrial membrane permeability [109]. SNAP donation of NO has been shown to amplify cell death in human neuroblastoma cells as a consequence of p53 upregulation [110,111]. SNAP is also associated with radiosensitization of hypoxic murine mammary carcinoma cells [112]. GSNO has, likewise, been proved to initiate apoptosis in colon tumor lines [113]. At hyperelevated concentrations, GSNO halted proliferation and initiated apoptosis in several colon malignancies without consequence related to their cyclooxygenase-2 (COX-2) expression and activities [114]. The peroxynitrite generator, SIN-1, initiates apoptosis in cancer cells via oxidative DNA damage and activation of caspase-3-like proteases [115]. SIN-1 also restricts the activity of arylamine *N*-acetyltranferase isoform 1 (NAT1) in breast cancer cells [116] and activates apoptotic cascades in dopaminergic neuroblastoma cells [117].

Diazeniumdiolates (NONOates) have been exhaustively investigated for their potential in the clinical management of cancer [118]. Diethylamine NONOate/AM (NONO-AM, produced from DEA/NO), for example, was shown to induce apoptosis, in a dose-dependent manner, in both F10 and MDA-MB-231 (bone metastatic breast cancer) cell lines [119]. Similarly, NO released from PAPA ((Z)-1-[*N*-(3-ammonio-propyl)-*N*-(*n*-propyl) amino]diazen-1-ium-1,2diolate)/NO was proved to effectively radiosensitize hypoxic murine mammary carcinoma cells [120] and to initiate apoptosis in human colorectal adenocarcinoma cells [121]. However, NO generated from SPER/NO (spermine-nitric oxide complex hydrate) initiated elevated expression of heme oxygenase-1 (HO-1) resulting in cytoprotection of rat glioma cells challenged with cadmium-induced cytotoxicity [122] and in breast cancer cells challenged with light-induced chain peroxidation [123]. A more recent study illustrated only modest benefit to incorporating SPER/NO as an adjuvant to cisplatin treatments for resistant ovarian cancers [124]. DETA/NO (diethylenetriamine/nitric oxide) has been shown to induce cytostasis and cell-cycle arrest in human breast adenocarcinoma lines (MDA-MB-231 [125]) and (MDA-MB-468 [126,127]) resulting from dephosphorylation of extracellular signal-regulated kinase 1/2 (ERK1/2) [127]. However, the impact of DETA/NO on cisplatin-resistant ovarian cancer cells was shown to be only modestly significant [124].

There is growing consensus that the seemingly dichotomous impacts of NO donors (cytotoxic vs. cytoprotective) are the result of variability in cell-specific bearings and/or NO concentrations. That is, most NO donors lack target specificity, adequate half-lives, and controlled-release kinetics. In response to these limitations, many researchers are turning to NO-hybrid prodrugs. Such prospective therapeutics can deliver NO in a controlled and, often, site-specific manner. Shielded NO donors and donors integrated into macromolecules are being studied as more stable means for tightly controlled delivery of NO. For example, JS-K (O^2-(2,4-dinitrophenyl) 1-[(4-ethoxycarbonyl)piperazin-1-yl]diazen-1-ium-1,2-diolate) is a diazeniumdiolate-based prodrug with a well-defined NO release profile [128]. In cells, JS-K is metabolized by glutathione by way of dearylation of the diazeniumdiolate facilitated by a nucleophilic aromatic substitution [128]. Hydrolysis of the resulting CEP/NO anion yields NO. The consequent reduction of glutathione and precipitous change in redox potential initiate apoptosis. Thus, JS-K is being studied as a cytotoxic therapeutic for nonsmall-cell lung cancer and has been proved to induce apoptosis in multiple myeloma cells [129,130]. Likewise, site-specific glyco-*S*-nitrosothiols (sugar-SNAP) have proved effective in the specific targeting and destruction of prostate cancer cells where an increased uptake of the NO donor resulted from enhanced transport of saccharides into the cells, relative to SNAP, alone [131].

Prodrugs prepared using polysaccharide-based dextran thiomers with covalent incorporation of a NO donor [132] protect the donors against the fleeting and unrestrained release of NO. This results in the quantitatively predictable release of NO under physiological conditions [132]. The incorporation of polysaccharides proves to be a better alternative than monosaccharides for dose delivery applications by enhancing microcirculatory flow through diminished blood viscosity and reduction of erythrocytic aggregation by virtue of their colloidal nature. Fig. 3A provides a representative NO release profile for a *S*-nitrosated dextran-cysteine derivative. Fig. 3B provides an example of the impacts that such donors have on the proliferation of breast cancer cells [133,134]. This dextran-cysteine derivative releases 0.119 ± 0.049 mmol NO/g in approximately 2 h [133,134].

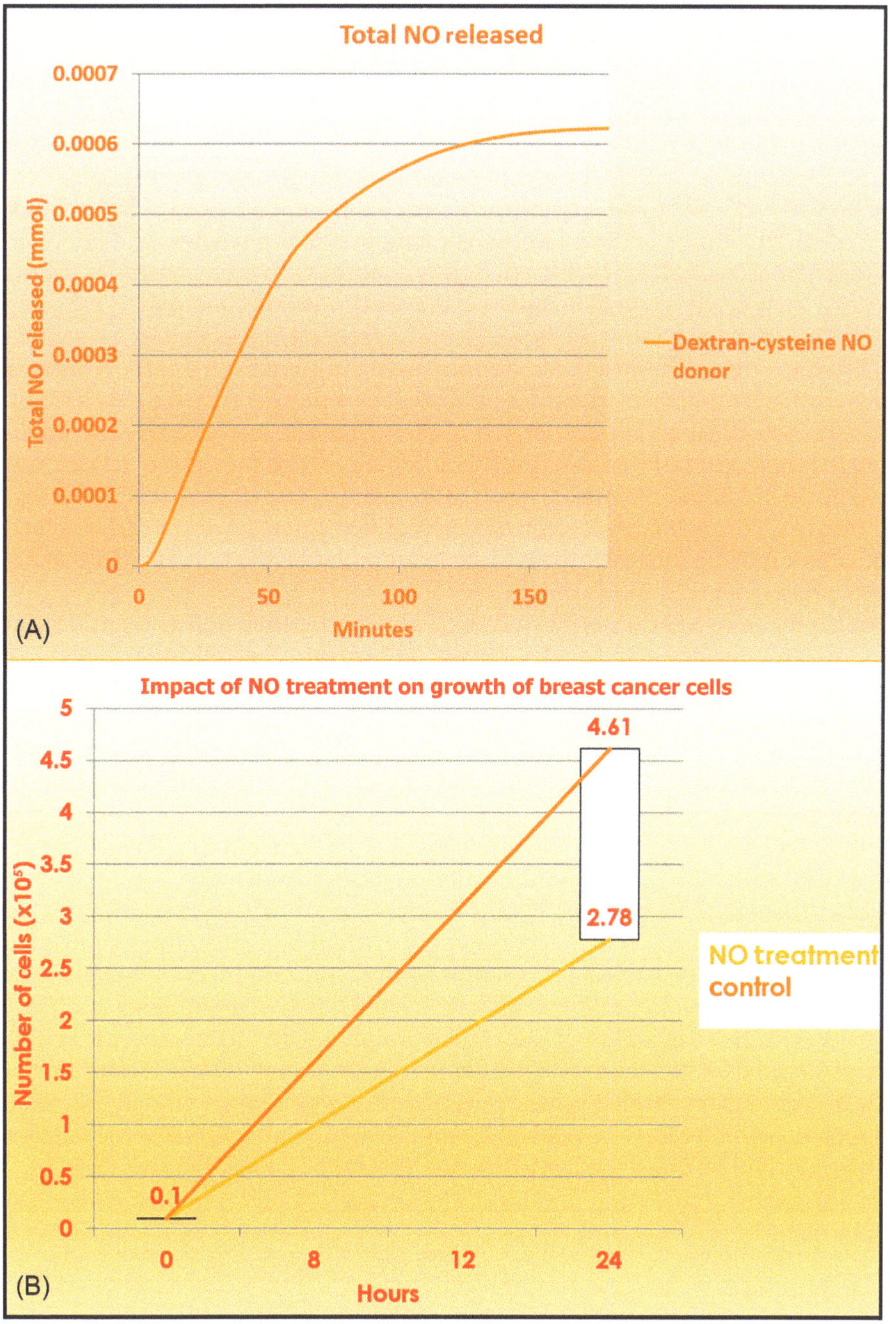

FIG. 3 (A) Representative NO release profile for an *S*-nitrosated dextran-cysteine [133,134]. (B) Impact of a dextran-cysteine NO donor on the proliferation of malignant breast cancer cells [133,134].

CONCLUSION

A simple review of the literature reveals that the idea of employing gene therapies with NO cancer applications is daunting. Issues related to safety, reliability, and efficacy along with complications associated with tissue microenvironments and longevity of expression top a growing list of concerns that will need to be addressed. Clearly, the greatest successes for the induction of NOS using gene therapeutic mechanisms have been achieved in the areas of cardiovascular therapeutics and radiosensitization of malignancies. The use of gene therapeutics for chemosensitization is still a fledgling concept with many obstacles to circumvent. Nevertheless, early studies reveal that gene therapeutic introductions of iNOS have great potential. The field of gene therapy is primed to make great advances in the coming decade that will likely lead to dozens of new clinical applications. Building from this foundation and from the lessons learned by pioneers in the use of NOS-generating gene therapeutics for cardiovascular and radiosensitization applications, it is likely that safe and effective therapeutics with chemosensitization applications will not be far behind. In the meantime, it is far more likely that therapeutics based on the introduction of exogenous NO donors will make the greatest impacts toward chemosensitization for the clinical management of cancer. However, in the long term, it is probable that well-defined gene therapeutics will overtake the use of exogenous donors as a preferred mode of treatment.

Although a range of NO donors have been evaluated for their ability to induce chemosensitization, toxicity profiles associated with the NO carrier, the cell uptake, and the lack of proper controlled-release kinetics have presented major obstacles to their clinical applications. For instance, although NO-releasing diazeniumdiolated silica nanoparticles significantly slowed the proliferation of ovarian tumor cells, the nondiazeniumdiolated silica nanoparticles (AHAP3-silica) resulted in crippling cytotoxicity for normal cells due to the primary amines on the silica surfaces [135]. NO-releasing prodrugs using noncytotoxic *S*-nitrosothiol NO donors in a controlled release may circumvent this issue, but the variable impacts of NO in changing tumor states continue to be a challenge.

The potential of NO donors as chemosensitizing adjuvant therapeutics has been well established. However, it has become increasingly evident that such donors must be tailored to accommodate controlled release at relatively high concentrations. Of equal importance, given the disparate physiological outcomes associated with NO treatments in different types of malignant cells, there is a glaring deficiency in data related to the specific cell types that may be best targeted by NO therapy. This dearth of any kind of library related to the impacts of specific NO donors on varying cancer cells underscores a serious void in the existing body of NO-donor-related literature. A deliberate approach to catalog cell-specific outcomes to fill this void would represent a major advance for NO-donor-based therapeutics.

Acknowledgments

MAB (1060548; 1460038), MAE (1060548; 1460038), and IA (1460038) thankfully acknowledge support from the National Science Foundation.

Conflict of Interest: No potential conflicts of interest were disclosed.

References

[1] Szakacs G, Paterson JK, Ludwig JA, Booth-Genthe C, Gottesman MM. Targeting multidrug resistance in cancer. Nat Rev Drug Discov 2006;5:219–34.

[2] O'Reilly EA, Gubbins L, Sharma S, Tully R, Guang MHZ, Weiner-Gorzel K, et al. The fate of chemoresistance in triple negative breast cancer (TNBC). BBA Clin 2015;3:257–75.

[3] Pan S-T, Li Z-L, He Z-X, Qiu J-X, Zhou S-F. Molecular mechanisms for tumour resistance to chemotherapy. Clin Exp Pharmacol Physiol 2016;43:723–37.

[4] Kachalaki S, Ebrahimi M, Mohamed Khosroshahi L, Mohammadinejad S, Baradaran B. Cancer chemoresistance; biochemical and molecular aspects: a brief overview. Eur J Pharm Sci 2016;89:20–30.

[5] Weyerbrock A, Baumer B, Papazoglou A. Growth inhibition and chemosensitization of exogenous nitric oxide released from NONOates in glioma cells in vitro. J Neurosurg 2009;110:128–36.

[6] Kogias E, Osterberg N, Baumer B, Psarras N, Koentges C, Papazoglou A, et al. Growth-inhibitory and chemosensitizing effects of the glutathione-*S*-transferase-π-activated nitric oxide donor PABA/NO in malignant gliomas. Int J Cancer 2012;130:1184–94.

[7] Altieri R, Fontanella M, Agnoletti A, Panciani PP, Spena G, Crobeddu E, et al. Role of nitric oxide in glioblastoma therapy: another step to resolve the terrible puzzle? Transl Med UniSa 2015;12:54–9.

[8] Coulter JA, McCarthy HO, Xiang J, Roedl W, Wagner E, Robson T, et al. Nitric oxide—a novel therapeutic for cancer. Nitric Oxide 2008;19:192–8.

[9] Bonavida B, Garban H. Nitric oxide-mediated sensitization of resistant tumor cells to apoptosis by chemo-immunotherapeutics. Redox Biol 2015;6:486–94.

[10] Bonavida B, Baritaki S, Huerta-Yepez S, Vega MI, Chatterjee D, Yeung K. Novel therapeutic applications of nitric oxide donors in cancer: roles in chemo- and immunosensitization to apoptosis and inhibition of metastases. Nitric Oxide 2008;19:152–7.

[11] Hibbs Jr. J, Taintor R, Vavrin Z. Macrophage cytotoxicity: role for L-arginine deiminase and imino nitrogen oxidation to nitrite. Science 1987;235:473–6.

[12] Hibbs Jr. JB, Taintor RR, Vavrin Z, Rachlin EM. Nitric oxide: a cytotoxic activated macrophage effector molecule. Biochem Biophys Res Commun 1988;157:87–94.

[13] Wink D, Hines H, Cheng R, Switzer C, Flores-Santana W, Vitek M, et al. Nitric oxide and the redox mechanisms in the immune response. J Leukoc Biol 2011;89:873–91.

[14] Binder C, Schulz M, Hiddemann W, Oellerich M. Induction of inducible nitric oxide synthase is an essential part of tumor necrosis factor-alpha-induced apoptosis. Lab Invest 1999;79:1703–12.

[15] Farias-Eisner R, Sherman MP, Aeberhard E, Chaudhuri G. Nitric oxide is an important mediator for tumoricidal activity in vivo. Proc Natl Acad Sci U S A 1994;91:9407–11.

[16] Sarti P, Forte E, Mastronicola D, Giuffrè A, Arese M. Cytochrome c oxidase and nitric oxide in action: molecular mechanisms and pathophysiological implications. Biochim Biophys Acta Bioenerg 2012;1817:610–9.

[17] Boyd CS, Cadenas E. Nitric oxide and cell signaling pathways in mitochondrial-dependent apoptosis. Biol Chem 2002;383(3–4):411–23.

[18] Lane P, Hao G, Gross SS. S-nitrosylation is emerging as a specific and fundamental posttranslational protein modification: head-to-head comparison with O-phosphorylation. Sci STKE 2001;2001:re1.

[19] Medeiros TN, Jarigese DM, Edwards MA, Brown MA. Impacts of S-nitrosylation in cancer. In: Bonavida B, editor. Nitric oxide and cancer: pathogenesis and therapy. Cham: Springer International Publishing; 2015. p. 87–95.

[20] Foster MW, Hess DT, Stamler JS. Protein S-nitrosylation in health and disease: a current perspective. Trends Mol Med 2009;15(9):391–404.

[21] Stamler JS, Simon DI, Osborne JA, Mullins ME, Jaraki O, Michel T, et al. S-nitrosylation of proteins with nitric oxide: synthesis and characterization of biologically active compounds. Proc Natl Acad Sci U S A 1992;89:444–8.

[22] Jaffrey SR, Erdjument-Bromage H, Ferris CD, Tempst P, Snyder SH. Protein S-nitrosylation: a physiological signal for neuronal nitric oxide. Nat Cell Biol 2001;3:193–7.

[23] Martínez-Ruiz A, Lamas S. Two decades of new concepts in nitric oxide signaling: from the discovery of a gas messenger to the mediation of nonenzymatic posttranslational modifications. IUBMB Life 2009;61:91–8.

[24] Benhar M, Forrester MT, Hess DT, Stamler JS. Regulated protein denitrosylation by cytosolic and mitochondrial thioredoxins. Science 2008;320:1050–4.

[25] Liu L, Yan Y, Zeng M, Zhang J, Hanes MA, Ahearn G, et al. Essential roles of S-nitrosothiols in vascular homeostasis and endotoxic shock. Cell 2004;116:617–28.

[26] Trujillo M, Alvarez MN, Peluffo G, Freeman BA, Radi R. Xanthine oxidase-mediated decomposition of S-nitrosothiols. J Biol Chem 1998;273:7828–34.

[27] Mannick JB. Regulation of apoptosis by protein S-nitrosylation. Amino Acids 2007;32:523–6.

[28] Benhar M, Stamler JS. A central role for S-nitrosylation in apoptosis. Nat Cell Biol 2005;7:645–6.

[29] Melino G, Bernassola F, Knight R, Corasaniti M, Nistico G, Finazzi-Agro A. S-nitrosylation regulates apoptosis. Nature 1997;388:432–3.

[30] Consuelo G, Faustino M. Lipid rafts and Fas/CD95 signaling in cancer chemotherapy. Recent Pat Anticancer Drug Discov 2011;6:274–83.

[31] Scheel-Toellner D, Wang K, Singh R, Majeed S, Raza K, Curnow SJ, et al. The death-inducing signalling complex is recruited to lipid rafts in Fas-induced apoptosis. Biochem Biophys Res Commun 2002;297:876–9.

[32] Singh S, Gupta AK. Nitric oxide: role in tumour biology and iNOS/NO-based anticancer therapies. Cancer Chemother Pharmacol 2011;67:1211–24.

[33] Circu ML, Aw TY. Reactive oxygen species, cellular redox systems and apoptosis. Free Radic Biol Med 2010;48:749–62.

[34] Sha Y, Marshall HE. S-nitrosylation in the regulation of gene transcription. Biochim Biophys Acta Gen Subj 2012;1820:701–11.

[35] Schonhoff CM, Daou M-C, Jones SN, Schiffer CA, Ross AH. Nitric oxide-mediated inhibition of Hdm2–p53 binding. Biochemistry 2002;41:13570–4.

[36] Hara MR, Cascio MB, Sawa A. GAPDH as a sensor of NO stress. Biochim Biophys Acta (BBA) - Mol Basis Dis 2006;1762:502–9.

[37] Tarze A, Deniaud A, Le Bras M, Maillier E, Molle D, Larochette N, et al. GAPDH, a novel regulator of the pro-apoptotic mitochondrial membrane permeabilization. Oncogene 2006;26:2606–20.

[38] Diers AR, Broniowska KA, Hogg N. Nitrosative stress and redox-cycling agents synergize to cause mitochondrial dysfunction and cell death in endothelial cells. Redox Biol 2013;1:1–7.

[39] Bonavida B, Baritaki S. Dual role of NO donors in the reversal of tumor cell resistance and EMT: downregulation of the NF-κB/Snail/YY1/RKIP circuitry. Nitric Oxide 2011;24:1–7.

[40] Aranda E, Lopez-Pedrera C, Haba-Rodriguez JRDL, Rodriguez-Ariza A. Nitric oxide and cancer: the emerging role of S-nitrosylation. Curr Mol Med 2012;12:50–67.

[41] Marshall HE, Hess DT, Stamler JS. S-nitrosylation: physiological regulation of NF-κB. Proc Natl Acad Sci U S A 2004;101:8841–2.

[42] Marshall HE, Stamler JS. Inhibition of NF-κB by S-nitrosylation. Biochemistry 2001;40:1688–93.

[43] Wang Z. Protein S-nitrosylation and cancer. Cancer Lett 2012;320:123–9.

[44] Lavrik IN, Krammer PH. Regulation of CD95/Fas signaling at the DISC. Cell Death Differ 2012;19:36–41.

[45] Guan X. Cancer metastases: challenges and opportunities. Acta Pharm Sin B 2015;5:402–18.

[46] Elenbaas B, Dobbelstein M, Roth J, Shenk T, Levine AJ. The MDM2 oncoprotein binds specifically to RNA through its RING finger domain. Mol Med 1996;2:439–51.

[47] Böttger A, Böttger V, Garcia-Echeverria C, Chène P, Hochkeppel H-K, Sampson W, et al. Molecular characterization of the hdm2-p53 interaction. J Mol Biol 1997;269:744–56.

[48] Slack A, Shohet JM. MDM2 as a critical effector of the MYCN oncogene in tumorigenesis. Cell Cycle 2005;4:857–60.

[49] Tristan C, Shahani N, Sedlak TW, Sawa A. The diverse functions of GAPDH: views from different subcellular compartments. Cell Signal 2011;23:317–23.

[50] Millman MS, Omachi A. The role of oxidized nicotinamide adenine dinucleotide in fluoride inhibition of active sodium transport in human erythrocytes. J Gen Physiol 1972;60:337–50.

[51] Sirover MA. Emerging new functions of the glycolytic protein, glyceraldehyde-3-phosphate dehydrogenase, in mammalian cells. Life Sci 1996;58:2271–7.

[52] Zala D, Hinckelmann M-V, Yu H, Lyra da Cunha Marcel M, Liot G, Cordelières FP, et al. Vesicular glycolysis provides on-board energy for fast axonal transport. Cell 2013;152:479–91.

[53] Tisdale EJ, Talati NK, Artalejo CR, Shisheva A. GAPDH binds Akt to facilitate cargo transport in the early secretory pathway. Exp Cell Res 2016;349:310–9.

[54] Molina y Vedia L, McDonald B, Reep B, Brüne B, Di Silvio M, Billiar TR, et al. Nitric oxide-induced S-nitrosylation of glyceraldehyde-3-phosphate dehydrogenase inhibits enzymatic activity and increases endogenous ADP-ribosylation. J Biol Chem 1992;267:24929–32.

[55] Bernhard B, Susanne M. Protein thiol modification of glyceraldehyde-3-phosphate dehydrogenase and caspase-3 by nitric oxide. Curr Protein Pept Sci 2001;2:61–72.
[56] Hara MR, Thomas B, Cascio MB, Bae B-I, Hester LD, Dawson VL, et al. Neuroprotection by pharmacologic blockade of the GAPDH death cascade. Proc Natl Acad Sci U S A 2006;103:3887–9.
[57] Hara MR, Snyder SH. Nitric oxide-GAPDH-siah: a novel cell death cascade. Cell Mol Neurobiol 2006;26:525–36.
[58] Ghosh G, Wang VY-F, Huang D-B, Fusco A. NF-κB regulation: lessons from structures. Immunol Rev 2012;246:36–58.
[59] Lenardo MJ, Baltimore D. NF-κB: a pleiotropic mediator of inducible and tissue-specific gene control. Cell 1989;58:227–9.
[60] Ahmed A, Williams B, Hannigan G. Transcriptional activation of inflammatory genes: mechanistic insight into selectivity and diversity. Biomolecules 2015;5:3087.
[61] Beg AA, Baldwin AS. The I kappa B proteins: multifunctional regulators of Rel/NF-kappa B transcription factors. Genes Dev 1993;7:2064–70.
[62] Coulter JA, Page NL, Worthington J, Robson T, Hirst DG, McCarthy HO. Transcriptional regulation of inducible nitric oxide synthase gene therapy: targeting early stage and advanced prostate cancer. J Gene Med 2010;12:755–65.
[63] Adams C, McCarthy HO, Coulter JA, Worthington J, Murphy C, Robson T, et al. Nitric oxide synthase gene therapy enhances the toxicity of cisplatin in cancer cells. J Gene Med 2009;11:160–8.
[64] Twentyman PR, Brown JM, Gray JW, Franko AJ, Scoles MA, Kallman RF. A new mouse tumor model system (RIF-1) for comparison of end-point studies. J Natl Cancer Inst 1980;64:595–604.
[65] Alimirah F, Chen J, Basrawala Z, Xin H, Choubey D. DU-145 and PC-3 human prostate cancer cell lines express androgen receptor: implications for the androgen receptor functions and regulation. FEBS Lett 2006;580:2294–300.
[66] Martínez-Maqueda D, Miralles B, Recio I. HT29 cell line. In: Verhoeckx K, Cotter P, López-Expósito I, Kleiveland C, Lea T, Mackie A, et al., editors. The impact of food bioactives on health: in vitro and ex vivo models. Cham: Springer International Publishing; 2015. p. 113–24.
[67] Worthingon J, Robson T, O'Keeffe M, Hirst D. Tumour cell radiosensitization using constitutive (CMV) and radiation inducible (WAF1) promoters to drive the iNOS gene: a novel suicide gene therapy. Gene Ther 2002;9:263–9.
[68] Cook T, Wang Z, Alber S, Liu K, Watkins SC, Vodovotz Y, et al. Nitric oxide and ionizing radiation synergistically promote apoptosis and growth inhibition of cancer by activating p53. Cancer Res 2004;64:8015–21.
[69] Wang Z, Cook T, Alber S, Liu K, Kovesdi I, Watkins SK, et al. Adenoviral gene transfer of the human inducible nitric oxide synthase gene enhances the radiation response of human colorectal cancer associated with alterations in tumor vascularity. Cancer Res 2004;64:1386–95.
[70] Worthington J, Robson T, Scott S, Hirst D. Evaluation of a synthetic CArG promoter for nitric oxide synthase gene therapy of cancer. Gene Ther 2005;12:1417–23.
[71] McCarthy HO, Worthington J, Barrett E, Cosimo E, Boyd M, Mairs RJ, et al. p21(WAF1)-mediated transcriptional targeting of inducible nitric oxide synthase gene therapy sensitizes tumours to fractionated radiotherapy. Gene Ther 2006;14:246–55.
[72] Coulter JA, McCarthy HO, Worthington J, Robson T, Scott S, Hirst DG. The radiation-inducible pE9 promoter driving inducible nitric oxide synthase radiosensitizes hypoxic tumour cells to radiation. Gene Ther 2008;15:495–503.
[73] Teupe C, Richter S, Fisslthaler B, Randriamboavonjy V, Ihling C, Fleming I, et al. Vascular gene transfer of phosphomimetic endothelial nitric oxide synthase (S1177D) using ultrasound-enhanced destruction of plasmid-loaded microbubbles improves vasoreactivity. Circulation 2002;105:1104–9.
[74] Cable DG, Pompili VJ, O'Brien T, Schaff HV. Recombinant gene transfer of endothelial nitric oxide synthase augments coronary artery relaxations during hypoxia. Circulation 1999;100:II-335–9.
[75] Cable DG, O'Brien T, Kullo IJ, Schwartz RS, Schaff HV, Pompili VJ. Expression and function of a recombinant endothelial nitric oxide synthase gene in porcine coronary arteries. Cardiovasc Res 1997;35:553–9.
[76] Cable DG, Caccitolo JA, Caplice N, O'Brien T, Simari RD, Daly RC, et al. The role of gene therapy for intimal hyperplasia of bypass grafts. Circulation 1999;100:II-392–6.
[77] Kullo IJ, Mozes G, Schwartz RS, Gloviczki P, Crotty TB, Barber DA, et al. Adventitial gene transfer of recombinant endothelial nitric oxide synthase to rabbit carotid arteries alters vascular reactivity. Circulation 1997;96:2254–61.

[78] Kullo IJ, Mozes G, Schwartz RS, Gloviczki P, Tsutsui M, Katusic ZS, et al. Enhanced endothelium-dependent relaxations after gene transfer of recombinant endothelial nitric oxide synthase to rabbit carotid arteries. Hypertension 1997;30:314–20.

[79] Hayashi T, Sumi D, Juliet PAR, Matsui-Hirai H, Asai-Tanaka Y, Kano H, et al. Gene transfer of endothelial NO synthase, but not eNOS, plus inducible NOS regressed atherosclerosis in rabbits. Cardiovasc Res 2004;61:339–51.

[80] von der Leyen HE, Gibbons GH, Morishita R, Lewis NP, Zhang L, Nakajima M, et al. Gene therapy inhibiting neointimal vascular lesion: in vivo transfer of endothelial cell nitric oxide synthase gene. Proc Natl Acad Sci U S A 1995;92:1137–41.

[81] Janssens S, Flaherty D, Nong Z, Varenne O, van Pelt N, Haustermans C, et al. Human endothelial nitric oxide synthase gene transfer inhibits vascular smooth muscle cell proliferation and neointima formation after balloon injury in rats. Circulation 1998;97:1274–81.

[82] Varenne O, Pislaru S, Gillijns H, Van Pelt N, Gerard RD, Zoldhelyi P, et al. Local adenovirus-mediated transfer of human endothelial nitric oxide synthase reduces luminal narrowing after coronary angioplasty in pigs. Circulation 1998;98:919–26.

[83] Cooney R, Hynes SO, Sharif F, Howard L, O'Brien T. Effect of gene delivery of NOS isoforms on intimal hyperplasia and endothelial regeneration after balloon injury. Gene Ther 2006;14:396–404.

[84] Wang K, Kessler PD, Zhou Z, Penn MS, Forudi F, Zhou X, et al. Local adenoviral-mediated inducible nitric oxide synthase gene transfer inhibits neointimal formation in the porcine coronary stented model. Mol Ther 2003;7:597–603.

[85] Sharif F, Hynes SO, Cooney R, Howard L, McMahon J, Daly K, et al. Gene-eluting stents: adenovirus-mediated delivery of eNOS to the blood vessel wall accelerates re-endothelialization and inhibits restenosis. Mol Ther 2008;16(10):1674–80.

[86] Fishbein I, Alferiev I, Bakay M, Stachelek SJ, Sobolewski P, Lai M, et al. Local delivery of gene vectors from bare-metal stents by use of a biodegradable synthetic complex inhibits in-stent restenosis in rat carotid arteries. Circulation 2008;117:2096–103.

[87] Li Q, Guo Y, Tan W, Ou Q, Wu W-J, Sturza D, et al. Cardioprotection afforded by inducible nitric oxide synthase gene therapy is mediated by cyclooxygenase-2 via a nuclear factor-κB-dependent pathway. Circulation 2007;116:1577–84.

[88] Li Q, Guo Y, Tan W, Stein AB, Dawn B, Wu W-J, et al. Gene therapy with iNOS provides long-term protection against myocardial infarction without adverse functional consequences. Am J Physiol Heart Circ Physiol 2006;290:H584–9.

[89] Dawson TA, Li D, Woodward T, Barber Z, Wang L, Paterson DJ. Cardiac cholinergic NO-cGMP signaling following acute myocardial infarction and nNOS gene transfer. Am J Physiol Heart Circ Physiol 2008;295: H990–8.

[90] Smith Jr. RS, Agata J, Xia C-F, Chao L, Chao J. Human endothelial nitric oxide synthase gene delivery protects against cardiac remodeling and reduces oxidative stress after myocardial infarction. Life Sci 2005;76:2457–71.

[91] West NEJ, Qian H, Guzik TJ, Black E, Cai S, George SE, et al. Nitric oxide synthase (nNOS) gene transfer modifies venous bypass graft remodeling. Effects on vascular smooth muscle cell differentiation and superoxide production. Circulation 2001;104:1526–32.

[92] Ohta S, Komori K, Yonemitsu Y, Onohara T, Matsumoto T, Sugimachi K. Intraluminal gene transfer of endothelial cell-nitric oxide synthase suppresses intimal hyperplasia of vein grafts in cholesterol-fed rabbit: a limited biological effect as a result of the loss of medial smooth muscle cells. Surgery 2002;131:644–53.

[93] Pfeiffer T, Wallich M, Sandmann W, Schrader J, Gödecke A. Lipoplex gene transfer of inducible nitric oxide synthase inhibits the reactive intimal hyperplasia after expanded polytetrafluoroethylene bypass grafting. J Vasc Surg 2006;43:1021–7.

[94] Ylä-Herttuala S, Markkanen JE, Rissanen TT. Gene therapy for ischemic cardiovascular diseases: some lessons learned from the first clinical trials. Trends Cardiovasc Med 2004;14:295–300.

[95] Namba T, Koike H, Murakami K, Aoki M, Makino H, Hashiya N, et al. Angiogenesis induced by endothelial nitric oxide synthase gene through vascular endothelial growth factor expression in a rat hindlimb ischemia model. Circulation 2003;108:2250–7.

[96] Onoue H, Tsutsui M, Smith L, Stelter A, O'Brien T, Katusic ZS. Expression and function of recombinant endothelial nitric oxide synthase gene in canine basilar artery after experimental subarachnoid hemorrhage. Stroke 1998;29:1959–66.

[97] Tsutsui M, Onoue H, Iida Y, Smith L, O'Brien T, Katusic ZS. Effects of recombinant eNOS gene expression on reactivity of small cerebral arteries. Am J Physiol Heart Circ Physiol 2000;278:H420–7.

[98] Khurana VG, Smith LA, Weiler DA, Springett MJ, Parisi JE, Meyer FB, et al. Adenovirus-mediated gene transfer to human cerebral arteries. J Cereb Blood Flow Metab 2000;20:1360–71.

[99] Khurana VG, Smith LA, Baker TA, Eguchi D, O'Brien T, Katusic ZS. Protective vasomotor effects of in vivo recombinant endothelial nitric oxide synthase gene expression in a canine model of cerebral vasospasm. Stroke 2002;33:782–9.

[100] Janssens SP, Bloch KD, Nong Z, Gerard RD, Zoldhelyi P, Collen D. Adenoviral-mediated transfer of the human endothelial nitric oxide synthase gene reduces acute hypoxic pulmonary vasoconstriction in rats. J Clin Investig 1996;98:317–24.

[101] Budts W, Pokreisz P, Nong Z, Van Pelt N, Gillijns H, Gerard R, et al. Aerosol gene transfer with inducible nitric oxide synthase reduces hypoxic pulmonary hypertension and pulmonary vascular remodeling in rats. Circulation 2000;102:2880–5.

[102] Zhang F, Wu S, Lu X, Wang M, Liu M. Gene transfer of endothelial nitric oxide synthase attenuates flow-induced pulmonary hypertension in rabbits. Ann Thorac Surg 2008;85:581–5.

[103] Luo J-D, Wang Y-Y, Fu W-L, Wu J, Chen AF. Gene therapy of endothelial nitric oxide synthase and manganese superoxide dismutase restores delayed wound healing in type 1 diabetic mice. Circulation 2004;110:2484–93.

[104] Huerta S, Chilka S, Bonavida B. Nitric oxide donors: novel cancer therapeutics. Int J Oncol 2008;33:909–27.

[105] Bonavida B, Baritaki S, Huerta-Yepez S, Vega MI, Jazirehi AR, Berenson J. Nitric oxide donors are a new class of anti-cancer therapeutics for the reversal of resistance and inhibition of metastasis nitric oxide (NO) and cancer. In: Bonavida B, editor. Nitric oxide and cancer: prognosis, prevention and therapy. New York, NY: Springer; 2010. p. 459–77.

[106] Millet A, Bettaieb A, Renaud F, Prevotat L, Hammann A, Solary E, et al. Influence of the nitric oxide donor glyceryl trinitrate on apoptotic pathways in human colon cancer cells. Gastroenterology 2002;123:235–46.

[107] Postovit L-M, Adams MA, Lash GE, Heaton JPW, Graham CH. Nitric oxide-mediated regulation of hypoxia-induced B16F10 melanoma metastasis. Int J Cancer 2004;108:47–53.

[108] Frederiksen LJ, Sullivan R, Maxwell LR, Macdonald-Goodfellow SK, Adams MA, Bennett BM, et al. Chemosensitization of cancer in vitro and in vivo by nitric oxide signaling. Clin Cancer Res 2007;13:2199–206.

[109] Wang F, Zhang R, Xia T, Hsu E, Cai Y, Gu Z, et al. Inhibitory effects of nitric oxide on invasion of human cancer cells. Cancer Lett 2007;257:274–82.

[110] Kitamura Y, Kamoshima W, Shimohama S, Nomura Y, Taniguchi T. Nitric oxide donor-induced p53-sensitive cell death is enhanced by Bcl-2 reduction in human neuroblastoma cells. Neurochem Int 1998;32:93–102.

[111] Kamoshima W, Kitamura Y, Nomura Y, Taniguchi T. Possible involvement of ADP-ribosylation of particular enzymes in cell death induced by nitric oxide-donors in human neuroblastoma cells. Neurochem Int 1997;30:305–11.

[112] Janssens MY, Verovski VN, Van den Berge DL, Monsaert C, Storme GA. Radiosensitization of hypoxic tumour cells by S-nitroso-N-acetylpenicillamine implicates a bioreductive mechanism of nitric oxide generation. Br J Cancer 1999;79:1085–9.

[113] Jeon HK, Choi Su, Jung NP. Association of the ERK1/2 and p38 kinase pathways with nitric oxide-induced apoptosis and cell cycle arrest in colon cancer cells. Cell Biol Toxicol 2005;21:115–25.

[114] Liu Q, Chan STF, Mahendran R. Nitric oxide induces cyclooxygenase expression and inhibits cell growth in colon cancer cell lines. Carcinogenesis 2003;24:637–42.

[115] Inoue S, Kawanishi S. Oxidative DNA damage induced by simultaneous generation of nitric oxide and superoxide. FEBS Lett 1995;371:86–8.

[116] Dairou J, Atmane N, Rodrigues-Lima F, Dupret J-M. Peroxynitrite irreversibly inactivates the human xenobioticmetabolizing enzyme arylamine N-acetyltransferase 1 (NAT1) in human breast cancer cells. J Biol Chem 2004;279:7708–14.

[117] Oh-hashi K, Maruyama W, Yi H, Takahashi T, Naoi M, K-i Isobe. Mitogen-activated protein kinase pathway mediates peroxynitrite-induced apoptosis in human dopaminergic neuroblastoma SH-SY5Y cells. Biochem Biophys Res Commun 1999;263:504–9.

[118] Keefer LK. Fifty years of diazeniumdiolate research. From laboratory curiosity to broad-spectrum biomedical advances. ACS Chem Biol 2011;6:1147–55.

[119] Simeone A-M, Colella S, Krahe R, Johnson MM, Mora E, Tari AM. N-(4-Hydroxyphenyl)retinamide and nitric oxide pro-drugs exhibit apoptotic and anti-invasive effects against bone metastatic breast cancer cells. Carcinogenesis 2006;27:568–77.

[120] Janssens MY, Van den Berge DL, Verovski VN, Monsaert C, Storme GA. Activation of inducible nitric oxide synthase results in nitric oxide-mediated radiosensitization of hypoxic EMT-6 tumor cells. Cancer Res 1998;58:5646–8.

[121] Jarry A, Charrier L, Bou-Hanna C, Devilder M-C, Crussaire V, Denis MG, et al. Position in cell cycle controls the sensitivity of colon cancer cells to nitric oxide-dependent programmed cell death. Cancer Res 2004;64:4227–34.

[122] Srisook K, Jung N-H, Kim B-R, Cha S-H, Kim H-S, Cha Y-N. Heme oxygenase-1-mediated partial cytoprotective effect by NO on cadmium-induced cytotoxicity in C6 rat glioma cells. Toxicol In Vitro 2005;19:31–9.

[123] Niziolek M, Korytowski W, Girotti AW. Chain-breaking antioxidant and cytoprotective action of nitric oxide on photodynamically stressed tumor cells. Photochem Photobiol 2003;78:262–70.

[124] Kielbik M, Szulc-Kielbik I, Nowak M, Sulowska Z, Klink M. Evaluation of nitric oxide donors impact on cisplatin resistance in various ovarian cancer cell lines. Toxicol In Vitro 2016;36:26–37.

[125] Pervin S, Singh R, Chaudhuri G. Nitric oxide-induced cytostasis and cell cycle arrest of a human breast cancer cell line (MDA-MB-231): potential role of cyclin D1. Proc Natl Acad Sci U S A 2001;98:3583–8.

[126] Pervin S, Singh R, Freije WA, Chaudhuri G. MKP-1-induced dephosphorylation of extracellular signal-regulated kinase is essential for triggering nitric oxide-induced apoptosis in human breast cancer cell lines. Cancer Res 2003;63:8853–60.

[127] Pervin S, Singh R, Gau C-L, Edamatsu H, Tamanoi F, Chaudhuri G. Potentiation of nitric oxide-induced apoptosis of MDA-MB-468 cells by farnesyltransferase inhibitor. Cancer Res 2001;61:4701–6.

[128] Maciag AE, Holland RJ, Robert Cheng YS, Rodriguez LG, Saavedra JE, Anderson LM, et al. Nitric oxide-releasing prodrug triggers cancer cell death through deregulation of cellular redox balance. Redox Biol 2013;1:115–24.

[129] Maciag AE, Chakrapani H, Saavedra JE, Morris NL, Holland RJ, Kosak KM, et al. The nitric oxide prodrug JS-K is effective against non-small-cell lung cancer cells in vitro and in vivo: involvement of reactive oxygen species. J Pharmacol Exp Ther 2011;336:313–20.

[130] Kiziltepe T, Hideshima T, Ishitsuka K, Ocio EM, Raje N, Catley L, et al. JS-K, a GST-activated nitric oxide generator, induces DNA double-strand breaks, activates DNA damage response pathways, and induces apoptosis in vitro and in vivo in human multiple myeloma cells. Blood 2007;110:709–18.

[131] Hou Y, Wang J, Andreana PR, Cantauria G, Tarasia S, Sharp L, et al. Targeting nitric oxide to cancer cells: cytotoxicity studies of glyco-S-nitrosothiols. Bioorg Med Chem Lett 1999;9:2255–8.

[132] Damodaran VB, Place LW, Kipper MJ, Reynolds MM. Enzymatically degradable nitric oxide releasing S-nitrosated dextran thiomers for biomedical applications. J Mater Chem 2012;22:23038–48.

[133] Reynolds MM, Witzeling SD, Damodaran VB, Medeiros TN, Knodle RD, Edwards MA, et al. Applications for nitric oxide in halting proliferation of tumor cells. Biochem Biophys Res Commun 2013;431:647–51.

[134] Reynolds MM, Witzeling SD, Damodaran VB, Jarigese DM, Edwards MA, Knodle RD, et al. Application of a nitric oxide-releasing pro-drug for halting growth of human breast and canine mammary carcinoma cells. J Vet Sci Med 2012;1:6–12.

[135] Stevens EV, Carpenter AW, Shin JH, Liu J, Der CJ, Schoenfisch MH. Nitric oxide-releasing silica nanoparticle inhibition of ovarian cancer cell growth. Mol Pharm 2010;7:775–85.

CHAPTER

2

Nitric Oxide Donors Sensitize Resistant Cancer Cells to Apoptosis Induced by Chemotherapy: Molecular Mechanisms of Sensitization

Benjamin Bonavida

University of California at Los Angeles, Los Angeles, CA, United States

Abstract

The role of nitric oxide (NO) in cancer has been the subject of many contradictory reports. Both NO-mediated protumorigenic effects and antitumorigenic effects have been reported. It has been proposed that the dual contradictory effects by NO may be due to both the levels and the tumor type. High levels of NO are shown to be antitumorigenic in several cancer cell lines examined. We and others have reported that high levels of NO donors sensitized chemotherapy-resistant cell lines to drug-induced cell death and apoptosis. Analyses were performed to delineate the underlying potential mechanisms of NO-mediated chemosensitization to various chemotherapeutic drugs. Treatment of resistant cell lines with NO donors (e.g., DETANONOate) resulted in the inhibition of the constitutively hyperactivated proliferative/antiapoptotic NF-κB signaling pathway. The NF-κB inhibition was the result of S-nitrosylation of p50 and p65 polypeptide chains of NF-κB. The inhibition of NF-κB by NO, thus, resulted in also the inhibition downstream of several antiapoptotic gene products. Further analyses identified in cancer cells a dysregulated NF-κB/Snail/YY1/RKIP/PTEN loop that was shown to regulate cell survival, cell growth, EMT/metastasis, and resistance to both chemo- and immunotherapeutic drugs. Treatment with the NO donors resulted in the inhibition of NF-κB, Snail, and Yin Yang 1 (YY1), an induction of the repressed Raf-1-kinase inhibitor protein (RKIP) and PTEN expressions. Hence, the treatment of drug-resistant cancer cells with NO resulted in the disruption of this loop on one hand and sensitization to drug-induced apoptosis on the other hand. We, then, examined the individual role of each of the gene products of the loop in the regulation of chemoresistance following NO treatment and their role in the reversal of resistance. The role of YY1 in chemosensitization was shown to result in both the inhibition of YY1 transcription (due to NF-κB inhibition) and the S-nitrosylation of YY1 that prevented its DNA-binding activity. Treatment of drug-resistant tumor cells with siRNA YY1 sensitized the cells to drug-induced apoptosis, mimicking the treatment with NO donors. The inhibition of Snail by NO was the result of both the inhibition of NF-κB and the inhibition of YY1 that also regulate Snail transcription. The role of Snail in the regulation of resistance was demonstrated by treatment of the cancer cells with siRNA Snail and mimicking NO-mediated

http://dx.doi.org/10.1016/B978-0-12-811020-1.00002-8

chemosensitization. The induction of RKIP expression by NO treatment resulted from the inhibition of its repressor Snail. The direct role of RKIP in chemosensitization was shown following overexpression of RKIP in drug-resistant cells and resulting in chemosensitization similar to NO. Likewise, the induction of PTEN by NO was the result, in part, of the inhibition of its repressor YY1 by NO. The induction of PTEN inhibited the activity of the PI3K/AKT survival/antiapoptotic pathway and sensitized the cells to drug-induced apoptosis.

Overall, treatment of drug-resistant cells with NO interfered with the dysregulated survival antiapoptotic NF-κB/Snail/YY1/RKIP/PTEN loop, and each modified gene product expression in the loop was involved in the regulation of chemoresistance. The sensitization of drug-resistant tumor cells by NO in vitro was corroborated in vivo in mice bearing human tumor xenografts treated with a combination of an NO donor and a chemotherapeutic drug. This treatment resulted in the inhibition of tumor growth and tumor tissues derived from the treated mice examined ex vivo showed the modification of the dysregulated loop that was observed in vitro with the cell lines. It is of interest, therefore, that the above findings in vivo and in vitro with tumor cell lines need to be validated in tumor specimens from patients treated with NO donors and chemotherapy. In addition, studies are warranted to examine the clinical significance of a number of NO donors in vivo in patients that have tumor-specific chemosensitizing activities and minimal toxic side effects. It is also suggested to design NO-drug conjugates that are specifically targeted to tumor cells via antibody/ligand incorporated into the NO-drug complex.

Abbreviations

ABCs ATP-binding cassettes
EMT epithelial mesenchymal transition
GTN glyceryl trinitrate
IAP inhibitor of apoptosis
MDR multidrug resistance
NO nitric oxide
NOS nitric oxide synthase
PEM premetexel
RKIP Raf-1-kinase inhibitor protein
TME tumor microenvironment
YY1 Yin Yang 1

INTRODUCTION

It is well recognized that cancer is not a single disease entity, but it consists of a large number of distinct diseases with both common and unique properties. Therefore, most current conventional anticancer therapeutics are minimally effective against all cancers and are primarily restricted to certain cancers and, most likely, more restricted on particular stages of a single cancer. Thus, although significant objective clinical responses have been achieved, there have been seldom cures. The major challenges faced to date in cancer treatments are the inherent and adaptive resistances of cancer cells to particular treatments and the development of cross-resistance to different unrelated therapeutics.

While the initial development and the application of various anticancer nonimmune therapeutics (e.g., chemotherapy, hormonal therapy, and radiation) have not been tumor-specific; however, they share with tumor-specific immunotherapies a common denominator by inducing cell death by apoptosis and/or necrosis. Thus, the resistance of cancer cells to drug-induced apoptosis is a hallmark mechanism of resistance.

MECHANISMS OF TUMOR CELL RESISTANCE TO CHEMOTHERAPEUTICS

Clinically, the response rates of patients with tumor progression or relapse are dismal. Deciphering the underlying mechanisms of resistance can lead to the development of novel therapeutics. Also, the rate of resistance to targeted agents has been experienced in many patients and continues to increase. Importantly, the malignant variants that relapsed after the initial treatment are not only resistant to the initial treatment but also cross resistant to other therapeutics used alone or in combination. Garraway and Jänne (2012) described four main categories of resistance to targeted therapeutics, namely, (1) secondary genetic alterations in drug targets (e.g., mutation in the drug contact residues, altered conformation of the target, amplification, and alternative splice forms), (2) bypass mechanisms (activation of paralleled signaling pathways), (3) alterations in upstream or downstream effectors, and (4) pathway-independent (epithelial mesenchymal transition (EMT), changes in TME, and altered angiogenesis) [1].

Tumor cells develop various mechanisms to escape the antiproliferative/cytotoxic stimuli induced by chemotherapeutic drugs. One good example is the constitutively hyperactivated pathways that govern tumorigenesis, angiogenesis, inflammation, immunomodulation, metastasis, and drug resistance [2–4].

While chemotherapy remains one of the most common modes of treatment for various cancers, whether used alone as a single drug or in combination, its clinical effectiveness is limited due to the resistant phenotype of cancers in general. Both innate and adaptive chemoresistance have been characterized. The innate or intrinsic resistance is the resistance before treatment due to the presence of a subset of resistant cancer stem cells (CSCs) and of various resistant factors in the tumor cells that make them resistant to chemotherapy. The adaptive or acquired resistance develops following treatment as a result of various mechanisms such as mutations, overexpression of the therapeutic targets, and alternative compensatory signaling pathways [5]. Also, resistance can be developed through drug-induced selective pressure of a preexisting minor subpopulation of cells that was present in the heterogeneous tumor cell population [6].

Several advances have been made using molecular (genomic and proteomic) and functional analytic methods to identify novel gene products in signaling pathways that inhibit the responsiveness of tumor cells to a specific drug treatment. In addition, the adaptation of high-throughput techniques in combination with bioinformatic approaches have enabled to interrogate clinical samples and molecular signatures and genotypes that predict responses to specific drugs. Furthermore, such above approaches have been beneficial and allow the development of new therapeutic targets to overcome resistance.

Among the various mechanisms of drug resistance that have been documented experimentally and clinically, they include the following:

Drug Efflux

Drug efflux is mediated by the ATP-binding cassette (ABC) transporter family of transmembrane proteins that regulate the flux across the plasma membrane for a variety of drugs that are structurally unrelated. Among the 49 members of the ABC family, only 3 have been studied that regulate the multidrug resistance (MDR) and consist of multidrug resistance protein 1 (MDR1; *p*-glycoprotein; ABCB1), MDR-associated protein 1 (MRP1; ABCC1), and breast-cancer-resistant protein (BCRP; ABCG2). These overlapping efflux systems promote the elimination of various hydrophobic drugs including taxanes, topoisomerase inhibitors, and antimetabolites [5,7].

Alterations in Drug Metabolism

Mechanisms of drug inactivation or the lack of drug activation that are specific for every drug have also been observed for platinum [8], for 5-fluorouracil and methotrexate [9,10], and for epigenetic silencing for capecitabine [7,11].

Mutation of Drug Targets

Alterations of drug targets by mutations or expression levels can lead to drug resistance [12–14]. Chemotherapeutic drugs induce DNA damage, and the cell response is under repair mechanisms. The role of DNA damage repair mechanisms in drug resistance has been reviewed by Bell et al. [15].

Activation of Survival Pathways

Several studies reported the activation of survival pathways that are involved in drug resistance [16–18].

Inactivation of Death-Signaling Pathways

Resistance to drug-induced apoptosis is a hallmark of cancer and is responsible, in large part, for drug resistance. A small number of antiapoptotic proteins has been implicated in the resistance [19,20]. The most relevant ones are the Bcl2 family of proteins, the inhibitors of apoptosis (IAPs), and the caspase inhibitor FLIP. Amplification, mutation, chromosomal translocation, and overexpression of these above gene products have been associated with various malignancies and linked to resistance to chemotherapy [21,22].

Epigenetic Changes

Cancer drug resistance can also be manifested as a result of epigenomics by methylation, acetylation, and altered levels of microRNAs leading to alterations in both upstream and downstream effectors involved in drug toxicity [23–25]. An example of the role of epigenetic mechanisms in drug resistance was reported by Borley and Brown [26]. Patients with epithelial ovarian cancer are resistant to chemotherapies and usually succumb to the disease. In their review, the authors report on the roles of DNA methylation, histone modification, and microRNAs and their associations with chemotherapy resistance in ovarian cancer.

Influence of the Tumor Microenvironment

In solid tumors, the tumor microenvironment (TME) consists of extracellular matrix, cancer-associated fibroblasts, inflammatory cells, blood vessels, and immune cells. Histological analysis of the TME in hematologic malignancies revealed that it consists of bone marrow stromal cells, bone marrow endothelial cells, osteoclasts, osteoblasts, macrophages, and T cells. These various factors protect the tumor cells for any drug-induced effect [23,24,27].

Presence of CSCs

In the majority of tumors, there exists a small subpopulation of CSCs that have the capabilities of self-renewal, dedifferentiation, and tumor heterogeneity and with an inherent resistance to chemotherapy. The CSCS subset remains a challenge to current treatments with chemotherapies [28,29]. Given the properties of CSCS, significant efforts have been placed to develop new drugs targeting CSCS, and many of these have progressed into clinical trials [30].

Molecular and Genetic Heterogeneity

Most tumors studied were documented to be heterogeneous. The recent development of rapid analyses of individual cancer genomes have advanced the field and demonstrated intertumor heterogeneity shared among tumors of the same histological type. In addition, evidence also supports the intratumor genetic heterogeneity found within individual tumor biopsies. This intratumor genetic heterogeneity evolves during the disease's course. Of interest, is the finding that cancer therapeutics may augment intratumor heterogeneity and resulting in a poor clinical response to chemotherapies through the development of various resistance mechanisms [6].

OVERCOMING RESISTANCE TO CHEMOTHERAPY

Several mechanisms have been reported to overcome chemotherapy resistance [22]. Briefly below, some strategies that have been used to overcome resistance.

Inhibitors of Drug Efflux

The development of inhibitors of drug efflux whereby they were effective in in vitro studies; however, they were clinically disappointing. For example, the MDR1 inhibitors zosuquidar and tariquidar are of high potency and specific. However, they did not provide any clinical benefits in breast cancer patients [31,32].

Drug Activation and Inactivation

One example is platinum drugs that need to be inactivated by glutathione [8] or the absence of required enzymes to convert the antimetabolites (e.g., 5-FU and methotrexate for their active forms) [9,10].

Alteration in Drug Targets

These can occur by mutations or overexpression [22].

DNA Damage Repair

The use of inhibitors to DNA repair mechanisms is a potential therapeutic target. Such agents have been developed such as inhibitors of the single-stranded break DNA repair enzyme poly(ADP-ribose) polymerase 1 [33].

Downstream Resistance Mechanisms

The process of transformation results in dysfunctional apoptotic pathways, a hallmark of cancer [34]. Pharmacological inhibitors directed against antiapoptotic Bcl2 family members have been developed. For example, ABT-737 or its orally bioavailable form ABT-263 are inhibitors for Bcl2, Bcl-xL, and BCLW and promote the proapoptotic proteins Bax and Bak [35]. Also, obatoclax has high affinity for anti-BCL2 family members and signals for apoptosis [36]. In addition, Smac mimetic drugs inhibit XIAP and cIAPs and promote the activation of caspases 3, 7, 8, and 9 and sensitize various types of tumors to chemotherapy [37].

Inhibition of Prosurvival Signaling/Pathways

Most tumors exhibit constitutively hyperactivated survival/antiapoptotic pathways that render the tumor cells highly resistant to chemotherapy. Thus, several chemical inhibitors have been developed and have been approved clinically targeting these pathways and include growth factor signaling, the Raf-MEK-ERK pathway, the NF-κB pathway, and the PI3K/AKT pathway. However, while these inhibitors showed significant clinical responses, there is the emergence of drug resistance (initially responsive cancers have led to only a marginal benefit). For example, inhibitors of the RAS/ERK pathway with the clinically approved Raf and MEK inhibitors have improved the clinical outcome of B-Raf mutated melanoma; however, drug resistance emerges and limits their efficacies. Several drug-resistant mechanisms have been discovered, including the expression of drug-resistant Raf isoforms, alterations of downstream factors, overexpression of upstream components of the cascade, and activation of alternative pathways [38,39]. The PI3K-AKT-mTOR pathway plays a major role in tumor cell growth, proliferation, and survival. There are many mTOR inhibitors used clinically; however, resistance to mTOR inhibitors have occurred [40]. Colorectal cancers overexpress EGFR and VEGF, and those have been associated with a poor outcome. The use of targeted therapies, such as EGFR tyrosine kinase inhibitors, EGFR monoclonal antibodies, and VEGF inhibitors, has improved survival in these colorectal cancer patients.

However, despite the efficacy of those inhibitors, resistance develops invariably within 1–2 years. The resistant mechanisms include resistance of drug targets and activation of alternative survival pathways [41]. Likewise, inhibition of the NF-κB pathway has been evaluated as they have also been effective targeting CSCS; here, also resistance does occur [42].

NITRIC OXIDE AND CANCER

General Properties

Nitric oxide (NO) is an ubiquitous molecule that has been reported to exert a large number of biological effects [43]. The free radical NO is an uncharged molecule that contains an impaired electron in its outermost orbital, allowing it to mediate several chemical reactions and acting as a weak oxidant (electron donor) or as an antioxidant (electron acceptor). NO reacts with inorganic molecules (oxygen, superoxides, and transition metals), DNA (pyrimidine bases), prosthetic groups (like heme), and proteins, leading to S-nitrosylation of thiol groups, nitration of tyrosine residues, or disruption of metal sulfide clusters such as zinc finger domains or iron sulfide complexes [44]. In addition, NO can function as an antioxidant against reactive oxygen species (ROS) such as hydrogen peroxide and superoxide. NO reacts with ROS such as superoxide to form peroxynitrite ($ONOO^-$). NO reacts with the thiol side chain of cysteine residues with proteins and peptides leading to S-nitrosylation and plays a significant role in the ubiquitous role of NO in cellular signal transduction [45].

L-Arginine is the amino acid essential for the biosynthesis of NO. The oxidation of L-arginine is catalyzed by a family of enzymes such as nitric oxide synthases (NOS) [46]. Three major isoforms of NOS have been characterized, namely, inducible NOS (NOS II/iNOS), endothelial NOS (NOSIII/eNOS), and neuronal NOS (NOSI/nNOS). Both NOSI and NOSIII are constitutively expressed and are calcium/calmodulin-dependent and require several cofactors for their enzymatic activities [47]. The calcium/calmodulin-independent iNOS is found in various cell types and in a variety of cancer cells. iNOS is induced by cytokines and microbial products and is responsible for the generation of high levels of NO sustained over long periods of time [43].

NO is involved in the regulation of fundamental, physiological processes such as vasodilation, neurotransmission, inflammation, and cell death via posttranslational modification of proteins. This is achieved via NO binding to soluble guanylyl cyclase (sGC)-coupled by cytosolic receptors and resulting in conformational changes that trigger sGC activity and the consequent generation of cGMP from GTP and, thus, activating various downstream targets (protein kinases, phosphodiesterases, and ion channels) [43]. NO signaling can also be cGMP-dependent and cGMP-independent [48,49].

NO regulates mitochondrial respiration through binding to cytochrome *c* oxidase, the terminal enzyme in the mitochondrial electron transport chain [50]. In addition, NO also plays a role in programmed cell death or apoptosis [51]. Depending on the concentration of NO, high levels promote apoptosis, while low levels can result in the resistance to apoptosis [43,52,53].

NO in Health and Diseases

NO is involved in various aspects of inflammation and immunity. Several cell types (fibroblasts, endothelial, epithelial cells, keratinocytes, chondrocytes, monocytes, macrophages, antigen-presenting cells, and NK cells) all express NO [54]. NO is also involved in cardiovascular regulation such as vascular tone (vasodilation), vascular structure (inhibition of smooth muscle cell proliferation), and cell-cell interactions on blood vessels [55]. NO also regulates the CNS [56]. In addition, NO is also involved in a variety of diseases of the immune system, including chronic asthma, rheumatoid arthritis, and autoimmunity [57].

NO Signaling

NO signaling encompasses three molecular categories, namely, (1) S-nitrosylation of thiol groups, (2) peroxynitrite generation and nitrotyrosine formation, and (3) donation of electrons through transition metals (copper, zinc, and iron) [58,59]. NO can modulate hypoxia-induced phenotypes by mimicking oxygen. It can inhibit hypoxic upregulation of genes (*EPO*, *ET1*, and *VGF*), and NO donors decrease the hypoxic accumulation of HIF-1-α [60–62]. NO mimetics inhibit hypoxia-induced resistance to doxorubicin and 5-FU in a variety of tumor cell types [63,64].

PRO AND ANTI TUMOROGENIC ACTIVITIES

The activities of NO on tumor cells are variable and cover both sides of the spectrum as antineoplastic and proneoplastic. The overall effect of NO depends on the TME, the type of tumor cells exposed to NO, the redox state of the reaction, the intracellular concentration, and the duration of NO exposure. The dual roles of NO in cancer have been reported [65,66].

The Protumorigenic Effect of NO

Depending on levels, NO can induce DNA damage alone or through its interaction with oxygen and superoxide, and initiate carcinogenesis [67]. DNA single-strand breaks in DNA results in the formation of NO-mediated basic sites (as a result of DNA deamination and/or via direct effect of peroxynitrite) [68]. NO can also cause DNA damage indirectly through the generation of nitrosamines that are prominent carcinogens [69].

Most growing tumors in vivo survive in hypoxic conditions, and under hypoxic conditions, many genes are induced, such as iNOS that promote tumor cell growth, survival, and metastasis [59,63,70,71].

Tumor progression can be significantly influenced by NO. Solid tumor cells and host cells express active NOS isoforms. Thomsen and Miles [72] reported that NOS activity was higher in malignant tissues compared with benign tissues, and there was a correlation between tumor progression and NOS activity. Also, NOS plays a direct role in metastasis [67]. NO was reported to stimulate angiogenesis through its role in triggering proliferation and migration of endothelial cells [73,74].

The Antitumorigenic Activity of NO

Activated macrophages synthesize nitrite and nitrate that led to tumor cell cytotoxicity [75]. High levels of NO exert antitumor effects by inhibiting cell growth and induction of apoptosis and also sensitized resistant tumor cells to various cytotoxic drugs. Xie et al. [76] reported in a murine melanoma model that overexpression of NOS II resulted in poor tumor growth and survival of the mice. The mechanisms of NO-mediated apoptosis in cancer cells include the accumulation of p53, the inhibition of mitochondrial respiration, the alteration of Bcl2 family members' expression, the activation of caspase signaling, and the DNA damage [57,77,78].

NO DONORS IN CANCER

Several NO donors in cancer have been reviewed by Huerta et al. [79]. The major classes of NO donors in cancer include organic nitrates, metal complexes (sodium nitroprusside), *S*-nitrosothiols, syndonimines, diazeniumdiolates (NONOates), and NO-drug hybrids.

Organic nitrates are a class of NO donors in clinical use [80] and include glyceryl trinitrate (GTN) and isosorbide (ISDN). GTN has been used for over a century for the relaxation of coronary vessels to improve chest pain [81]. In mice, treatment with GTN inhibited tumor-induced metastasis in the lungs of nude mice [59]. Under hypoxic conditions, GTN sensitized prostate cancer cells to doxorubicin [64]. These same authors also reported the potential role of GTN as a sensitizing agent in vivo by transdermal application in a murine model of prostate cancer. Likewise, nitroglycerin (NTG) patches have a strong effect in patients with SCLC in combination with chemotherapeutic drugs [82,83].

Metal NO Complexes Such As SNP

NO release requires mediation or a thiol-mediated reaction [80,84]. T-lymphoma cells incubated with SNP resulted in the induction of apoptosis as a consequence of the inhibition of NF-κB and Bcl2 [85].

S-Nitrosothiols (RSNO)

These donors are typically unstable; however, two relatively stable compounds are *S*-nitroso-*N*-acetylpenicillamine (SNAP) and *S*-nitrosoglutathione (GSNO) [80]. Treatment with SNAP induced potentiation in apoptosis in neuroblastoma tumor cells as a result of the augmentation of p53 [86]. GSNO also induced apoptosis in colon cancer cell lines [87].

Sydnonimines

The most extensively studied compound in this class is 3-morpholinosydnonimine (SIN-1) [80]. SIN-1 exhibited antitumor activity in many cell lines as a consequence of NO-mediated inhibition of HIF-1-α [60].

NONOates

Under optimal conditions, this compound generates up to 2 mol of NO per 1 mol of NONOate [80]. The compounds in this group consist of (a) sodium-α-oxytrinitrite (Angele's salt), (b) DEA/NO, (c) PAPA/NO, (d) SPER/NO, (e) PROLI/NO, (f) MAHMA/NO, and (g) DETA/NO. The antitumor activities of these compounds have been reviewed elsewhere [79]. Additional antitumor activity by DETA/NO will be described in more detail below.

DETA/NO

DETA/NO sensitized Chinese hamster V79 lung fibroblasts and head and neck squamous cell carcinoma to cisplatin-mediated cytotoxicity [88,89]. In an in vivo B16 melanoma, DETA/NO in combination with cisplatin inhibited tumor growth [90].

NO-Drug Hybrids

The NO/NSAIDs compounds have been reviewed by Rigas and Kashfi [91]. In human ovarian cancer cells, the NO aspirin compound, NO-4016, sensitized the tumor cells to cisplatin-mediated cytotoxicity [92].

NO AS A CHEMOINHIBITOR

Melanoma tissues examined demonstrated that they expressed high levels of iNOS and nitrotyrosine [93]. These tumor cells generated high levels (nM) of NO intracellularly. Treatment of the tumor cells with inhibitors of either iNOS or NO sensitized the cells to CDDP apoptosis. The inhibition of S-nitrosylation enhanced the apoptotic effect mediated by CDDP. These investigators suggested that NO inhibited caspase 3 by S-nitrosylation and also inhibited prohydrolase (an enzyme that targets HIF1-α) and resulting in the inhibition of resistance. In another report by Sinha et al. [94], treatment of tumor cells with NO in combination with VP-16 resulted in the inhibition of VP-16-induced cell death. The interaction of NO with VP16 generated nitroso-VP16 that diminished VP16 topoisomerase-dependent DNA cleavage and inhibited cytotoxicity. Wongvaromb et al. [95] reported that long-term treatment (7–14 days) of lung cancer cells with NO rendered the cells resistant to various chemotherapeutic drugs. There was an increased survival via overexpression of caveolin-1 and Bcl2 and AKT activity. The removal of NO reversed the resistance.

NO AS A CHEMOSENSITIZER AND MECHANISMS OF CHEMOSENSITIZATION

Evig et al. [96] reported the treatment of MCF-7 breast cancer cells to an aqueous solution of NO delivered (as bolus) prior to treatment to doxorubicin resulted in tumor cell survival of only 5% as compared with survival of 40% with doxorubicin alone. Riganti et al. [97] reported that treatment of the doxorubicin-resistant epithelial colon cancer cell line (HT29-d) with NO donors (SNAP, GSNO, and SNP) or the induction of NO synthesis (atorvastatin) resulted in the reversal of resistance. This effect was mediated, in part, by tyrosine nitration of MDR-associated protein and the inhibition of drug efflux.

Studies in vitro demonstrated that treatment of cisplatin drug-resistant tumor cells with the NO donor DETA/NO resulted in the induction of apoptosis by cisplatin in a synergistic manner [98]. These findings extended prior studies on chemosensitization [99,100]. Analysis of the chemosensitization revealed that the treatment with DETA/NO inhibited the constitutively activated prosurvival/antiapoptotic signaling pathway mediated by NF-κB. The inhibition of NF-κB resulted downstream in the inhibition of the antiapoptotic resistant factors Yin Yang 1 (YY1) and Bcl-xL [101]. The inhibition of YY1 and Bcl-xL by treatment with siRNA sensitized the tumor cells to cisplatin apoptosis. Of interest, patient-derived human prostatic cancer tissues showed overexpression of both YY1 and Bcl-xL and the levels of expression increased as a function of tumor progression. The inhibition of NF-κB by DETA/NO was the result of NO-mediated nitrosylation of both p50 and p65 [102,103]. Of interest, treatment with siRNA YY1 or siRNA Bcl-xL resulted in the inhibition of both YY1 and Bcl-xL. The mechanism of inhibition of YY1 and Bcl-xL by either sRNA is not very clear. The in vitro findings with DETA/NO-induced chemosensitization to CDDP were corroborated in vivo in mice bearing human prostate tumor xenografts. Treatment of the mice with DETA/NO or CDDP alone had no antitumor effect. However, the combination treatment resulted in the inhibition of tumor growth. Tumor tissues from the treated mice showed the inhibition of both YY1 and Bcl-xL and, thus, corroborating the in vitro findings [104].

A study reported by Bratasz et al. [92] demonstrated that treatment of both sensitive and cisplatin-resistant ovarian cancer cell lines with the NO-releasing aspirin, NCX-4040 (2-(acetyloxy)benzoic acid 4-(nitrooxymethyl) phenyl ester), sensitized the cell lines to cisplatin-induced cell death, both in vitro and in vivo tumor xenografts models in mice. NCX-4040 belongs to a class of NO-releasing, nonsteroidal anti-inflammatory drugs (NO-NSAID). The NCX-4040 molecule contains an aspirin backbone attached to an ONO2 moiety linked via an aromatic molecular spacer. The NO moiety is required for its cancer activity [105,106]. The mechanism by which NCX-4040 mediates its effects is through the inhibition of thiol levels, as glutathione levels regulate cisplatin resistance. In addition, NCX-4040 inhibited EGFR and STAT-3 signaling and, hence, increased the levels of p53 and Bax and decreased the levels of Bcl-xL, survivin, and cyclin D1, resulting in the inhibition of proliferation and the induction of apoptosis by cisplatin.

A study reported by Kogias et al. [107] investigated the effect of an NO donor, PAPA/NO that required an enzymatic activation by glutathione-S-transferase (GST) to release NO [108]. Human glioma overexpressing GST was a model to examine the effect of PAPA/NO alone or in combination with chemotherapy. Both in vitro and in vivo models were examined. Treatment of glioma cells with carboplatin (CPT) or temozolomide (TMZ) inhibited cell proliferation and induced a significant synergy when combined with both TMZ and CPT. However, treatment with PAPA/NO intratumorally and intracranially inhibited tumor cell proliferation and induced apoptosis with no significant prolongation of survival. The inactivation of DNA repair enzymes by intracellular NO release might have been a factor in chemosensitization [109].

Reynolds et al. [110] have reviewed the potential role of NO as an anticancer therapeutic. In addition, NO interfered with the NF-κB/Snail/YY1/RKIP loop and also upregulated PTEN, leading to the inhibition of PI3K/AKT survival antiapoptotic pathway. Both Snail and YY1 negatively regulate PTEN transcription and expression. Inhibition of NF-κB, Snail, and YY1 by NO resulted in the derepression of PTEN and inhibition of antiapoptotic gene products regulated by the PI3K/AKT pathway [111].

Treatment of tumor cells with NO in combination with a multitargeted antifolate antineoplastic drug, premetexel (PEM), resulted in significant cytotoxicity. Furthermore, the combination of the NO donor, NOC-18, and GTN+EM inhibited tumor cell growth in vivo compared with treatment with PEM alone [112]. Gene transfection with the iNOS gene in combination with a low dose of CDDP resulted in tumor cell inhibition of proliferation, invasion, migration, and induction of apoptosis [113]. Colorectal cancer is resistant to a variety of chemotherapeutic drugs [114]. The colorectal tumor cell line SW480 and its metastatic variant SW620 were both derived from the same patient from a primary stage II and metastatic lymph node stage III, respectively [115]. Treatment of the drug-resistant SW620 tumor cells with DETA/NO sensitized the tumor cells to CDDP apoptosis. This treatment resulted in the upregulation of AIF expression that may be involved in the sensitization of the tumor cells to CDDP. Inhibition of AIF by siRNA AIF reversed DETA/NO sensitization to CDDP. The chemosensitization by DETA/NO and CDDP in vitro was corroborated in vivo in mice bearing SW620 tumor xenografts [116].

The NO donor JS-K generates NO by enzymatic activation with glutathione and GST. Glioma cells treated with JS-K resulted in the inhibition and proliferation and induction of apoptosis in vitro and the inhibition of tumor cell growth in mice [117].

The combination of DETA and farnesyl transferase inhibitors augmented apoptosis of breast cancer cells with minimal toxicity to normal breast cells [118]. Matthews et al. [63] reported the sensitization of drug-resistant breast cancer cells to 5-FU and doxorubicin-induced apoptosis. In a mouse model bearing human melanoma tumor xenografts, treatment intratumorally with DETA and i.p. with CDDP inhibited significantly tumor growth [90].

Cancer cells exhibit constitutively activated survival signaling pathways including NF-κB [119]. Downstream targets regulated by NF-κB include various antiapoptotic proteins such as YY1, Bcl-xL, and XIAP [120,121] and the metastasis inducer Snail [122]. Snail has been shown to repress the metastasis suppressor Raf-1-kinase inhibitor protein (RKIP) [123].

In most cancers, there is a dysregulation of the apoptotic signaling pathways that have been implicated in the pathogenesis, the progression, and the resistance to cytotoxic cells/drugs [124–126]. We have reported the presence of a dysregulated NF-κB/Snail/YY1/RKIP loop in many cancers that is involved in cell survival, proliferation, metastasis, and resistance [127–129].

Treatment of tumor cells with DETA/NO interferes with the dysregulated loop by inhibiting NF-κB and downstream targets YY1 and Snail while derepressing Snail-mediated induction of RKIP. Further, NO treatment reversed drug resistance of tumor cells when used in combination with chemotherapeutic drugs [127,130].

Every gene product of the loop, by itself, was shown to regulate the resistance to chemotherapy. Snail, a member of the Snail super family of zinc finger transcription factors, is transcriptionally regulated by NF-κB and YY1 [122,131]. Snail silencing sensitized drug-resistant tumor cells to apoptosis by chemotherapeutic drugs [127,132]. YY1 is a multifunctional DNA-binding protein that can activate, repress, or initiate transcription depending on the context in which it binds [133]. The role of YY1 expression in drug resistance was reported in which the inhibition of YY1 with siRNA YY1 sensitized the tumor cells to drug-induced apoptosis [134]. The RKIP is a member of the phosphatidylethanolamine-binding protein family. RKIP functions by inhibiting both the Raf/MEK/ERK and NFκ-B pathways [135–137]. RKIP overexpression reverses tumor cell resistance following treatment with chemotherapeutic drugs [127,134]. Both the inhibition of Snail by NO and treatment of tumor cells with sRNA Snail resulted in reversing drug resistance [102]. Treatment with DETA/NO resulted in the inhibition of YY1 transcription and its inactivation by S-nitrosylation [138]. Inhibition of YY1 sensitized the resistant tumor cells to CDDP-induced apoptosis [98]. Overexpression of tumor cells with RKIP resulted in the reversal of drug resistance and synergistic apoptotic activity following treatment with CDDP.

RRX-001 was designed to combine two structural entities: a dinitroazetidine and an α-bromoacetamide. Both are required for maximum activity. The bromoacetamide group serves to deplete glutathione, leading to increases in the ROS and RNS, and in addition, it conjugates to an endogenous carrier molecule, RBCi, by hemoglobin adduction. RRX-001 was effective in vitro and in vivo against multiple tumor cell lines and synergy with radiotherapy. The binding of RRX-001 to hemoglobin results in the release of NO. In a phase I clinical study, in patients with different tumors, treatment with RRX-001 resulted in a clinical benefit in 70% of the patients of different tumors. RRX-001 resensitized patients who failed prior therapies and a subset responded to subsequent chemotherapy and radiation post-RRX-001 [139–141].

Lung cancer is the leading cause of death from cancer worldwide [142]. The prognosis of patients with NSCLC is poor with a medium survival of 14–17 months and a 3-year survival of approximately 25% [143]. Treatment with radiotherapy has a minimal effect. One mechanism of tumor cell resistance to chemotherapy is the expression of HIF-1α, and it was reported that NO inhibits HIF-1-α by proteasomal degradation [144]. The authors hypothesized that NTG increases vascular flow and drug delivery in hypoxic tissues, and thus, its combination with chemotherapy may improve its cytotoxic effect [145,146]. The findings of this phase II nonrandomized clinical study demonstrated that the addition of NTG to induction chemotherapy and concurrent chemoradiotherapy in patients with NSCLC showed poor toxicity and improved clinical response.

Patients with stage II and III rectal cancer are treated with neoadjuvant chemoradiation (CRT). This treatment strategy leads to a pathological complete response (pCR) in 9%–37% of patients [147,148]. Yasuda et al. [82,83] in a pivotal phase II trial investigated the efficacy and safety of NTG plus vinorelbine and cisplatin in patients with previously untreated stages IIIB/IV NSCLC [82,83]. In this randomized trial, patients receiving NTG experienced a high response rate compared with patients who did not receive NTG. In rectal cancer, in a nonrandomized study, treatment with NTG patches in combination with standard CRT demonstrated safety and tolerability in the patients and poor toxicity. This study suggested a phase II randomized trial [149].

CONCLUDING REMARKS

In this review, I have discussed, in general, the role of NO in cancer, and I have presented potential molecular mechanisms by which NO interferes with the survival pathways in cancer cells and resulting in their modifications leading to the reversal of tumor cell resistance and their sensitization by the same or different chemotherapeutic drugs to which they were resistant. In addition, there was a discussion on the potential of additional therapeutic targets involved in the resistance of chemotherapy, for example, the role of NO-mediated modification of the dysregulated survival/antiapoptotic NF-κB/Snail/YY1/RKIP/PTEN loop and its role in chemoresistance [150]. It is also important to note that not only NO treatment reverses resistance but also its pleiotropic effects lead to its roles in the reversal of immune resistance and the inhibition of EMT and metastasis and its role in the inhibition of CSCs.

We have witnessed significant advances in the challenging and contradictory roles of NO in cancer and have resolved, in large part, the underlying mechanism of this controversy. We have also witnessed considerable advances in promoting the anticancer role mediated by NO and/or its metabolites in sensitizing drug-resistant tumor cells to cytotoxicity when used in combination. Several reports have conclusively demonstrated that NO donors can sensitize drug-resistant tumor cells to apoptosis when used in combination with a variety of chemotherapeutic drugs, both in vitro cell lines and in vivo mice bearing human tumor xenografts. These findings clearly established a new paradigm for pursuing the clinical application of NO donors in patients with resistant tumors when used in combination with chemotherapy. While a few clinical studies have demonstrated the therapeutic efficacy of NO donors in cancer patients, clearly, additional clinical and randomized studies need to be pursued for validation. Of similar interest, we have also witnessed the scientific interest in the development and application of new NO donors by biotech companies and the development of a battery of novel NO donors including NO-drug complexes and their potential applications. A new feature is being investigated, namely, the introduction of novel NO donors that are targeted specifically to the tumor via antibodies and/or ligands. Likewise, it is possible to generate immunotherapeutic agents such as genetically engineered tumor-specific cytotoxic T cells that can deliver toxic levels of NO to the tumor via their interactions with the tumor and also the development of CART cells that have been engineered to deliver toxic levels of NO.

Acknowledgments

The author acknowledges all of the investigators who have researched closely with him and have jointly published several studies dealing with the roles of NO in cancer, most of which were reviewed in this chapter. These investigators are doctors Hermes Garban, Fumiya Hongo, Sara Huerta-Yepez, Stavroula Baritaki, Sergio Huerta, and Valentina Rapozzi. The author also acknowledges the Jonsson Comprehensive Center at UCLA and prior support from the Department of Defense (DOD/US Army, DAMD-17-02-1-0023), the UCLA gene therapy program, and the UCLA SPORE in prostate cancer. The author also acknowledges the assistance of Leah Moyal in the preparation and formatting of this manuscript.

References

[1] Garraway LA, Jänne PA. Circumventing cancer drug resistance in the era of personalized medicine. Cancer Discov 2012;2(3):214–26.

[2] Siegfried Z, Bonomi S, Ghigna C, Karni R. Regulation of the Ras-MAPK and PI3K-mTOR signalling pathways by alternative splicing in cancer. Int J Cell Biol 2013;2013:568931.
[3] Chan CH, Jo U, Kohrman A, Rezaeian AH, Chou PC, Logothetis C, et al. Posttranslational regulation of Akt in human cancer. Cell Biosci 2014;4(1):59.
[4] Wicki A, Mandalà M, Massi D, Taverna D, Tang H, Hemmings BA, et al. Acquired resistance to clinical cancer therapy: a twist in physiological signaling. Physiol Rev 2016;96(3):805–29.
[5] Longley DB, Johnston PG. Molecular mechanisms of drug resistance. J Pathol 2005;205(2):275–92.
[6] Swanton C. Intratumor heterogeneity: evolution through space and time. Cancer Res 2012;72(19):4875–82.
[7] Gottesman MM, Fojo T, Bates SE. Multidrug resistance in cancer: role of ATP-dependent transporters. Nat Rev Cancer 2002;2(1):48–58.
[8] Meijer C, Mulder NH, Timmer-Bosscha H, Sluiter WJ, Meersma GJ, de Vries EG. Relationship of cellular glutathione to the cytotoxicity and resistance of seven platinum compounds. Cancer Res 1992;52(24):6885–9.
[9] Houghton JA, Houghton PJ. Elucidation of pathways of 5-fluorouracil metabolism in xenografts of human colorectal adenocarcinoma. Eur J Cancer Clin Oncol 1983;19(6):807–15.
[10] Schwartz PM, Moir RD, Hyde CM, Turek PJ, Handschumacher RE. Role of uridine phosphorylase in the anabolism of 5-fluorouracil. Biochem Pharmacol 1985;34(19):3585–9.
[11] Moutinho C, Martinez-Cardús A, Santos C, Navarro-Pérez V, Martínez-Balibrea E, Musulen E, et al. Epigenetic inactivation of the BRCA1 interactor SRBC and resistance to oxaliplatin in colorectal cancer. J Natl Cancer Inst 2014;106(1):djt322.
[12] Borst P, Elferink RO. Mammalian ABC transporters in health and disease. Annu Rev Biochem 2002;71:537–92.
[13] Fojo T, Bates S. Strategies for reversing drug resistance. Oncogene 2003;22(47):7512–23.
[14] Longley DB, Harkin DP, Johnston PG. 5-Fluorouracil: mechanisms of action and clinical strategies. Nat Rev Cancer 2003;3(5):330–8.
[15] Bell DW, Gore I, Okimoto RA, Godin-Heymann N, Sordella R, Mulloy R, et al. Inherited susceptibility to lung cancer may be associated with the T790M drug resistance mutation in EGFR. Nat Genet 2005;37(12):1315–6.
[16] Debatin KM, Krammer PH. Death receptors in chemotherapy and cancer. Oncogene 2004;23(16):2950–66.
[17] Sumitomo M, Asano T, Asakuma J, Horiguchi A, Hayakawa M. ZD1839 modulates paclitaxel response in renal cancer by blocking paclitaxel-induced activation of the epidermal growth factor receptor-extracellular signal-regulated kinase pathway. Clin Cancer Res 2004;10(2):794–801.
[18] Kishida O, Miyazaki Y, Murayama Y, Ogasa M, Miyazaki T, Yamamoto T, et al. Gefitinib ("Iressa", ZD1839) inhibits SN38-triggered EGF signals and IL-8 production in gastric cancer cells. Cancer Chemother Pharmacol 2005;55(4):393–403.
[19] Lowe SW, Cepero E, Evan G. Intrinsic tumour suppression. Nature 2004;432(7015):307–15.
[20] Letai AG. Diagnosing and exploiting cancer's addiction to blocks in apoptosis. Nat Rev Cancer 2008;8 (2):121–32.
[21] Sentman CL, Shutter JR, Hockenbery D, Kanagawa O, Korsmeyer SJ. Bcl-2 inhibits multiple forms of apoptosis but not negative selection in thymocytes. Cell 1991;67(5):879–88.
[22] Holohan C, Van Schaeybroeck S, Longley DB, Johnston PG. Cancer drug resistance: an evolving paradigm. Nat Rev Cancer 2013;13(10):714–26.
[23] Taylor ST, Hickman JA, Dive C. Epigenetic determinants of resistance to etoposide regulation of Bcl-XL and Bax by tumor microenvironmental factors. J Natl Cancer Inst 2000;92(1):18–23.
[24] Maier S, Dahlstroem C, Haefliger C, Plum A, Piepenbrock C. Identifying DNA methylation biomarkers of cancer drug response. Am J Pharm 2005;5(4):223–32.
[25] Rueff J, Rodrigues AS. Cancer drug resistance: a brief overview from a genetic viewpoint. Methods Mol Biol 2016;1395:1–18.
[26] Borley J, Brown R. Epigenetic mechanisms and therapeutic targets of chemotherapy resistance in epithelial ovarian cancer. Ann Med 2015;47(5):359–69.
[27] McMillin M, Grant S, Frampton G, Andry S, Brown A, DeMorrow S. Fractalkine suppression during hepatic encephalopathy promotes neuroinflammation in mice. J Neuroinflammation 2016;13(1):198.
[28] Valent P, Bonnet D, De Maria R, Lapidot T, Copland M, Melo JV, et al. Cancer stem cell definitions and terminology: the devil is in the details. Nat Rev Cancer 2012;12(11):767–75.
[29] Ranji P, Salmani Kesejini T, Saeedikhoo S, Alizadeh AM. Targeting cancer stem cell-specific markers and/or associated signaling pathways for overcoming cancer drug resistance. Tumour Biol 2016;37(10):13059–75.
[30] Marcucci F, Rumio C, Lefoulon F. Anti-cancer stem-like cell compounds in clinical development—an overview and critical appraisal. Front Oncol 2016;6:115.

[31] Pusztai L, Wagner P, Ibrahim N, Rivera E, Theriault R, Booser D, et al. Phase II study of tariquidar, a selective P-glycoprotein inhibitor, in patients with chemotherapy-resistant, advanced breast carcinoma. Cancer 2005;104(4):682–91.

[32] Ruff P, Vorobiof DA, Jordaan JP, Demetriou GS, Moodley SD, Nosworthy AL, et al. A randomized, placebo-controlled, double-blind phase 2 study of docetaxel compared to docetaxel plus zosuquidar (LY335979) in women with metastatic or locally recurrent breast cancer who have received one prior chemotherapy regimen. Cancer Chemother Pharmacol 2009;64(4):763–8.

[33] Farmer H, McCabe N, Lord CJ, Tutt AN, Johnson DA, Richardson TB, et al. Targeting the DNA repair defect in BRCA mutant cells as a therapeutic strategy. Nature 2005;434(7035):917–21.

[34] Hanahan D, Weinberg RA. The hallmarks of cancer. Cell 2000;100(1):57–70.

[35] van Delft MF, Wei AH, Mason KD, Vandenberg CJ, Chen L, Czabotar PE, et al. The BH3 mimetic ABT-737 targets selective Bcl-2 proteins and efficiently induces apoptosis via Bak/Bax if Mcl-1 is neutralized. Cancer Cell 2006;10(5):389–99.

[36] Konopleva M, Watt J, Contractor R, Tsao T, Harris D, Estrov Z, et al. Mechanisms of antileukemic activity of the novel Bcl-2 homology domain-3 mimetic GX15-070 (obatoclax). Cancer Res 2008;68(9):3413–20.

[37] Chen DJ, Huerta S. Smac mimetics as new cancer therapeutics. Anticancer Drugs 2009;20(8):646–58.

[38] El-Gamal M, Khan MA, Oh CH. Advanced melanoma: a progressive review of V600E-B-RAF kinase and drugs inhibiting it. Mini Rev Med Chem 2017;17(4):351–65.

[39] Maik-Rachline G, Seger R. The ERK cascade inhibitors: towards overcoming resistance. Drug Resist Updat 2016;25:1–12.

[40] Brotelle T, Bay JO. PI3K-AKT-mTOR pathway: description, therapeutic development, resistance, predictive/prognostic biomarkers and therapeutic applications for cancer. Bull Cancer 2016;103(1):18–29.

[41] Mirone G, Shukla A, Marfe G. Signaling mechanisms of resistance to EGFR- and anti-angiogenic inhibitors cancer. Crit Rev Oncol Hematol 2016;97:85–95.

[42] Vazquez-Santillan K, Melendez-Zajgla J, Jimenez-Hernandez L, Martínez-Ruiz G, Maldonado V. NF-κB signaling in cancer stem cells: a promising therapeutic target? Cell Oncol (Dordr) 2015;38(5):327–39.

[43] Blaise GA, Gauvin D, Gangal M, Authier S. Nitric oxide, cell signaling and cell death. Toxicology 2005;208(2):177–92.

[44] Bogdan C. Nitric oxide and the regulation of gene expression. Trends Cell Biol 2001;11(2):66–75.

[45] Hess DT, Matsumoto A, Kim SO, Marshall HE, Stamler JS. Protein S-nitrosylation: purview and parameters. Nat Rev Mol Cell Biol 2005;6(2):150–66.

[46] Koesling D, Russwurm M, Mergia E, Mullershausen F, Friebe A. Nitric oxide-sensitive guanylyl cyclase: structure and regulation. Neurochem Int 2004;45(6):813–9.

[47] Haynes V, Elfering S, Traaseth N, Giulivi C. Mitochondrial nitric-oxide synthase: enzyme expression, characterization, and regulation. J Bioenerg Biomembr 2004;36(4):341–6.

[48] Birschmann I, Walter U. Physiology and pathophysiology of vascular signaling controlled by guanosine 3′,5′-cyclic monophosphate-dependent protein kinase. Acta Biochim Pol 2004;51(2):397–404.

[49] Krumenacker JS, Hanafy KA, Murad F. Regulation of nitric oxide and soluble guanylyl cyclase. Brain Res Bull 2004;62(6):505–15.

[50] Moncada S, Erusalimsky JD. Does nitric oxide modulate mitochondrial energy generation and apoptosis? Nat Rev Mol Cell Biol 2002;3(3):214–20.

[51] Brune B. Nitric oxide: NO apoptosis or turning it ON? Cell Death Differ 2003;10(8):864–9.

[52] Zech B, Köhl R, von Knethen A, Brüne B. Nitric oxide donors inhibit formation of the Apaf-1/caspase-9 apoptosome and activation of caspases. Biochem J 2003;371(Pt 3):1055–64.

[53] Borst P, Rottenberg S. Cancer cell death by programmed necrosis? Drug Resist Updat 2004;7(6):321–4.

[54] Coleman JW. Nitric oxide in immunity and inflammation. Int Immunopharmacol 2001;1(8):1397–406.

[55] Cannon III RO. Role of nitric oxide in cardiovascular disease: focus on the endothelium. Clin Chem 1998;44(8 Pt 2):1809–19.

[56] Bicker G. STOP and GO with NO: nitric oxide as a regulator of cell motility in simple brains. Bioessays 2005;27(5):495–505 [Review].

[57] Bonavida B, Khineche S, Huerta-Yepez S, Garbán H. Therapeutic potential of nitric oxide in cancer. Drug Resist Updat 2006;9(3):157–73.

[58] Hanafy KA, Krumenacker JS, Murad F. NO, nitrotyrosine, and cyclic GMP in signal transduction. Med Sci Monit 2001;7(4):801–19.

[59] Postovit LM, Adams MA, Lash GE, Heaton JP, Graham CH. Nitric oxide-mediated regulation of hypoxia-induced B16F10 melanoma metastasis. Int J Cancer 2004;108(1):47–53.

[60] Sogawa K, Numayama-Tsuruta K, Ema M, Abe M, Abe H, Fujii-Kuriyama Y. Inhibition of hypoxia-inducible factor 1 activity by nitric oxide donors in hypoxia. Proc Natl Acad Sci U S A 1998;95(13):7368–73.

[61] Ghiso N, Rohan RM, Amano S, Garland R, Adamis AP. Suppression of hypoxia-associated vascular endothelial growth factor gene expression by nitric oxide via cGMP. Investig Ophthalmol Vis Sci 1999;40 (6):1033–9.

[62] Hagen T, Taylor CT, Lam F, Moncada S. Redistribution of intracellular oxygen in hypoxia by nitric oxide: effect on HIF1alpha. Science 2003;302(5652):1878–975.

[63] Matthews NE, Adams MA, Maxwell LR, Gofton TE, Graham CH. Nitric oxide-mediated regulation of chemosensitivity in cancer cells. J Natl Cancer Inst 2001;93(24):1879–85.

[64] Frederiksen LJ, Siemens DR, Heaton JP, Maxwell LR, Adams MA, Graham CH. Hypoxia induced resistance to doxorubicin in prostate cancer cells is inhibited by low concentrations of glyceryl trinitrate. J Urol 2003;170 (3):1003–7.

[65] Burke AJ, Sullivan FJ, Giles FJ, Glynn SA. The yin and yang of nitric oxide in cancer progression. Carcinogenesis 2013;34(3):503–12.

[66] Cheng H, Wang L, Mollica M, Re AT, Wu S, Zuo L. Nitric oxide in cancer metastasis. Cancer Lett 2014;353 (1):1–7.

[67] Lala PK, Chakraborty C. Role of nitric oxide in carcinogenesis and tumour progression. Lancet Oncol 2001; 2(3):149–56.

[68] Tamir S, Burney S, Tannenbaum SR. DNA damage by nitric oxide. Chem Res Toxicol 1996;9(5):821–7.

[69] Felley-Bosco E. Role of nitric oxide in genotoxicity: implication for carcinogenesis. Cancer Metastasis Rev 1998; 17(1):25–37.

[70] Chaudary N, Hill RP. Hypoxia and metastasis. Clin Cancer Res 2007;13(7):1947–9.

[71] Postovit LM, Sullivan R, Adams MA, Graham CH. Nitric oxide signalling and cellular adaptations to changes in oxygenation. Toxicology 2005;208(2):235–48.

[72] Thomsen LL, Miles DW. Role of nitric oxide in tumour progression: lessons from human tumours. Cancer Metastasis Rev 1998;17(1):107–18.

[73] Ambs S, Merriam WG, Ogunfusika MO, Bennett WP, Ishibe N, Hussain SP, et al. p53 and vascular endothelial growth factor regulate tumor growth of NOS2-expressing human carcinoma cells. Nat Med 1998;4 (12):1371–6.

[74] Jenkins DC, Charles IG, Thomsen LL, Moss DW, Holmes LS, Baylis SA, et al. Roles of nitric oxide in tumor growth. Proc Natl Acad Sci U S A 1995;92(10):4392–6.

[75] Hibbs Jr. JB, Taintor RR, Vavrin Z, Rachlin EM. Nitric oxide: a cytotoxic activated macrophage effector molecule. Biochem Biophys Res Commun 1988;157(1):87–94.

[76] Xie K, Fidler IJ. Therapy of cancer metastasis by activation of the inducible nitric oxide synthase. Cancer Metastasis Rev 1998;17(1):55–75.

[77] Alexandrova R, Mileva M, Zvetkova E. Nitric oxide and cancer (minireview). Exp Pathol Parasitol 2001;4:13–8.

[78] Umansky V, Schirrmacher V. Nitric oxide-induced apoptosis in tumor cells. Adv Cancer Res 2001;82:107–31.

[79] Huerta S, Chilka S, Bonavida B. Nitric oxide donors: novel cancer therapeutics. Int J Oncol 2008;33(5):909–27.

[80] Wang PG, Xian M, Tang X, Wu X, Wen Z, Cai T, et al. Nitric oxide donors: chemical activities and biological applications. Chem Rev 2002;102(4):1091–134.

[81] Parker JD, Parker JO. Nitrate therapy for stable angina pectoris. N Engl J Med 1998;338(8):520–31.

[82] Yasuda H, Nakayama K, Watanabe M, Suzuki S, Fuji H, Okinaga S, et al. Nitroglycerin treatment may enhance chemosensitivity to docetaxel and carboplatin in patients with lung adenocarcinoma. Clin Cancer Res 2006;12 (22):6748–57.

[83] Yasuda H, Yamaya M, Nakayama K, Sasaki T, Ebihara S, Kanda A, et al. Randomized phase II trial comparing nitroglycerin plus vinorelbine and cisplatin with vinorelbine and cisplatin alone in previously untreated stage IIIB/IV non-small-cell lung cancer. J Clin Oncol 2006;24(4):688–94.

[84] Wang EQ, Lee WI, Brazeau D, Fung HL. cDNA microarray analysis of vascular gene expression after nitric oxide donor infusions in rats: implications for nitrate tolerance mechanisms. AAPS PharmSci 2002;4(2):E10.

[85] Rishi L, Dhiman R, Raje M, Majumdar S. Nitric oxide induces apoptosis in cutaneous T cell lymphoma (HuT-78) by downregulating constitutive NF-kappaB. Biochim Biophys Acta 2007;1770(8):1230–9.

[86] Kitamura Y, Kamoshima W, Shimohama S, Nomura Y, Taniguchi T. Nitric oxide donor-induced p53-sensitive cell death is enhanced by Bcl-2 reduction in human neuroblastoma cells. Neurochem Int 1998;32(1):93–102.

[87] Jeon HK, Choi SU, Jung NP. Association of the ERK1/2 and p38 kinase pathways with nitric oxide-induced apoptosis and cell cycle arrest in colon cancer cells. Cell Biol Toxicol 2005;21(2):115–25.

[88] Wink DA, Cook JA, Christodoulou D, Krishna MC, Pacelli R, Kim S, et al. Nitric oxide and some nitric oxide donor compounds enhance the cytotoxicity of cisplatin. Nitric Oxide 1997;1(1):88–94.

[89] Azizzadeh B, Yip HT, Blackwell KE, Horvath S, Calcaterra TC, Buga GM, et al. Nitric oxide improves cisplatin cytotoxicity in head and neck squamous cell carcinoma. Laryngoscope 2001;111(11 Pt 1):1896–900.

[90] Perrotta C, Bizzozero L, Falcone S, Rovere-Querini P, Prinetti A, Schuchman EH, et al. Nitric oxide boosts chemoimmunotherapy via inhibition of acid sphingomyelinase in a mouse model of melanoma. Cancer Res 2007;67(16):7559–64.

[91] Rigas B, Kashfi K. Nitric-oxide-donating NSAIDs as agents for cancer prevention. Trends Mol Med 2004;10(7):324–30.

[92] Bratasz A, Weir NM, Parinandi NL, Zweier JL, Sridhar R, Ignarro LJ, et al. Reversal to cisplatin sensitivity in recurrent human ovarian cancer cells by NCX-4016, a nitro derivative of aspirin. Proc Natl Acad Sci U S A 2006;103(10):3914–9.

[93] Godoy LC, Anderson CT, Chowdhury R, Trudel LJ, Wogan GN. Endogenously produced nitric oxide mitigates sensitivity of melanoma cells to cisplatin. Proc Natl Acad Sci U S A 2012;109(50):20373–8.

[94] Sinha BK, Bhattacharjee S, Chatterjee S, Jiang J, Motten AG, Kumar A, et al. Role of nitric oxide in the chemistry and anticancer activity of etoposide (VP-16,213). Chem Res Toxicol 2013;26(3):379–87.

[95] Wongvaranon P, Pongrakhananon V, Chunhacha P, Chanvorachote P. Acquired resistance to chemotherapy in lung cancer cells mediated by prolonged nitric oxide exposure. Anticancer Res 2013;33(12):5433–44.

[96] Evig CB, Kelley EE, Weydert CJ, Chu Y, Buettner GR, Burns CP. Endogenous production and exogenous exposure to nitric oxide augment doxorubicin cytotoxicity for breast cancer cells but not cardiac myoblasts. Nitric Oxide 2004;10(3):119–29.

[97] Riganti C, Miraglia E, Viarisio D, Costamagna C, Pescarmona G, Ghigo D, et al. Nitric oxide reverts the resistance to doxorubicin in human colon cancer cells by inhibiting the drug efflux. Cancer Res 2005;65(2):516–25.

[98] Huerta-Yepez S, Baritaki S, Baay-Guzman G, Hernandez-Luna MA, Hernandez-Cueto A, Vega MI, et al. Contribution of either YY1 or BclXL-induced inhibition by the NO-donor DETANONOate in the reversal of drug resistance, both in vitro and in vivo. YY1 and BclXL are overexpressed in prostate cancer. Nitric Oxide 2013;29:17–24.

[99] Frederiksen LJ, Sullivan R, Maxwell LR, Macdonald-Goodfellow SK, Adams MA, Bennett BM, et al. Chemosensitization of cancer in vitro and in vivo by nitric oxide signaling. Clin Cancer Res 2007;13(7):2199–206.

[100] Sullivan R, Graham CH. Chemosensitization of cancer by nitric oxide. Curr Pharm Des 2008;14(11):1113–23.

[101] Vega MI, Jazirehi AR, Huerta-Yepez S, Bonavida B. Rituximab-induced inhibition of YY1 and Bcl-xL expression in Ramos non-Hodgkin's lymphoma cell line via inhibition of NF-kappa B activity: role of YY1 and Bcl-xL in Fas resistance and chemoresistance, respectively. J Immunol 2005;175(4):2174–83.

[102] Huerta-Yepez S, Vega M, Jazirehi A, Garban H, Hongo F, Cheng G, et al. Nitric oxide sensitizes prostate carcinoma cell lines to TRAIL-mediated apoptosis via inactivation of NF-kappa B and inhibition of Bcl-xl expression. Oncogene 2004;23(29):4993–5003.

[103] Marshall HE, Hess DT, Stamler JS. S-nitrosylation: physiological regulation of NF-kappaB. Proc Natl Acad Sci U S A 2004;101(24):8841–2.

[104] Huerta-Yepez S, Liu H, Baritaki S, Del Lourdes Cebrera-Muñoz M, Rivera-Pazos C, Maldonado-Valenzuela A, et al. Overexpression of Yin Yang 1 in bone marrow-derived human multiple myeloma and its clinical significance. Int J Oncol 2014;45(3):1184–92.

[105] Gao J, Kashfi K, Rigas B. In vitro metabolism of nitric oxide-donating aspirin: the effect of positional isomerism. J Pharmacol Exp Ther 2005;312(3):989–97.

[106] Fabbri F, Brigliadori G, Ulivi P, Tesei A, Vannini I, Rosetti M, et al. Pro-apoptotic effect of a nitric oxide-donating NSAID, NCX 4040, on bladder carcinoma cells. Apoptosis 2005;10(5):1095–103.

[107] Kogias E, Osterberg N, Baumer B, Psarras N, Koentges C, Papazoglou A, et al. Growth-inhibitory and chemosensitizing effects of the glutathione-S-transferase-π-activated nitric oxide donor PABA/NO in malignant gliomas. Int J Cancer 2012;130(5):1184–94.

[108] Calatozzolo C, Gelati M, Ciusani E, Sciacca FL, Pollo B, Cajola L, et al. Expression of drug resistance proteins Pgp, MRP1, MRP3, MRP5 and GST-pi in human glioma. J Neurooncol 2005;74(2):113–21.

[109] Laval F, Wink DA. Inhibition by nitric oxide of the repair protein, O6-methylguanine-DNA-methyltransferase. Carcinogenesis 1994;15(3):443–7.

[110] Reynolds MM, Witzeling SD, Damodaran VB, Medeiros TN, Knodle RD, Edwards MA, et al. Applications for nitric oxide in halting proliferation of tumor cells. Biochem Biophys Res Commun 2013;431(4):647–51.

[111] Bonavida B, Baritaki S. The novel role of Yin Yang 1 in the regulation of epithelial to mesenchymal transition in cancer via the dysregulated NF-κB/Snail/YY1/RKIP/PTEN circuitry. Crit Rev Oncog 2011;16(3–4):211–26.

[112] Nagai H, Yasuda H, Hatachi Y, Xue D, Sasaki T, Yamaya M, et al. Nitric oxide (NO) enhances pemetrexed cytotoxicity via NO-cGMP signaling in lung adenocarcinoma cells in vitro and in vivo. Int J Oncol 2012;41:24–30.

[113] Ye S, Yang W, Wang Y, Ou W, Ma Q, Yu C, et al. Cationic liposome-mediated nitric oxide synthase gene therapy enhances the antitumor effects of cisplatin in lung cancer. Int J Mol Med 2013;31(1):33–42.

[114] Jamieson ER, Lippard SJ. Structure, recognition, and processing of cisplatin-DNA adducts. Chem Rev 1999;99:2467–98.

[115] Huerta S, Harris DM, Jazirehi A, Bonavida B, Elashoff D, Livingston EH, et al. Gene expression profile of metastatic colon cancer cells resistant to cisplatin-induced apoptosis. Int J Oncol 2003;22:663–70.

[116] Huerta S, Heinzerling JH, Nguiano-Hernandez YM, Huerta-Yepez S, Lin J, Chen D, et al. Modification of gene products involved in resistance to apoptosis in metastatic colon cancer cells: roles of Fas, Apaf-1, NFkappaB, IAPs, Smac/DIABLO, and AIF. J Surg Res 2007;142:184–94.

[117] Weyerbrock A, Osterberg N, Psarras N, Baumer B, Kogias E, Werres A, et al. JS-K, a glutathione S-transferase-activated nitric oxide donor with antineoplastic activity in malignant gliomas. Neurosurgery 2012;70(2):497–510.

[118] Pervin S, Singh R, Gau CL, Edamatsu H, Tamanoi F, Chaudhuri G. Potentiation of nitric oxide-induced apoptosis of MDA-MB-468 cells by farnesyltransferase inhibitor: implications in breast cancer. Cancer Res 2001;61:4701–6.

[119] Lin Y, Bai L, Chen W, Xu S. The NF-kappaB activation pathways, emerging molecular targets for cancer prevention and therapy. Expert Opin Ther Targets 2010;14(1):45–55.

[120] Kleinberg L, Davidson B. Cell survival and apoptosis-related molecules in cancer cells in effusions: a comprehensive review. Diagn Cytopathol 2009;37(8):613–24.

[121] Wang H, Hertlein E, Bakkar N, Sun H, Acharyya S, Wang J, et al. NF-kappaB regulation of YY1 inhibits skeletal myogenesis through transcriptional silencing of myofibrillar genes. Mol Cell Biol 2007;27(12):4374–87.

[122] Julien S, Puig I, Caretti E, Bonaventure J, Nelles L, van Roy F, et al. Activation of NF-kappaB by Akt upregulates Snail expression and induces epithelium mesenchyme transition. Oncogene 2007;26(53):7445–56.

[123] Beach S, Tang H, Park S, Dhillon AS, Keller ET, Kolch W, et al. Snail is a repressor of RKIP transcription in metastatic prostate cancer cells. Oncogene 2008;27(15):2243–8.

[124] Reed JC. Dysregulation of apoptosis in cancer. J Clin Oncol 1999;17(9):2941–53.

[125] Tan DS, Gerlinger M, Teh BT, Swanton C. Anti-cancer drug resistance: understanding the mechanisms through the use of integrative genomics and functional RNA interference. Eur J Cancer 2010;46(12):2166–77.

[126] Singh A, Settleman J. EMT, cancer stem cells and drug resistance: an emerging axis of evil in the war on cancer. Oncogene 2010;29(34):4741–51.

[127] Baritaki S, Yeung K, Palladino M, Berenson J, Bonavida B. Pivotal roles of snail inhibition and RKIP induction by the proteasome inhibitor NPI-0052 in tumor cell chemoimmunosensitization. Cancer Res 2009;69(21):8376–85.

[128] Huerta-Yepez S, Vega M, Escoto-Chavez SE, Murdock B, Sakai T, Baritaki S, et al. Nitric oxide sensitizes tumor cells to TRAIL-induced apoptosis via inhibition of the DR5 transcription repressor Yin Yang 1. Nitric Oxide 2009;20(1):39–52.

[129] Bonavida B, Baritaki S, Huerta-Yepez S, Vega MI, Jazirehi AR, Berenson J. Nitric oxide donors are a new class of anti-cancer therapeutics for the reversal of resistance and inhibition of metastases. New York: Springer; 2010. p. 459–77.

[130] Baritaki S, Suzuki E, Umezawa K, Spandidos DA, Berenson J, Daniels TR, et al. Inhibition of Yin Yang 1-dependent repressor activity of DR5 transcription and expression by the novel proteasome inhibitor NPI-0052 contributes to its TRAIL-enhanced apoptosis in cancer cells. J Immunol 2008;180(9):6199–210.

[131] Palmer MB, Majumder P, Cooper JC, Yoon H, Wade PA, Boss JM. Yin yang 1 regulates the expression of snail through a distal enhancer. Mol Cancer Res 2009;7(2):221–9.

[132] Olmeda D, Moreno-Bueno G, Flores JM, Fabra A, Portillo F, Cano A. SNAI1 is required for tumor growth and lymph node metastasis of human breast carcinoma MDA-MB-231 cells. Cancer Res 2007;67(24):11721–31.

[133] Gordon S, Akopyan G, Garban H, Bonavida B. Transcription factor YY1: structure, function, and therapeutic implications in cancer biology. Oncogene 2006;25(8):1125–42.
[134] Baritaki S, Huerta-Yepez S, Sakai T, Spandidos DA, Bonavida B. Chemotherapeutic drugs sensitize cancer cells to TRAIL-mediated apoptosis: up-regulation of DR5 and inhibition of Yin Yang 1. Mol Cancer Ther 2007;6(4):1387–99.
[135] Yeung K, Seitz T, Li S, Janosch P, McFerran B, Kaiser C, et al. Suppression of Raf-1 kinase activity and MAP kinase signalling by RKIP. Nature 1999;401(6749):173–7.
[136] Yeung KC, Rose DW, Dhillon AS, Yaros D, Gustafsson M, Chatterjee D, et al. Raf kinase inhibitor protein interacts with NF-kappaB-inducing kinase and TAK1 and inhibits NF-kappaB activation. Mol Cell Biol 2001;21(21):7207–17.
[137] Tang H, Park S, Sun SC, Trumbly R, Ren G, Tsung E, et al. RKIP inhibits NF-kappaB in cancer cells by regulating upstream signaling components of the IkappaB kinase complex. FEBS Lett 2010;584(4):662–8.
[138] Hongo F, Garban H, Huerta-Yepez S, Vega M, Jazirehi AR, Mizutani Y, et al. Inhibition of the transcription factor Yin Yang 1 activity by S-nitrosation. Biochem Biophys Res Commun 2005;336(2):692–701.
[139] Reid T, Dad S, Korn R, Oronsky B, Knox S, Scicinski J. Two case reports of resensitization to previous chemotherapy with the novel hypoxia-activated hypomethylating anticancer agent RRx-001 in metastatic colorectal cancer patients. Case Rep Oncol 2014;7(1):79–85.
[140] Reid TR, Oronsky B, Infante JR, Burris HA, Scribner C, Knox SJ, et al. A phase 1 trial and pharmacokinetic study of RRx-001, a novel ROS-mediated pan-epginentic agent. J Clin Oncol 2014;32(suppl; abstr2578).
[141] Oronsky B, Oronsky N, Lybeck M, Fanger G, Scicinski Bonavida B. Targeting hyponitroxia in cancer therapy. Nitric Oxide and Cancer: Pathogenesis and Therapy. Springer; 2015;39–47.
[142] Jemal A, Bray F, Center MM, Ferlay J, Ward E, Forman D. Global cancer statistics. CA Cancer J Clin 2011;61(2):69–90.
[143] Bradley JD, Paulus R, Komaki R, Masters G, Blumenschein G, Schild S, et al. Standard-dose versus high-dose conformal radiotherapy with concurrent and consolidation carboplatin plus paclitaxel with or without cetuximab for patients with stage IIIA or IIIB non-small-cell lung cancer (RTOG 0617): a randomised, two-by-two factorial phase 3 study. Lancet Oncol 2015;16(2):187–99.
[144] Díez I, Calatayud S, Hernández C, Quintana E, O'Connor JE, Esplugues JV, et al. Nitric oxide, derived from inducible nitric oxide synthase, decreases hypoxia inducible factor-1alpha in macrophages during aspirin-induced mesenteric inflammation. Br J Pharmacol 2010;159(8):1636–45.
[145] Seki T, Fang J, Maeda H. Enhanced delivery of macromolecular antitumor drugs to tumors by nitroglycerin application. Cancer Sci 2009;100(12):2426–30.
[146] Liu YS, Chuang MT, Tsai YS, Tsai HM, Lin XZ. Nitroglycerine use in transcatheter arterial (chemo)embolization in patients with hepatocellular carcinoma and dual-energy CT assessment of Lipiodol retention. Eur Radiol 2012;22(10):2193–200.
[147] Carraro S, Roca EL, Cartelli C, Rafailovici L, Castillo Odena S, Wasserman E, et al. Radiochemotherapy with short daily infusion of low-dose oxaliplatin, leucovorin, and 5-FU in T3-T4 unresectable rectal cancer: a phase II IATTGI study. Int J Radiat Oncol Biol Phys 2002;54(2):397–402.
[148] Minsky BD, Röedel C, Valentini V. Combined modality therapy for rectal cancer. Cancer J 2010;16(3):253–61.
[149] Illum H, Wang DH, Dowell JE, Hittson WJ, Torrisi JR, Meyer J, et al. Phase I dose escalation trial of nitroglycerin in addition to 5-fluorouracil and radiation therapy for neoadjuvant treatment of operable rectal cancer. Surgery 2015;158(2):460–5.
[150] Bonavida B. Pivotal role of nitric oxide in chemo and immunosensitization of resistant tumor cells to apoptosis. Nitric Oxide and Cancer: Pathogenesis and Therapy. Springer; 2015;179–201.

CHAPTER

3

RRx-001 Reset: Chemoresensitization via NO-Mediated M1 Macrophage Repolarization

*Jan Scicinski**, Bryan Oronsky**, Scott Caroen**, Pedro Cabrales*[†], *Thomas A. Summers*[‡], *Tony R. Reid*[†], *Michelle M. Kim*[§], *Jane B. Trepel*[¶], *Min-Jung Lee*[¶], *Nacer D. Abrouk*[**], *Regina M. Day*[††], *Saheli Jha*[*,††], *Arnold Oronsky*[‡‡], *Corey A. Carter*[‡]

[*]EpicentRx, Inc., San Diego, CA United States

[†]University of California at San Diego (UCSD), San Diego, CA United States

[‡]Walter Reed National Military Medical Center, Bethesda, MD, United States

[§]University of Michigan, Ann Arbor, MI, United States

[¶]National Cancer Institute, National Institutes of Health, Bethesda, MD, United States

[**]Innovexe Inc., Mountain View, CA United States

[††]Uniformed Services University, Bethesda, MD, United States

[‡‡]InterWest Partners, Menlo Park, CA United States

Abstract

RRx-001 is a nitric oxide-mediated immunostimulatory clinical stage small molecule that arose as a result of a single collaborative enterprise between InterWest Partners, a venture capital company, and Stanford University, a biotechnology company and the defense industry; its discovery and development have been guided and driven by careful observations of both preclinical and clinical data with data-driven adaptation of approaches and strategies in real time. In this chapter, we focus on the chemo- and radiosensitization properties of RRx-001, describing initial observations and follow-up clinical trials on the application of RRx-001 as a prosensitizer, a hybrid word that refers to the protection of normal tissue with sensitization of refractory and resistant tumors to chemo- and radiotherapy.

Nitric Oxide (Donor/Induced) in Chemosensitization, Volume 1
http://dx.doi.org/10.1016/B978-0-12-811020-1.00003-X

Abbreviations

ARE	antioxidant response element
$AUC_{0-\infty}$	pharmacokinetic area under curve extrapolated to infinity
CEA	carcinoembryonic antigen, an indicator of tumor burden
CFI	chemotherapy-free interval
Cmax	pharmacokinetic maximum concentration
CT	computerized tomography
Cys	amino acid cysteine residue
DCE-MRI	dynamic contrast-enhanced MRI
FOLFIRI	drug combination—fluorouracil, leucovorin, irinotecan
GI	gastrointestinal
GSH	reduced glutathione
Hb	hemoglobin
MRI	magnetic resonance imaging
Nrf2	NF-E2-related factor
PS	phosphatidylserine
RBC	red blood cell
SCLC	small-cell lung cancer
TAMS	tumor-associated macrophages
TNAZ	trinitroazetidine
TNT	trinitrotoluene
WBRT	whole brain radiotherapy

INTRODUCTION

A reset button restores or reboots a device to its original state. If the cancer cell is akin to a kind of biologic computer with a digital DNA program code that operates on input data from the cell environment and generates output, for example, proliferation, invasiveness, progression, and drug resistance in response, then, in theory, it should be possible to design and develop a pharmacological reset, which aborts or, if this is not possible, reprograms the malignant code such that the resistant cancer cell returns to a less aggressive state and becomes retreatable with standard therapies. In oncology, resensitization, which refers to the restoration of the original sensitivity to chemotherapy, has been predominately associated with epigenetic inhibition and is a potential mechanism to extend survival for patients who are otherwise out of options and time.

With the estimated total deaths due to cancer at almost 600,000 in 2016, cancer accounts for nearly one of every four deaths and is the second most common cause of death in the United States [1]. Although the incidence of cancer has been gradually declining since the early 1990s and improvements in treatment such as the recent explosion in the use of immunooncology agents that activate the immune system, improvements in the treatment of cancer are still a major healthcare imperative.

Recent advances in the understanding of how tumor cells escape surveillance and eradication by the immune system have resulted in the development of immunooncology agents that activate the host immune system to recognize and mount an attack against cancer cells [2]. Although the results have been exciting and remarkable responses and extension of survival have been observed, it is now clear that not all tumor types are susceptible and not all patients respond. Further advances in the armamentarium against cancer are therefore required.

Of the anticancer agents that have been developed and marketed, many target relatively small patient populations where potential responders need to be identified through the careful analysis of specifically developed biomarkers. Consequently, conventional chemotherapy, which is broadly active, remains the fallback standard of care for many cancer types. However, the inevitable emergence of acquired resistance, together with extensive systemic toxicity that impacts the quality of life, accounts in large part for the failure of these so-called targeted agents. Moreover, once resistance develops, the reversal to a sensitive phenotype occurs rarely or not at all. Clearly, an agent that addresses both acquired resistance and extensive toxicity associated with chemotherapy could become a widely used treatment modality to enhance and extend the use of chemotherapy, thereby potentially benefiting a large number of patients with many different tumor types. *Prosensitization* is a hybrid word, coined to describe this type of activity comprising of the chemo*sensitization* of tumors together with concomitant chemo*pro*tection and radio*pro*tection of normal tissue.

In this chapter, we discuss RRx-001 as a potential tumor "reset button" in the context of a number of radiation, chemotherapy, and immunotherapy combination phase-2 studies designed to explore this activity in detail. Subjects with advanced, heavily treated, and drug-resistant disease may particularly benefit from the prosensitizing effects of RRx-001, which is conducive to the repeated reuse of previously effective first- and second-line therapies, hopefully adding time, but not toxicity, to the lives of cancer patients.

RRx-001: ORIGIN AND HISTORY

The immunostimulatory agent, RRx-001, is a novel first-in-class small molecule that possesses a variety of multiple mechanisms, vascular normalization, prosensitization, epigenetic inhibition, and macrophage polarization that converge to sensitize and resensitize tumors to conventional therapies. RRx-001 was discovered [3] through an unprecedented partnership between Stanford University, a major defense contractor, Orbital ATK, a farsighted Venture Partnership, InterWest Partners, and a fledgling Silicon Valley biotechnology company called EpicentRx to explore the potential of energetic compounds as radio- and chemosensitizers in the treatment of cancer.

Energetic molecules are chemical compounds with a high amount of stored chemical energy that is released when the chemical bonds are broken. The released highly reactive fragments, in the form of free radicals, were hypothesized to act as DNA repair inhibitors, similar to the action of oxygen in the context of radiotherapy [4]. The discovery and development of RRx-001, a close analog of trinitroazetidine (TNAZ), a melt-cast explosive explored as a potential replacement for TNT [5], are described in detail elsewhere [3,6]; however, biological testing of a library of energetic materials, followed by lead optimization, resulted in the identification of RRx-001, a dinitroazetidine derivative with single-agent, radio-, and chemosensitization activity without the drawback of bulk explosive properties (Fig. 1) [7].

RRx-001 represents a new class of pharmacophores that were rationally designed by combining two structural components: a dinitroazetidine, derived from TNAZ, and an α-bromoacetamide. The rationale for combining these structural units was to reduce the explosive potential of the molecule compared with TNAZ by substitution of the very labile nitramine with a functional group that would add mass while reducing the total number of nitrogen and oxygen atoms in the molecule and to increase biological activity while also enabling the

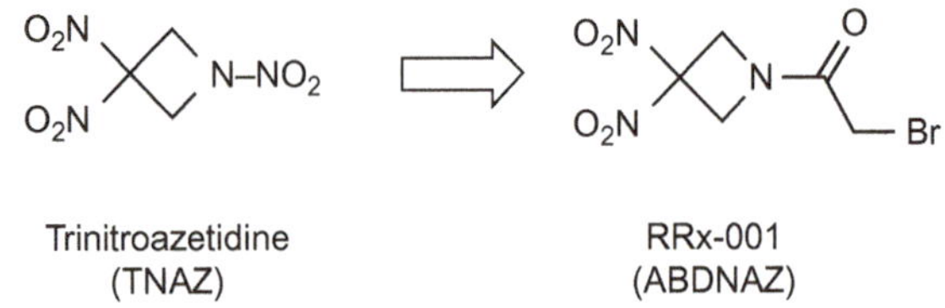

FIG. 1 Molecular structure of RRx-001 (also known as ABDNAZ or 1-bromoacetyl-3,3 dinitroazetidine) showing its derivation from trinitroazetidine (TNAZ).

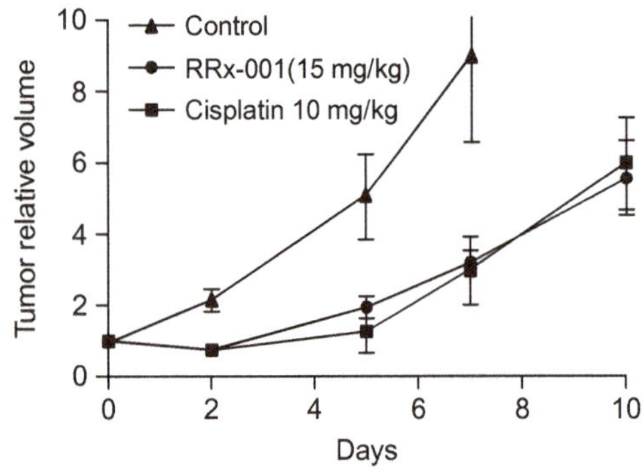

FIG. 2 RRx-001 single-dose antitumor activity in the mouse SCCVII tumor model compared with cisplatin.

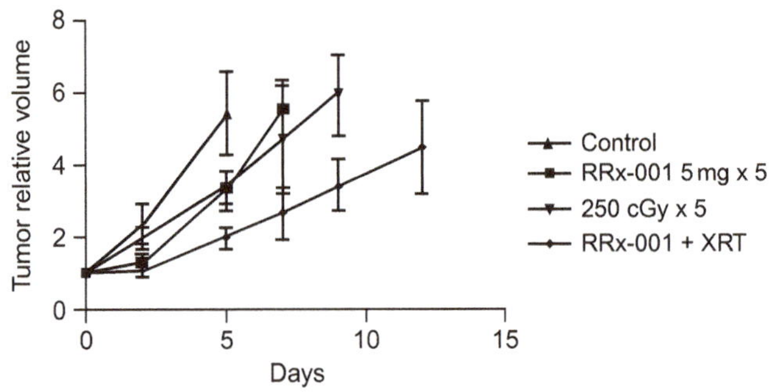

FIG. 3 Radiosensitizing effects of RRx-001 (5 mg × 5, QD) with whole body radiation (250 cGy × 5, QD) in the mouse SCCVII tumor model.

assembly of antibody and prodrug derivatives. Structure-activity relationship studies revealed that both of these structural units contribute to the overall activity of the molecule [6].

RRx-001 demonstrated activity both in vitro and in vivo in a number of different animal models of cancer, particularly in mouse syngeneic models where RRx-001 demonstrated equipotent activity with cisplatin with none of the associated toxicity and weight loss (Fig. 2). RRx-001 was found to be a potent radiosensitizer (Fig. 3), and no overt systemic toxicity as determined by weight loss or by changes in hematological, biochemical, or histopathologic markers was observed [7] (Figs. 2 and 3).

RRx-001: METABOLISM AND PHARMACOKINETICS

The α-bromoamide functionality of RRx-001 is a reactive, sulfhydryl-selective electrophile [8] that, on infusion, binds rapidly and irreversibly to reduced glutathione and the highly conserved cysteine-93 residue on the beta subunit of hemoglobin (Hb) of a subpopulation of red blood cells (RBCs) forming RRx-001-GSH and RRx-001-Hb metabolites, respectively [9]. The half-life of RRx-001 in whole blood is in the order of seconds, and this rapid binding to a subpopulation of RBCs contributes to the selectivity and low toxicity of the molecule. RRx-001 analogs containing α-chloro or α-iodoamide functionality in place of the α-bromoamide exhibited lower activity both in vivo and in vitro indicating the importance of relative electrophilic reactivity to the overall activity of the molecule [6]. Binding of RRx-001 to beta-Cys93 was confirmed using proteomic studies, which also showed an absence of binding to the amino terminus of the protein. Whole body quantitative autoradiography using ^{14}C-RRx-001 showed that the label was rapidly eliminated in the urine (36%) and feces (2%) while the remainder (62%) remained in the blood for the duration of the experiment (168 h), further confirming the binding to Hb [9].

The rapid metabolism and disposition characteristics of RRx-001 meant that it was not possible to follow the pharmacokinetics of RRx-001 directly; however, because of its rapid formation, the glutathione adduct [10], RRx-001-GSH, was chosen as the bioanalyte for both clinical and preclinical pharmacokinetic studies [11]. In the clinic, the pharmacokinetics before, during, and various time points post intravenous infusion on the first (day 1) and last day (day 22 or 50) of treatment were studied. The terminal half-life of RRx-001-GSH was found to be approximately 30 min, while changes in the calculated $AUC_{0-\infty}$ and pharmacokinetic maximum concentration (Cmax) were broadly dose-dependent with the exception of the highest dose (83 mg/m^2) (Fig. 4) [12].

Clinically, RRx-001-GSH is not responsible for the anticancer activity of RRx-001, which renders its measurement essentially moot for purposes of pharmacodynamic analyses. By contrast, data on the impact of RRx-001 on tumor blood flow, measured preclinically in mice using contrast-enhanced ultrasound [7] or immunohistochemical staining [13] of perfused blood vessels and hypoxic regions, are likely to be a more relevant pharmacodynamic indicator. Clinically, changes in tumor blood flow akin to vascular normalization (see detailed description below) were observed in the liver lesions of colorectal cancer patients using contrast-enhanced ultrasound and in patients [14] with brain metastases using dynamic contrast magnetic resonance imaging (MRI) techniques [15]. The duration of blood flow changes varied from no change to elevated flow compared with baseline for up to a week.

SAFETY AND TOLERABILITY

The safety and tolerability of RRx-001 was assessed in a phase-1 first-in-man clinical trial. RRx-001 was dose-escalated in a 3+3 design in six cohorts (10, 16.7, 24.6, 33, 55, and 83 mg/m^2). The starting dose was based on one-sixth of the no adverse event level (NOAEL) (equivalent to the highest nonseverely toxic dose) in dogs, the most sensitive species. A total of 25 patients were treated with RRx-001 [12]. In this study, most adverse events were grade 1 or 2 with grade 1 and 2 injection site pain as the most common treatment-related adverse event. This transient injection site pain, associated frequently with vasodilation, resolved spontaneously within seconds after the infusion was completed and was

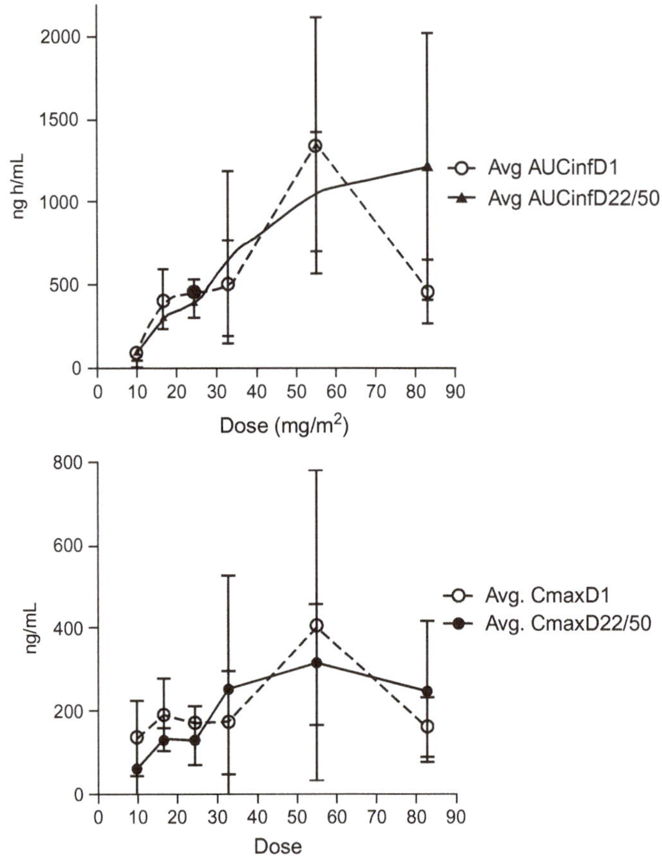

FIG. 4 Pharmacokinetics of RRx-001-GSH from the phase-1 clinical trial. AUC and Cmax for RRx-001-GSH were found to be broadly dose-dependent from 10 to 55 mg/m^2.

ameliorated by extending the infusion time and by pretreatment with steroids and opioids, as required.

Notably, RRx-001 was found to lack systemic toxicity at all doses, with no relevant treatment-related changes in laboratory values observed and with a marked absence of all of the toxicities typically observed with chemotherapeutic and even targeted agents such as alopecia, myelosuppression, diarrhea, mucositis, nausea, vomiting (Table 1). Therefore, as no dose-limiting toxicities were observed at any of the dose levels, a maximum tolerated dose was not reached, and the highest phase-1 dose of 83 mg/m^2 was determined to the be highest feasible dose as further dose escalation would have required prolonging the infusion beyond 8 h, requiring patients to be dosed in an inpatient setting.

In subsequent clinical trials, the incidence of local infusion pain was addressed by modifying the infusion procedure. In place of direct infusion, RRx-001 was coinfused with approximately 10 mL of autologous blood. Coadministration with blood was found to mitigate the pain on infusion while enabling much faster infusion rates and shorter patient clinical visits.

TABLE 1 Treatment-Emergent Adverse Events Attributed to RRx-001 in Decreasing Order of Frequency for 5% of Patients or More Found in the Phase-1 Clinical Trial

	10 mg/m^2 ($n = 6$)		16.7 mg/m^2 ($n = 3$)		24.6 mg/m^2 ($n = 3$)		33 mg/m^2 ($n = 4$)		55 mg/m^2 ($n = 3$)		83 mg/m^2 ($n = 6$)	
	Grades 1–2	Grade 3	Grades 1–2	Grade 3	Grades 1–2	Grade 3	Grades 1–2	Grade 3	Grades 1–2	Grade 3	Grades 1–2	Grade 3
Infusion site pain	4 (67%)	0	3 (100%)	0	1 (33%)	0	3 (75%)	1 (25%)	3 (100%)	0	6 (100%)	0
Arm swelling or edema	0	0	1 (33%)	0	0	0	1 (25%)	0	2 (67%)	0	4 (67%)	0
Vein hardening	0	0	0	0	0	0	1 (25%)	0	1 (33%)	0	5 (83%)	0
Dyspnea or wheezing[a]	1 (17%)	0	0	0	0	0	0	0	1 (33%)	0	3 (50%)	0
Mouth tingling or burning	0	0	0	0	0	0	0	0	2 (67%)	0	2 (33%)	0
Anxiety	0	0	0	0	0	0	0	0	1 (33%)	1 (33%)	1 (17%)	0
Chest discomfort	1 (17%)	0	1 (33%)	0	0	0	0	0	0	0	1 (17%)	0
Cough	0	0	0	0	0	0	0	0	1 (33%)	0	2 (33%)	0
Fatigue	0	0	0	0	0	0	1 (25%)	0	0	0	2 (33%)	0
Skin discoloration	0	0	0	0	0	0	1 (25%)	0	0	0	2 (33%)	0
Flushing[b]	1 (17%)	0	0	0	0	0	0	0	0	0	1 (17%)	0
Throat discomfort	0	0	0	0	0	0	0	0	0	0	2 (33%)	0
Vasodilation	1 (17%)	0	0	0	0	0	1 (25%)	0	0	0	0	0
Vein occlusion or DVT	0	0	0	0	0	0	0	0	0	0	2 (33%)	0
Weakness	0	0	0	0	0	0	0	0	2 (67%)	0	0	0

[a] *Terms wheezing, dyspnea, and dyspnea/wheezing were combined.*
[b] *Terms flushing and intermittent flushing were combined.*
Data are number of patients (%). DVT, deep vein thrombosis. No grade 4 adverse events were observed.
From Reid T, Oronsky B, Scicinski J, Scribner CL, Knox SJ, Ning S, et al. Safety and activity of RRx-001 in patients with advanced cancer: a first-in-human, open-label, dose-escalation phase 1 study. Lancet Oncol 2015;16(9):1133–42.

IMMUNE-MEDIATED MECHANISM OF ACTION OF RRx-001

The anticancer activity of RRx-001 arises from the initial interactions within the RBC (Fig. 5). As described above, RRx-001 increases the oxidative and nitrosative stress of the small subset of RBCs that it binds to via the beta-Cys 93 Hb residue. This highly conserved residue [16,17] is implicated in nitric oxide transport and in the modulation of oxygen affinity. The affinity of RRx-001 for this residue displaces nitric oxide, which leads to the previously described infusion-related pain and vasodilation. The oxidative and nitrosative stress of these RRx-001 bound RBCs increase under hypoxia due to the conversion of nitrite to nitric oxide by deoxyhemoglobin [18], which is highly potentiated by RRx-001 (Fig. 6) [19].

A consequence of the increase in RBC oxidative stress is membrane scrambling with the externalization of phosphatidylserine (PS) [20] and the display of other adhesion molecules on the RBC surface. Other RBC changes, such as an increase in rigidity, contribute to intratumoral microvascular occlusion, mediated by the surface presence of other molecules such as CD36, CD47, CD71, and ICAM-1. The expression of these adhesion factors in a small subset of RBCs that have been primed by RRx-001 results in their tendency to aggregate in the aberrant tumor vasculature leading to microvascular clogging and stasis.

The highly localized PS "eat me" signal of the RRx-001 bound RBCs sounds a "dinner bell" for engulfing tumor-associated macrophages (TAMS), which are subsequently converted from an M2 quiescent, protumor, phenotype to an activated M1-like phenotype [20–22] (Fig. 7).

The transfer of oxidative and nitrosative stress from the RBC to the tumor *via* TAMS is supported by the increase of intratumoral free radicals measured in vitro [7,23] and by

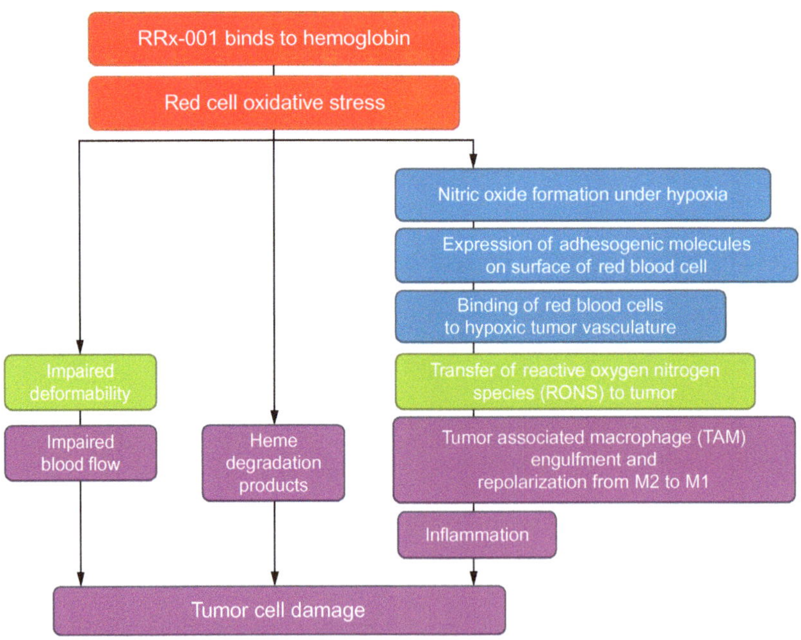

FIG. 5 Sequence of red-blood-cell-based events that occur after infusion of RRx-001.

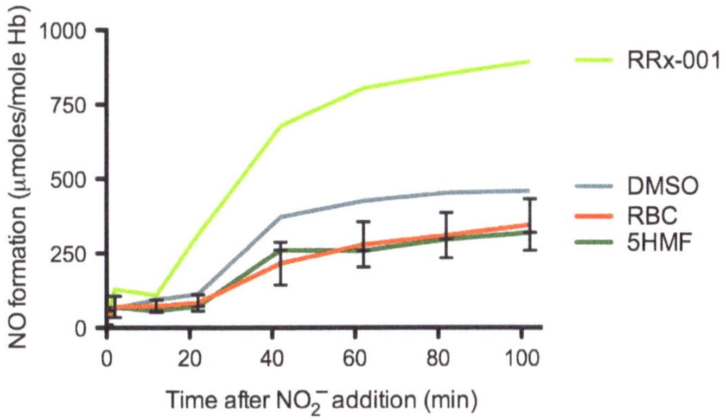

FIG. 6 Potentiation of hypoxia-mediated NO generation through hemoglobin nitrite reductase activity by RRx-001.

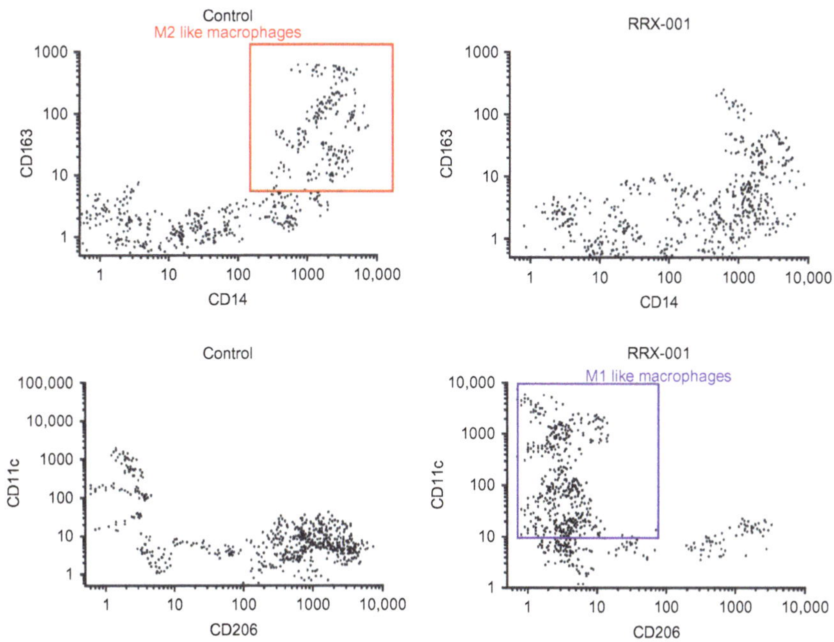

FIG. 7 Administration of RRx-001 results in an increase of M1-like macrophages together with a decrease in M2-like macrophages.

MRI studies that show the oxidation of tumor thiols in vivo [24]. With respect to higher-than-normal oxidative stress levels in tumors, a fine line exists between enough to drive proliferation signaling pathways and too much resulting in irreversible damage, senescence, and cell death [25]. Through direct and indirect action, RRx-001 generates a toxic mixture of oxygen and nitrogen-based free radicals such as peroxynitrite, nitrogen dioxide, and dinitrogen

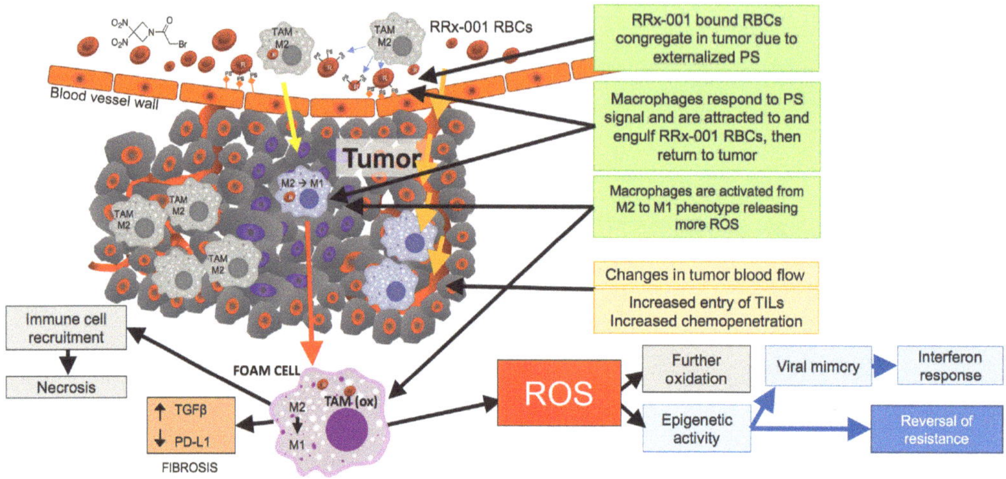

FIG. 8 RRx-001 immune modulated activity.

trioxide [26,27] that react with and cause damage to organelles and macromolecules, ultimately leading to cell death [28,29].

Free-radical generation is also responsible for the inhibition of DNA methyltransferases (DNMTs) and histone deacetylases (HDACs) [30,31] resulting in changes in gene expression in critical enzymes that may affect chemo- and radiosensitivity such as p53 [23].

As shown in Fig. 8, the release of free radicals from the activated TAMs induces fibrosis with an associated increase in TGFβ. Analysis of pre- and postdose tumor biopsy samples of responding patients on the QUADRUPLE THREAT study showed the presence of TAMs and of higher levels of TGFβ [32], suggesting that these features could be used to identify potential responders to RRx-001.

RRx-001: PROSENSITIZATION

As outlined above, a prosensitizing agent that reverses the acquired resistance and mitigates the extensive toxicity associated with chemotherapy represents the next best thing to a cure: conversion of an inexorably fatal disease to a more chronic and manageable one. The prosensitization activity of RRx-001 was observed both in preclinical experiments and in clinical trials.

Preclinically, chemoprotection against cisplatin, doxorubicin, and irinotecan toxicities was studied in various mouse models; for example, treatment of mice with RRx-001 before administration of cisplatin resulted in a lower incidence of myelosuppression (Fig. 9) chromosomal breaks, gaps, and rearrangements (Fig. 10) compared with cisplatin alone. In addition, cisplatin-induced nephrotoxicity, as evidenced by significantly elevated levels of the blood urea nitrogen, the serum creatinine, and the kidney-to-body-weight ratio, was significantly

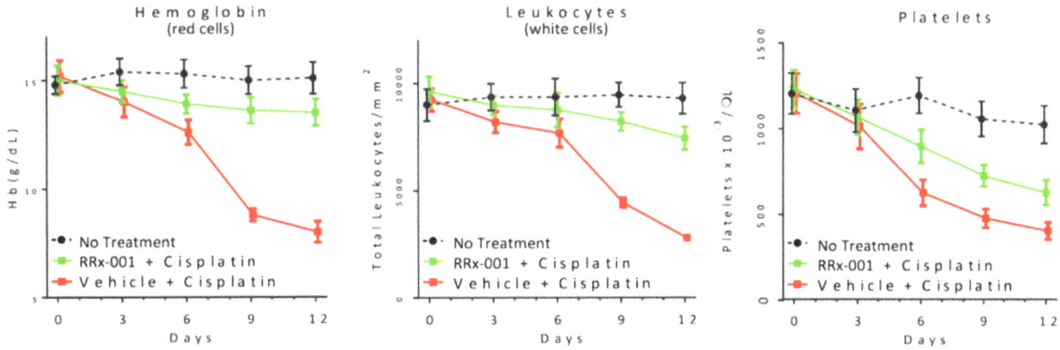

FIG. 9 Chemoprotection of RRx-001 against cisplatin-induced myelosuppression. Hemoglobin concentration, platelets, and WBC counts were significantly less decreased in mice pretreated with RRx-001.

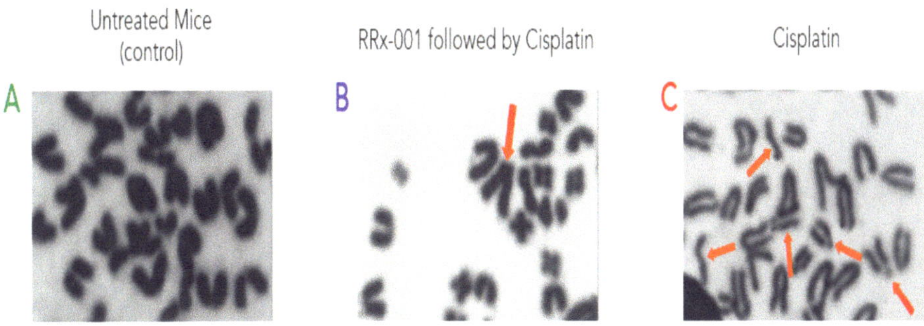

FIG. 10 Preclinical evidence of chemoprotection. RRx-001 treated mice prior to treatment with cisplatin. The *red arrows* mark gaps, breaks, or rearrangements in metaphase chromosomes.

attenuated by pretreatment with RRx-001. Importantly, while RRx-001 attenuated the toxic effects of cisplatin, the activity of cisplatin post RRx-001 treatment was enhanced compared with cisplatin alone (Fig. 11) [33].

The anthracycline antibiotics, while used extensively for the treatment of many cancers, including ovarian, bladder, lung, and breast cancer [34], are associated with irreversible degenerative cardiomyopathy that leads to congestive heart failure [35] in 10% of treated patients, thus limiting their utility [36]. Pretreatment with RRx-001 prior to doxorubicin was found to partially attenuate the cardiotoxic effects of doxorubicin in a similar way to the positive control, candesartan (Fig. 12) [33].

RRx-001 was also found to have radiosensitizing properties to tumors while sparing normal tissue [7]. In the gastrointestinal (GI) crypt cell assay, administration of RRx-001 to mice at 10 mg/kg before total body irradiation resulted in a protective effect to the GI stem cells (Fig. 13), while, in a separate experiment, RRx-001 was also shown to be a radiosensitizer (Fig. 14).

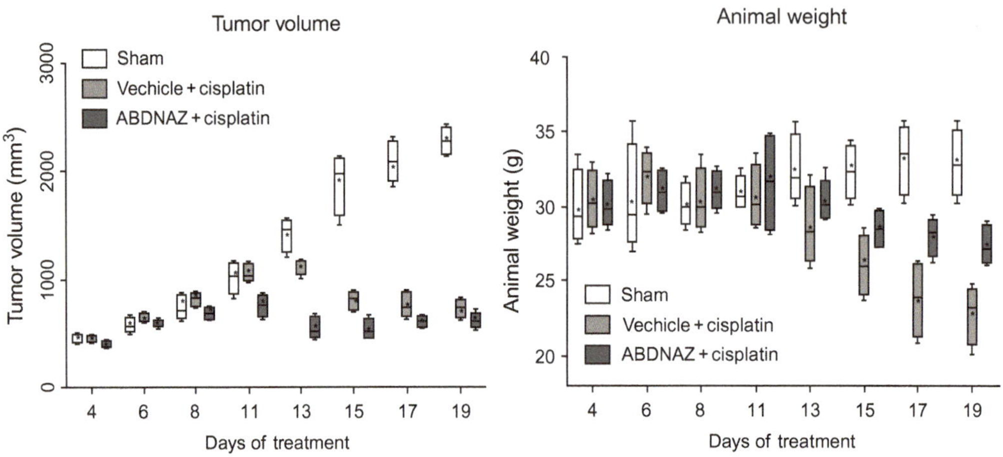

FIG. 11 RRx-001 (ABDNAZ) chemosensitization of tumors with cisplatin *(left panel)* with reduced body weight loss in cisplatin + RRx-001 (ABDNAZ)-treated mice *(right panel)*.

FIG. 12 Attenuation of doxorubicin cardiotoxicity by RRx-001: left ventricular pressure versus volume.

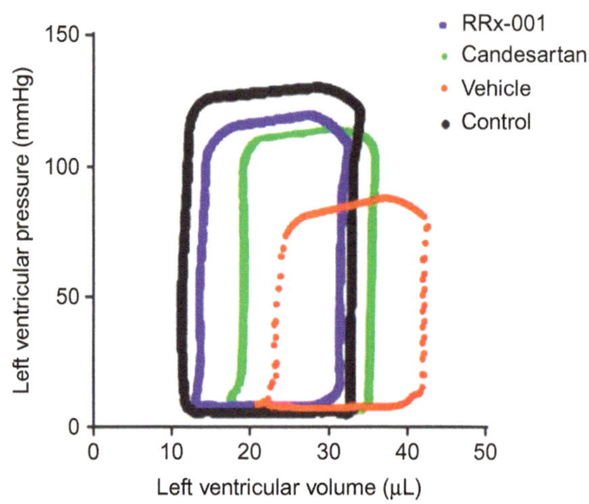

Evidence of chemoprotection was also observed in the clinic. Analysis of interim safety data in resensitization trials of RRx-001, where RRx-001 is dosed to resensitize refractory or resistant tumors to chemotherapy, suggested a much lower incidence of chemotherapy-induced toxicity than anticipated. In the "ROCKET" (NCT02096354) study, a trial in third- or fourth-line colorectal cancer in patients that had previously received and failed irinotecan-based therapies such as FOLFIRI, before being dosed with RRx-001 and then rechallenged with irinotecan, the incidence of diarrhea during irinotecan administration was much lower than that reported on the irinotecan package insert. Fig. 15 shows the incidence of treatment emergent adverse events in the GI disorder system organ class for RRx-001, irinotecan post RRx-001, and regorafenib, the study comparator drug.

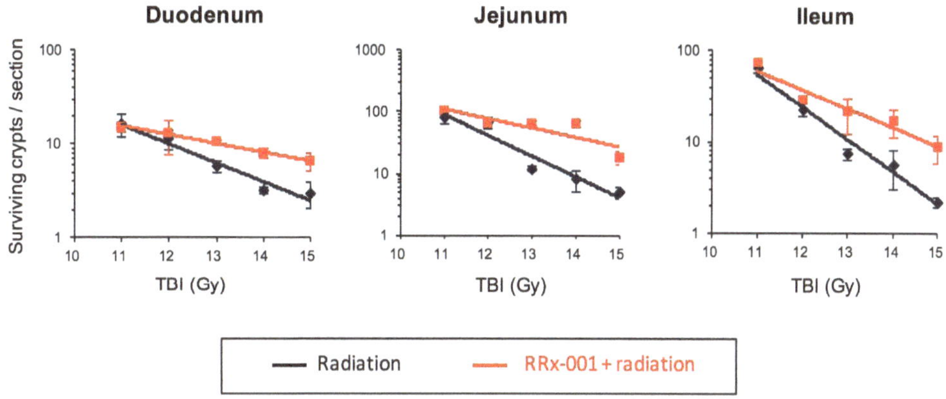

FIG. 13 Intestinal crypt stem cell assay. Normal C3H mice were irradiated with total body irradiation with or without preadministration of 10 mg/kg RRx-001. Three mice/dose point. Eight section per intestine (mean ± SD).

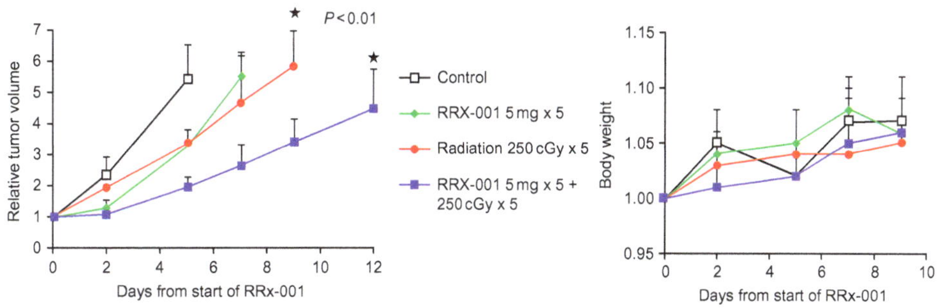

FIG. 14 Radiosensitizing effects of RRx-001 in the SCCVII mouse syngeneic tumor model.

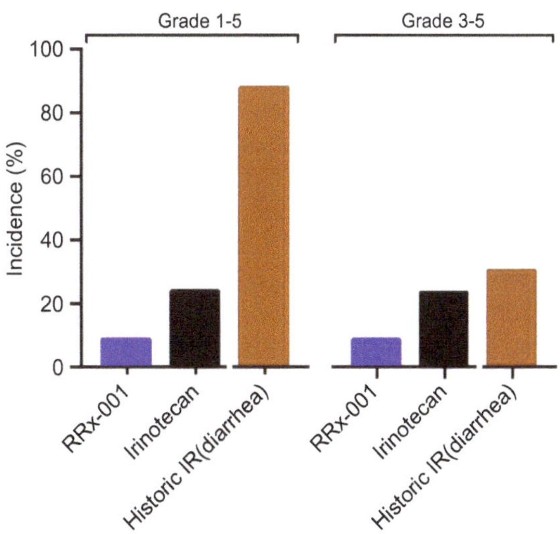

FIG. 15 Incidence of any adverse event (related or not related) by gastrointestinal system organ class comparing RRx-001 and irinotecan for grades 3–5 and all grades. Historic irinotecan is plotted next to the study data for comparison.

While comparisons with data not generated within the same study should be made with caution, the incidence of any GI adverse event for patients receiving irinotecan (180 mg/m^2) post RRx-001 appeared to be much lower than that for patients experiencing diarrhea > 24 h after administration of single-agent irinotecan (125 mg/m^2) weekly in the second-line colorectal cancer setting as described in the CAMPTOSAR package insert [37].

We hypothesize that these radio-and chemoprotective effects arise from the RRx-001-induced activation, in normal tissue, of NF-E2-related factor (Nrf2) [38], a key component of the antioxidant response element (ARE). Activation of Nrf2 and ARE by "prooxidant-antioxidant" molecules has previously been described as a potential mechanism of protection against both radiation damage [39] and chemotherapy-induced toxicity [40–42].

RRx-001: PROSENSITIZATION OBSERVED IN CLINICAL TRIALS

The second aspect of prosensitization is resensitization to chemo- and radiotherapy. Resensitization to previously failed therapy was first observed in the first-in-man phase-1 clinical study [43]. The trial enrolled a typical "all failed" phase-1 patient population and dose-escalated RRx-001 through six dose cohorts. Patients were continued on RRx-001 until toxicity or disease progression [12]. In this study, 10 of 12 (83%) patients who had completed RRx-001 received subsequent chemotherapy and responded well with respect to survival (median survival approximately 20 months). Four patients who were resistant to chemotherapy prior to enrollment were rechallenged to previously effective therapies after completing RRx-001 treatment and were found to respond well with a median overall survival of 18 months (Fig. 16).

The concept of resensitization to previously failed or refractory therapies was explored further in a series of phase-2 studies. The main two studies are ROCKET (NCT02096354) in colorectal cancer with patients receiving RRx-001 until progression followed by reintroduction of second-line irinotecan against a regorafenib comparator arm and QUADRUPLE THREAT (NCT02489903) in refractory/resistant small-cell lung cancer (SCLC), EGFR+ non-SCLC, high-grade neuroendocrine, and resistant/refractory gynecologic cancers including ovarian and malignant mixed Müllerian tumors.

In the ROCKET study, a number of instances of resensitization to irinotecan-based therapy in histologically confirmed metastatic refractory colorectal cancer patients after once-weekly treatment with RRx-001 until progression were observed. A majority of RRx-001 patients that completed the priming portion of the study were able to receive irinotecan-based therapy, either as irinotecan alone or as the FOLFIRI combination with or without bevacizumab, and a significant number of these patients demonstrated resensitization as determined by a decrease of carcinoembryonic antigen (CEA) after restarting irinotecan (Fig. 17).

At the first interim analysis, comparison of the median overall survival between the RRx-001 and regorafenib arms showed an advantage of approximately 2 months in favor of the RRx-001.

The QUADRUPLE THREAT clinical study focuses on the resensitization of tumors that are typically treated with platinum-based therapies such as lung, ovarian, and neuroendocrine tumors. On enrollment, patients receive RRx-001 once weekly until progression; then, first-line platinum doublets, corresponding to that typically indicated for the specific tumor type,

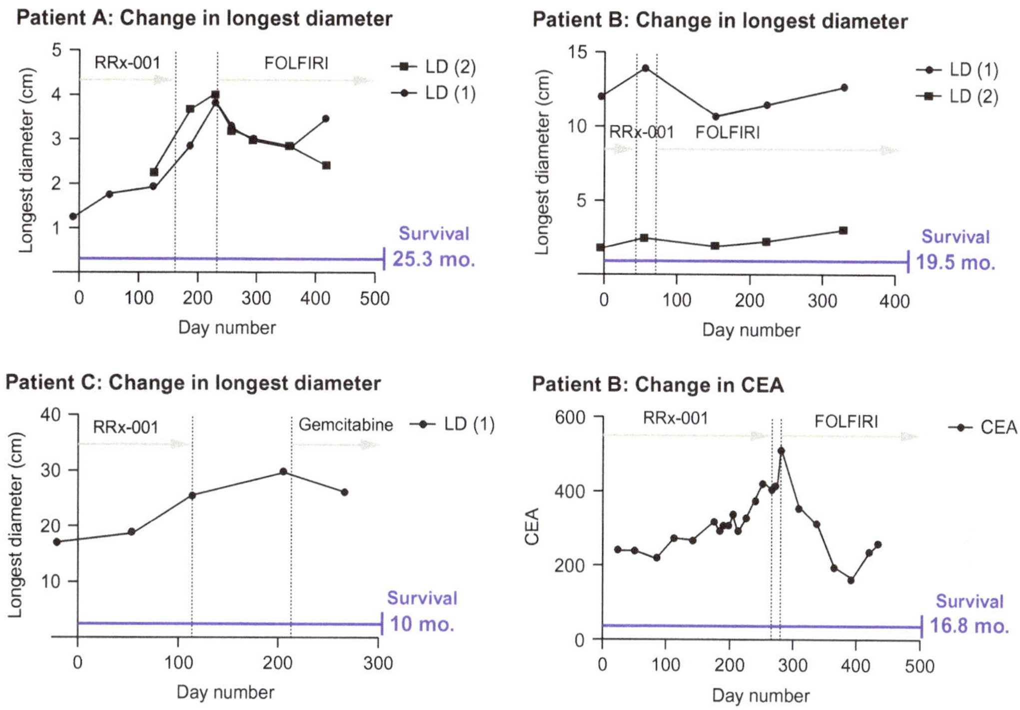

FIG. 16 Resensitization to previously failed therapies: overall survival, change in tumor diameter (patients A–C) and CEA levels, a marker of disease burden in colorectal cancer (patient D). Each graph shows time from first dose of RRx-001, indicating duration of RRx-001 and subsequent chemotherapy administration.

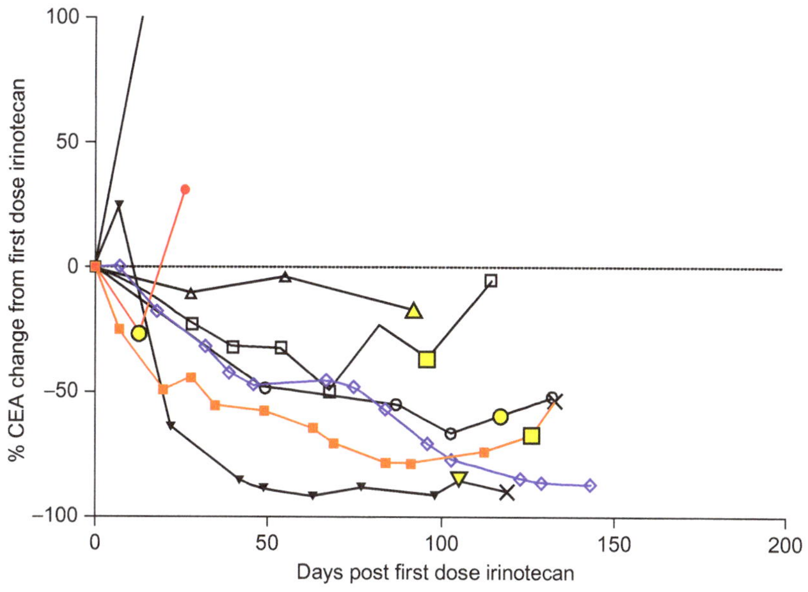

FIG. 17 Resensitization in the ROCKET study: percent change in CEA in representative patients after restarting irinotecan post RRx-001. Day 0 = fist irinotecan dose post RRx-001. *Yellow markers* show last dose of irinotecan.

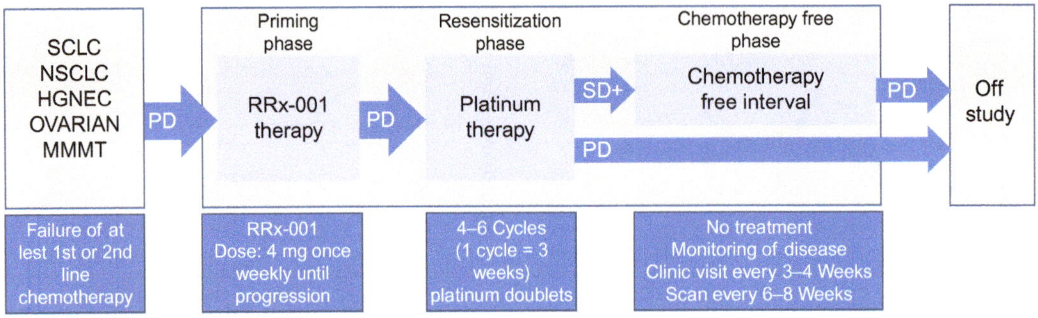

FIG. 18 QUADRUPLE THREAT study design.

are reintroduced. After completion of the required cycles of platinum-doublet therapy, patients with a stable or better response enter a chemotherapy-free interval (CFI) allowing the tumor, with the pressure of chemotherapy removed, to revert to a more chemosensitive, less energetically costly phenotype, that would enable further treatment (Fig. 18).

Although the trial is still in progress, to date, four of nine (44%) evaluable patients have a CFI longer than 3 months, suggesting of the transformation tumor to a more chemosensitive tumor phenotype by RRx-001. In addition, six of nine (71%) evaluable patients have experienced resensitization as evidenced by durable partial responses.

The following two case reports describe responses to resensitization by SCLC patients. Extensive stage SCLC patients are characterized as sensitive, resistant, or refractory based on their response to platinum therapy. Although a number of case reports on resensitization by RRx-001 have been published [15,44–53], these two reports are of particular of interest as they highlight a change in the disease from an aggressive and resistant/refractory phenotype to a sensitive one (Fig. 19).

In the first case [45], a 49-year-old Hispanic male ex-smoker with extensive stage refractory-relapsed SCLC with spinal and mediastinal metastases and with a malignant pleural effusion enrolled on the trial with a history of extensive platinum-doublet treatment in which he received a full six cycles of carboplatin-etoposide as first-line therapy and achieved a durable partial response. Subsequent relapse 8 months later was treated with cisplatin, etoposide, and belinostat; however, the patient progressed while on therapy suggesting that his disease had become refractory to platinum. On trial treatment with RRx-001 resulted in a stable response with the computerized tomography (CT) scan showing a reduction in the pleural effusion (Fig. 20).

At 10 weeks, the patient was discontinued from RRx-001 due to symptomatic progression despite radiographically stable disease and was rechallenged with platinum doublets per protocol. After 12 weeks of therapy, a routine CT scan showed a partial response (Fig. 21), which after the patient completed six cycles of platinum therapy (approximately 20 weeks) allows entry into the CFI that lasted 7 months. Survival is now 19 months from enrollment and is ongoing.

A similar case [53] describes an 82-year-old white male patient with platinum-resistant human papillomavirus (HPV)-positive SCLC of the base of the tongue metastatic to the liver (clinical stages T3, N0, and M1). On initial diagnosis, the patient completed six cycles of

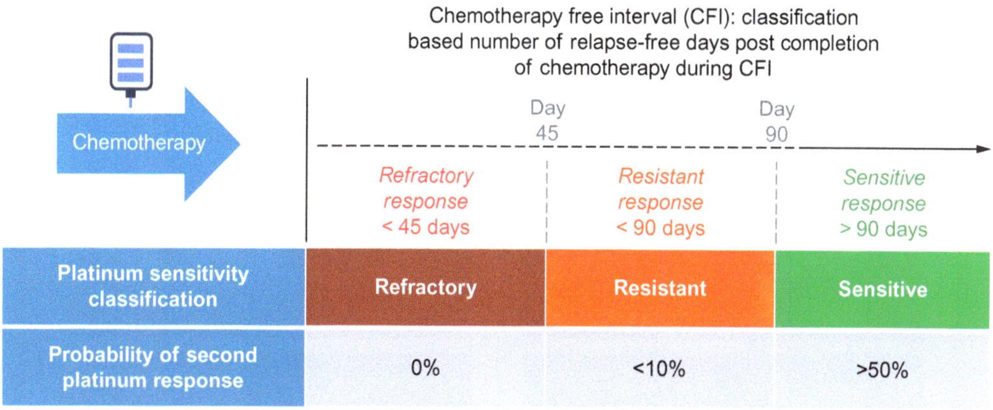

FIG. 19 Disease classification for small-cell lung cancer. Platinum sensitivity is classified as refractory, resistant, or sensitive, according to the time relapsed during a chemotherapy-free interval since finishing first-line treatment. Probability of retreatment response is shown for each group of patients.

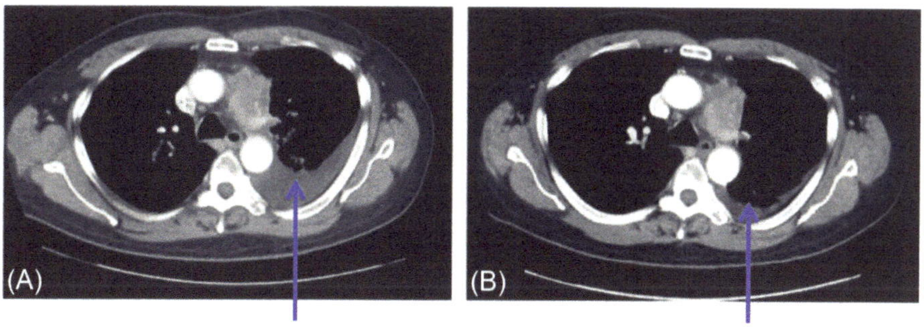

FIG. 20 CT scans demonstrating a left-sided pleural effusion *(blue arrows)* (A) before treatment with RRx-001 and (B) its near complete resolution after 6 weeks.

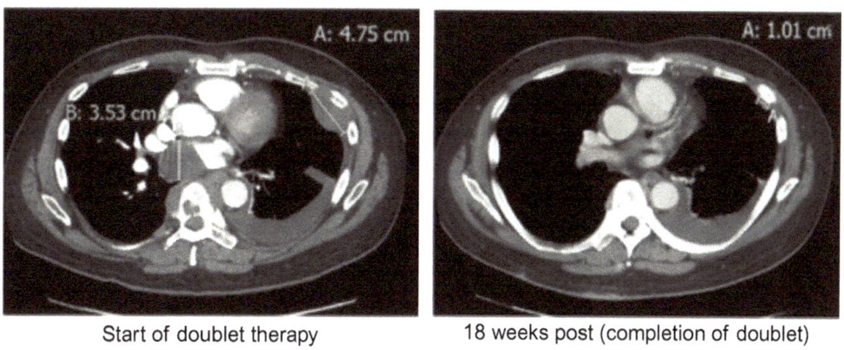

FIG. 21 Patient computed tomography (CT) scans during doublet therapy. *(Left)* At the start of doublet therapy; CT scan shows a 3.53 cm anterior mediastinal lesion (labeled B) and a metastasis measuring 4.75 cm invading the chest wall and the ribs (labeled A). *(Right)* CT scan at the end of 4 cycles (12 weeks) of carboplatin/etoposide. The mediastinal mass (B) disappeared, and the chest wall metastasis (A) significantly decreased in size.

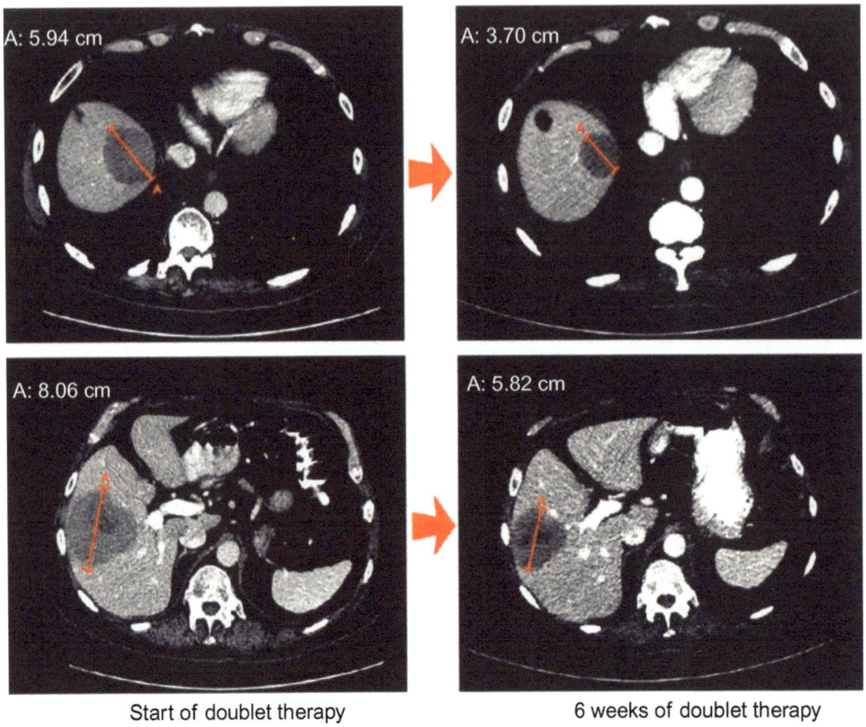

FIG. 22 CT scan performed at the start of doublet therapy *(left)* demonstrating liver metastases with the maximum diameters of 5.94 and 8.06 cm that showed significant shrinkage to 3.70 and 5.82 cm *(right)*.

etoposide-carboplatin; however, he progressed within a month of finishing the treatment, meeting the criteria of resistant disease. The patient enrolled on the QUADRUPLE THREAT study and received 12 weeks of RRx-001. Although a routine CT scan at 6 weeks indicated stability, the 12-week scan indicated progressive disease, and the patient was restarted on carboplatin-etoposide per protocol, completing the required six cycles of treatment. CT scans showed a stable disease at 6 weeks and a partial response at 12 weeks (Fig. 22). Interestingly, the fatigue and loss of appetite that the patient reported on enrollment was greatly alleviated during treatment with RRx-001 with the patient reporting normalization of energy and appetite and improvement in general well-being. Survival to date is 13 months and is ongoing.

These case reports are of particular interest because, unlike low- and intermediate-grade neuroendocrine tumors that are typically indolent, SCLC is an aggressive tumor type with poor prognosis and chemoresistance and rapid growth resulting in poor survival. The natural history of SCLC is characterized by a very rapid doubling time (25–160 days), metastases that are widespread and that manifest early in the disease history and the development of multidrug chemoresistance. Untreated patients have a median survival of 3 months for limited, localized disease and 1.5 months for extensive disease [54,55]. The conversion of a very aggressive disease into a more indolent phenotype that can be effectively treated with previously refractory or resistant therapy represents a new paradigm in the treatment of cancer.

RRx-001 AS A RADIOSENSITIZER

Nitric oxide exerts a strong radiosensitizing effect through similar direct mechanisms to oxygen, by fixing in place DNA radiation damage [56] and also through an indirect mechanism that include changes in tumor perfusion [57]. While RRx-001 acts as a local, hypoxia-mediated NO donor, through reduction of nitrite to nitric oxide, further mechanisms that arise from RBC oxidation—iron- and heme-rich exosome shedding, for example, lead to tumor oxidative and nitrosative stress and therefore further contribute to the radiosensitizing effects [58,59].

Melanoma brain metastases are a common but increasingly prevalent complication [60–63]. While melanoma brain metastases are commonly treated with whole brain radiotherapy (WBRT) for multiple brain metastases or with stereotactic radiosurgery (SRS) for nonresectable single metastases or oligometastases [15], the relative radioresistance of melanoma brain metastases means that the treatment goal is often limited to disease control to conserve neurological function and thus, potentially, performance status and survival.

The radiosensitizing effects of RRx-001 were explored in a phase-1/phase-2 clinical trial combining WBRT with RRx-001 in patients with brain metastases arising from any tumor type. The first patient enrolled on the BRAINSTORM (NCT02215512) study was a 62-year-old white male with brain metastases from primary melanoma, enrolling with decreased vision and central and right-sided visual field deficit due to large masses in the left and right occipital lobes. Per protocol, the patient underwent dynamic contrast-enhanced brain MRI (DCE-MRI) pre- and postsingle first dose of RRx-001 before receiving fractionated radiotherapy (30 Gy in 10 fractions over 2 weeks), together with RRx-001 (5 mg/m^2) twice weekly during radiation treatment only. DCE-MRI showed a decrease in k^{trans} suggesting a reduction in vessel leakiness post RRx-001. With 2 days of starting WBRT+ RRx-001, the patient reported an improvement in visual symptoms and, by the end of treatment, a near resolution of visual field deficit [64]. Four weeks after WBRT+ RRx-001 contrast MRI revealed a decrease in the size of the largest left and right occipital masses and imaging 4 months after treatment showed a continued response (Fig. 23).

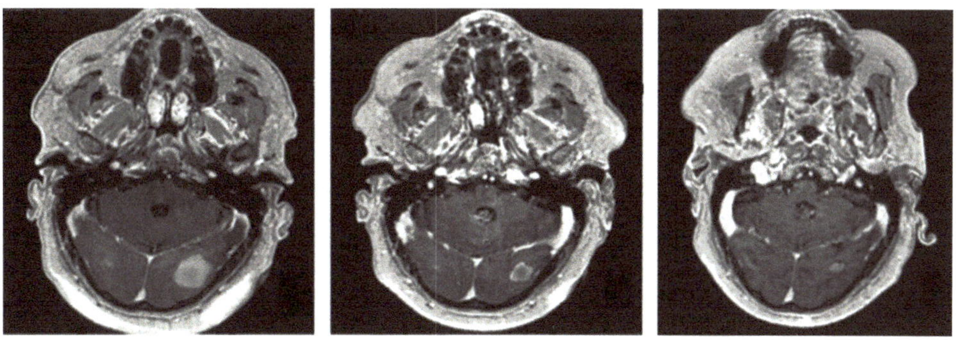

FIG. 23 Gadolinium-enhanced contrast MRI imaging: left, representative image of the patient's left occipital lobe at screening; middle, marked response at 4 weeks; right, image after 4 months showing a continued response.

CONCLUSIONS AND OUTLOOK

RRx-001 represents a novel aerospace-derived compound class with first-in-class biological activity and mechanisms of action. The selective tumor toxicity of RRx-001 is secondary to its red-blood-cell-mediated delivery and internalization by TAMS in the manner of a biological Trojan horse, which, in turn, influences multiple downstream neoplastic pathways and cellular activities. The potential to provide terminal patients that are otherwise out of options with a second chance at first-line chemotherapy treatment represents an unorthodox new paradigm where the typical linear progression through multiple lines of therapy is replaced with the reuse of previously effective agents, thus turning the clock back on cancer. The promise of this RRx-001 reset will hopefully add time, but not toxicity to the lives of cancer patients.

Acknowledgments

The authors would like to acknowledge collaboration with Stanford University and Orbital ATK on the discovery of RRx-001.

References

[1] American Cancer Society. Cancer facts & figures 2013. Atlanta: American Cancer Society; 2016.
[2] Chen L, Han X. Anti-PD-1/PD-L1 therapy of human cancer: past, present, and future. J Clin Invest 2015;125(9):3384–91.
[3] Oronsky B, Scicinski J, Ning S, Peehl D, Oronsky A, Cabrales P, et al. Rockets, radiosensitizers, and RRx-001: an origin story part I. Discov Med 2016;21(115):173–80.
[4] Oronsky BT, Knox SJ, Scicinski J. Six degrees of separation: the oxygen effect in the development of radiosensitizers. Transl Oncol 2011;4(4):189–98.
[5] Watt D, Cliff M. TNAZ based melt-cast explosives: technology review and AMRL research directions. Melbourne, Australia: DSTO-TR-0702, DSTO Aeronautical and Maritime Research Laboratory; 1998.
[6] Scicinski J, Oronsky B, Ning S, Fanger G, Knox S, Bednarski M. Discovery and development of RRx-001, a novel nitric oxide and ROS mediated epigenetic modulator. In: Bonavida B, editor. Nitric oxide and cancer: pathogenesis and therapy. Switzerland: Springer International Publishing; 2015. p. 259–77.
[7] Ning S, Bednarski M, Oronsky B, Scicinski J, Saul G, Knox SJ. Dinitroazetidines are a novel class of anticancer agents and hypoxia-activated radiation sensitizers developed from highly energetic materials. Cancer Res 2012;72(10):2600–8.
[8] Oronsky BT, Reid T, Knox SJ, Scicinski JJ. The scarlet letter of alkylation: a mini review of selective alkylating agents. Transl Oncol 2012;5(4):226–9.
[9] Scicinski J, Oronsky B, Taylor M, Luo G, Musick T, Marini J, et al. Preclinical evaluation of the metabolism and disposition of RRx-001, a novel investigative anticancer agent. Drug Metab Dispos 2012;40(9):1810–6.
[10] Wen B, Fitch WL. Analytical strategies for the screening and evaluation of chemically reactive drug metabolites. Expert Opin Drug Metab Toxicol 2009;5(1):39–55.
[11] Scicinski J, Oronsky B, Cooper V, Taylor M, Alexander M, Hadar R, et al. Development of methods for the bioanalysis of RRx-001 and metabolites. Bioanalysis 2014;6(7):947–56.
[12] Reid T, Oronsky B, Scicinski J, Scribner CL, Knox SJ, Ning S, et al. Safety and activity of RRx-001 in patients with advanced cancer: a first-in-human, open-label, dose-escalation phase 1 study. Lancet Oncol 2015;16(9):1133–42.
[13] Oronsky B, Scicinski J, Cabrales P, Minchinton A. RRx-001, an epigenetic-based radio- and chemosensitizer, has vascular normalizing effects on SCCVII and U87 tumors. Clin Epigenetics 2016;8:53.
[14] EpicentRx, Unpublished data.
[15] Kim MM, Parmar H, Cao Y, Pramanik P, Schipper M, Hayman J, et al. Whole brain radiotherapy and RRx-001: two partial responses in radioresistant melanoma brain metastases from a phase I/II clinical trial: a TITE-CRM phase I/II clinical trial. Transl Oncol 2016;9(2):108–13.

[16] Allen BW, Piantadosi CA. How do red blood cells cause hypoxic vasodilation? The SNO-hemoglobin paradigm. Am J Physiol Heart Circ Physiol 2006;291(4):H1507–12.
[17] Gladwin MT, Ognibene FP, Pannell LK, Nichols JS, Pease-Fye ME, Shelhamer JH, et al. Relative role of heme nitrosylation and beta-cysteine 93 nitrosation in the transport and metabolism of nitric oxide by hemoglobin in the human circulation. Proc Natl Acad Sci U S A 2000;97(18):9943–8.
[18] Fens MH, Larkin SK, Oronsky B, Scicinski J, Morris CR, Kuypers FA. The capacity of red blood cells to reduce nitrite determines nitric oxide generation under hypoxic conditions. PLoS One 2014;9(7):e101626.
[19] Fens MH, Cabrales P, Scicinski J, Larkin SK, Suh JH, Kuypers FA, et al. Targeting tumor hypoxia with the epigenetic anticancer agent, RRx-001: a superagonist of nitric oxide generation. Med Oncol 2016;33(8):85.
[20] Cabrales P, Scicinski J, Reid T, Kuypers F, Larkin S, Fens M, et al. A look inside the mechanistic black box: Are red blood cells the critical effectors of RRx-001 cytotoxicity? Med Oncol 2016;33(7):63.
[21] Cabrales P, Reid T, Oronsky B, Scicinski J. RRx-001 an EXO-based epigenetic anti-cancer agent in phase 2 clinical trials. In: ISEV2014 educational event, San Diego; 2014.
[22] Fens MH, van Wijk R, Andringa G, van Rooijen KL, Dijstelbloem HM, Rasmussen JT, et al. A role for activated endothelial cells in red blood cell clearance: implications for vasopathology. Haematologica 2012;97 (4):500–8.
[23] Das DS, Ray A, Das A, Song Y, Tian Z, Oronsky B, et al. A novel hypoxia-selective epigenetic agent RRx-001 triggers apoptosis and overcomes drug resistance in multiple myeloma cells. Leukemia 2016;30(11):2187–97.
[24] Raghunand N, Scicinski J, Oronsky B, Guntle G, Bruckheimer E, Korn R. RRx-001 oxidation of redox sensitive protein thiols in tumors measured by Gd-LC7-SH enhanced MRI in preclinical tumor models. In: AACR 105th annual meeting, San Diego, April 5–9; 2014 [Abstract 2068].
[25] Schulze A, Harris AL. How cancer metabolism is tuned for proliferation and vulnerable to disruption. Nature 2012;491(7424):364–73.
[26] Hirst DG, Robson T. Nitrosative stress in cancer therapy. Front Biosci 2007;12:3406–18.
[27] Hirst DG, Robson T. Nitrosative stress as a mediator of apoptosis: implications for cancer therapy. Curr Pharm Des 2010;16(1):45–55.
[28] van der Vliet A, Eiserich JP, Cross CE. Nitric oxide: a pro-inflammatory mediator in lung disease? Respir Res 2000;1(2):67–72.
[29] Szabo C, Ischiropoulos H, Radi R. Peroxynitrite: biochemistry, pathophysiology and development of therapeutics. Nat Rev Drug Discov 2007;6(8):662–80.
[30] Zhao H, Ning S, Scicinski J, Oronsky B, Knox S, Peehl DM. Abstract 3515: RRx-001: a double action systemically non-toxic epigenetic agent for cancer therapy. Cancer Res 2015;75(15 Suppl.):3515.
[31] Das DS, Tian Z, Ray A, Ravillah D, Song Y, Richardson PG, et al. Anti-myeloma activity of a novel free radical inducer Rrx-001. Blood 2014;124:4712.
[32] EpicentRx, Inc., Unpublished data.
[33] Oronsky B, Paulmurugan R, Foygel K, Scicinski J, Knox SJ, Peehl D, et al. RRx-001: a systemically non-toxic M2-to-M1 macrophage stimulating and prosensitizing agent in phase II clinical trials. Expert Opin Investig Drugs 2017;26(1):109–19.
[34] Rivankar S. An overview of doxorubicin formulations in cancer therapy. J Cancer Res Ther 2014;10(4):853–8.
[35] Vejpongsa P, Yeh ET. Prevention of anthracycline-induced cardiotoxicity: challenges and opportunities. J Am Coll Cardiol 2014;64(9):938–45.
[36] Octavia Y, Tocchetti CG, Gabrielson KL, Janssens S, Crijns HJ, Moens AL. Doxorubicin-induced cardiomyopathy: from molecular mechanisms to therapeutic strategies. J Mol Cell Cardiol 2012;52(6):1213–25.
[37] CAMPTOSAR Package Insert. http://www.rxlist.com/camptosar-inj-drug/side-effects-interactions.htm [Accessed 1 February 2017].
[38] Ning S, Sekar TV, Scicinski J, Oronsky B, Peehl DM, Knox SJ, et al. Nrf2 activity as a potential biomarker for the pan-epigenetic anticancer agent, RRx-001. Oncotarget 2015;6(25):21547–56.
[39] Gambhir L. 1,4-Naphthoquinone, a pro-oxidant, ameliorated radiation induced gastro-intestinal injury through perturbation of cellular redox and activation of Nrf2 pathway. Drug Discov Ther 2016;10(2):93–102.
[40] Chen W, Li S, Li J, Zhou W, Wu S, Xu S, et al. Artemisitene activates the Nrf2-dependent antioxidant response and protects against bleomycin-induced lung injury. FASEB J 2016;30(7):2500–10.
[41] Xu S, Chen W, Xie Q, Xu Y. Obacunone activates the Nrf2-dependent antioxidant responses. Protein Cell 2016;7(9):684–8.
[42] Poulose SM, Harris ED, Patil BS. Antiproliferative effects of citrus limonoids against human neuroblastoma and colonic adenocarcinoma cells. Nutr Cancer 2006;56(1):103–12.

[43] Reid T, Dad S, Korn R, Oronsky B, Knox S, Scicinski J. Two case reports of resensitization to previous chemotherapy with the novel hypoxia-activated hypomethylating anticancer agent RRx-001 in metastatic colorectal cancer patients. Case Rep Oncol 2014;7(1):79–85.
[44] Carter CA, Degesys A, Oronsky B, Scicinski J, Caroen SZ, Oronsky AL, et al. Flushing out carcinoid syndrome: beneficial effect of the anticancer epigenetic agent RRx-001 in a patient with a treatment-refractory neuroendocrine tumor. Case Rep Oncol 2015;8(3):461–5.
[45] Brzezniak C, Oronsky B, Scicinski J, Caroen S, Cabrales P, Dean Abrouk N, et al. Conversion of platinum-etoposide-resistant to sensitive SCLC after treatment with the epi-immunotherapeutic RRx-001: a case report. Oncol Res Treat 2016;39(11):720–3.
[46] Brzezniak C, Schmitz BA, Peterson PG, Degesys A, Oronsky BT, Scicinski JJ, et al. RRx-001-induced tumor necrosis and immune cell infiltration in an EGFR mutation-positive NSCLC with resistance to EGFR tyrosine kinase inhibitors: a case report. Case Rep Oncol 2016;9(1):45–50.
[47] Carter CA, Browning R, Oronsky BT, Scicinski JJ, Brzezniak C. The case of a zebra that was misdiagnosed as a horse: pulmonary tumor thrombotic microangiopathy, a new paraneoplastic syndrome, mimicking PD-1-induced pneumonitis. Case Rep Oncol 2016;9(1):68–75.
[48] Carter CA, Oronsky B, Caroen S, Scicinski J, Cabrales P, Degesys A, et al. Partial response to carboplatin in an RRx-001 pretreated patient with EGFR-inhibitor-resistance and T790M-negative NSCLC. Respir Med Case Rep 2016;18:62–5.
[49] Carter CA, Oronsky B, Caroen S, Scicinski J, Degesys A, Cabrales P, et al. Partial response in an RRx-001-primed patient with refractory small-cell lung cancer after a third introduction of platinum doublets. Case Rep Oncol 2016;9(2):285–9.
[50] Carter CA, Oronsky BT, Caroen SZ, Scicinski JJ, Cabrales P, Reid T, et al. Partial response to platinum doublets in refractory EGFR-positive non-small cell lung cancer patients after RRx-001: evidence of episensitization. Case Rep Oncol 2016;9(1):62–7.
[51] Carter CA, Oronsky BT, Caroen SZ, Scicinski JJ, Degesys A, Kim MM, et al. RRx-001 in refractory small-cell lung carcinoma: a case report of a partial response after a third reintroduction of platinum doublets. Case Rep Oncol 2016;9(1):171–6.
[52] Carter CA, Schmitz B, Peterson PG, Quinn M, Degesys A, Jenkins J, et al. Immune reactivity and pseudoprogression or tumor flare in a serially biopsied neuroendocrine patient treated with the epigenetic agent RRx-001. Case Rep Oncol 2016;9(1):164–70.
[53] Oronsky B, Caroen S, Zeman K, Quinn M, Brzezniak C, Scicinski J, et al. A partial response to reintroduced chemotherapy in a resistant small cell lung cancer patient after priming with RRx-001. Clin Med Insights Oncol 2016;10:105–8.
[54] Abrams J, Doyle LA, Aisner J. Staging, prognostic factors, and special considerations in small cell lung cancer. Semin Oncol 1988;15(3):261–77.
[55] Aisner J. Extensive-disease small-cell lung cancer: the thrill of victory; the agony of defeat. J Clin Oncol 1996;14(2):658–65.
[56] De Ridder M, Verellen D, Verovski V, Storme G. Hypoxic tumor cell radiosensitization through nitric oxide. Nitric Oxide 2008;19(2):164–9.
[57] Scicinski J, Oronsky B, Ning S, Knox S, Peehl D, Kim MM, et al. NO to cancer: the complex and multifaceted role of nitric oxide and the epigenetic nitric oxide donor, RRx-001. Redox Biol 2015;6:1–8.
[58] Oronsky B, Oronsky N, Fanger G, Oronsky A, Lybeck M, Lybeck H, et al. A review of two promising radiosensitizers in brain metastases: Rrx-001 and 2-deoxyglucose. J Cancer Sci Ther 2015;07:137–41.
[59] Oronsky B, Scicinski J, Ning S, Peehl D, Oronsky A, Cabrales P, et al. RRx-001, a novel dinitroazetidine radiosensitizer. Invest New Drugs 2016;34(3):371–7.
[60] Sampson JH, Carter Jr. JH, Friedman AH, Seigler HF. Demographics, prognosis, and therapy in 702 patients with brain metastases from malignant melanoma. J Neurosurg 1998;88(1):11–20.
[61] Fife KM, Colman MH, Stevens GN, Firth IC, Moon D, Shannon KF, et al. Determinants of outcome in melanoma patients with cerebral metastases. J Clin Oncol 2004;22(7):1293–300.
[62] Davies MA, Liu P, McIntyre S, Kim KB, Papadopoulos N, Hwu WJ, et al. Prognostic factors for survival in melanoma patients with brain metastases. Cancer 2011;117(8):1687–96.
[63] Meier S, Baumert BG, Maier T, Wellis G, Burg G, Seifert B, et al. Survival and prognostic factors in patients with brain metastases from malignant melanoma. Onkologie 2004;27(2):145–9.
[64] Kim MM, Parmar H, Cao Y, Knox SJ, Oronsky B, Scicinski J, et al. Concurrent whole brain radiotherapy and RRx-001 for melanoma brain metastases. Neuro Oncol 2016;18(3):455–6.

CHAPTER

4

Nitric Oxide and Nitric Oxide Donors in Preclinical Studies of Breast and Prostate Cancer

Pablo Garrido, Elaine M. Walsh, Sharon A. Glynn

National University of Ireland Galway, Galway, Ireland

Abstract

Inflammation has been proposed as one of the hallmarks of cancer and therefore one of the main constituents of tumor development. Nitric oxide (NO) is a highly reactive molecule that exerts pleiotropic roles depending on the tissue type, dose, and duration of exposure. The role of NO in cancer progression is particularly controversial. While low-to-moderate doses have been shown to stimulate a more aggressive and invasive phenotype, high doses have been associated with induced cytotoxicity and cytostasis. In this review, we analyze the role of NO in different therapeutic approaches for breast and prostate cancers. In this regard, we discuss the results of different studies showing the efficiency of NO-donating drugs as a single-agent therapy inducing cytostasis, increased oxidative stress or autophagy-mediated cell death. NO has been widely studied not only as a sensitizer of the conventional chemotherapy strategies by reducing the hypoxia-mediated drug resistance or inhibiting commonly hyperactivated survival pathways in cancer as NK-κB but also as conjugated forms of chemotherapeutic agents that result in an increased efficacy and reduced toxicity. However, because of the pleiotropic effects, the effect of NO on cancer progression and its employment as a therapeutic target is still highly debated, and more preclinical and clinical studies are needed to clarify the role of NO in the different types and stages of cancer.

Abbreviations

ASA	acetylsalicylic acid
BRCA	breast cancer susceptibility protein
DEA/NO-aspirin	O_2-(acetylsalicyloyloxymethyl)-1-(*N,N*-diethylamino)-diazen-1-ium-1,2-diolate
DETANONOate	(*Z*)-1-[2-(2-aminoethyl)-*N*-(2-ammonio-ethyl)amino]diazen-1-ium-1, 2-diolate
eNOS	endothelial nitric oxide synthase
GTN	glyceryl trinitrate
iNOS	inducible nitric oxide synthase
IPA/NO-aspirin	O_2-(acetylsalicyloyloxymethyl)-1-(*N*-isopropylamino)-diazen-1-ium-1,2-diolate
JS-K	O_2-(2,4-dinitrophenyl) 1-[(4-ethoxycarbonyl)piperazin-1-yl]diazen-1-ium-1,2-diolate

Nitric Oxide (Donor/Induced) in Chemosensitization, Volume 1
http://dx.doi.org/10.1016/B978-0-12-811020-1.00004-1

L-NAME	L-NG-nitroarginine methyl ester
MAPK	mitogen-activated protein kinases
NF-kB	nuclear factor kappa-light-chain-enhancer of activated B cells
NitDox	nitrooxy-Dox
nNOS	neuronal nitric oxide synthase
NO	nitric oxide
NOS	nitric oxide synthase
NSAIDs	nonsteroidal antiinflammatory drugs
$O_2^{\bullet-}$	superoxide
ONOO−	peroxynitrite
PARP	poly ADP-ribose polymerase
PDT	photodynamic therapy
RNS	reactive nitrogen species
ROS	reactive oxygen species
TNF	tumor necrosis factor
TRAIL	TNF-related apoptosis-inducing ligand
YY1	Yin-Yang 1

INTRODUCTION

Nitric oxide (NO) is a highly reactive molecule with multiple functions within biological systems. At room temperature, NO is an unstable gas with a half-life that varies depending on conditions. In vivo, the half-life does not exceed a few seconds [1]. In the human body, NO has pleiotropic roles that depend not only on the tissue in which it is located but also on a dose-dependent manner. NO is synthesized by three different nitric oxide synthase (NOS) enzymes, each the product of a separate gene. The enzymatic activity of the NOS is regulated by different factors and with each producing varying levels of NO. NO is synthesized by the NOS enzymes through the hydroxylation of L-arginine to N^{ω}-hydroxy-L-arginine. This is subsequently oxidized to produce L-citrulline and NO [2]. The neuronal isoform (nNOS) regulates cell signaling in the nervous system and thus controls central blood pressure, synaptic plasticity, and vasodilatation via peripheral nitrergic nerves [3,4]. The endothelial isoform (eNOS) is constitutively expressed in the endothelial cells and relates to vasodilation in the peripheral system with numerous vasoprotective effects [5,6]. Both nNOS and eNOS produce considerably low levels of NO in a calcium-dependent manner. The third isoform is inducible NOS (iNOS) that can be expressed by many different cell types. iNOS plays an important role in immune response regulation and wound healing, through the production of high levels of NO in response to infectious agents or cytokines [2,7]. When NO is produced by any of these enzymes, it functions by differing mechanisms, including the activation of the only known NO receptor (soluble guanylyl cyclase), interaction with transition metal-containing proteins or interaction with proteins without attachment of the NO group and by posttranslational modification, such as S-nitrosylation, that regulates the activity of the target proteins [8,9].

NO AND CANCER

The role of NO in cancer progression is controversial. Paradoxically, there are several studies showing the role of NO in both cancer promotion and protection against cancer [10–12]. Low concentrations of NO have been shown to cause increased growth factor signaling activation, increased cell migration and invasion, and tumor-promoting inflammatory mechanisms both in vivo and in vitro [12–17]. Accordingly, in malignant melanoma, patients expressing high levels of iNOS were associated with poorer survival outcomes compared with those that did not express iNOS [18]. Furthermore, there are preclinical and clinical data suggesting that iNOS overexpression, and therefore increased NO levels, triggers inflammation in the tumor microenvironment [19]. iNOS has been linked with poorer survival in estrogen receptor (ER)-negative [13,15,17] and triple-negative breast cancer [20,21]. In prostate cancer, an association of iNOS was found with tumor stage, high Gleason score, DNA aneuploidy, and perineural infiltration [22]. Expression of iNOS in prostate cancer may also be implicated in the development of androgen-independent prostate cancer, as NO inhibits the androgen receptor activation and proliferation of AR-positive prostate cancer cells [23], via S-nitrosylation and inactivation of the AR at the C601 residue present in the DNA-binding domain of the AR, thus inhibiting its ability to respond to androgen [24]. In contrast to these studies showing an association of iNOS with tumor progression, high doses of NO and its derivatives, nitrite, nitrate, *S*-nitrosothiols, nitrosamines, nitrous oxide, and *S*-nitrosoglutathione, have also been shown to be cytotoxic and induce cytostasis by cyclin D1 inhibition and triggering mitochondrial-mediated apoptosis through the activation of Bax in breast cancer cells [12,25,26].

NO DONORS AS A THERAPEUTIC AGENT

Among all chemical NO donors, the most commonly used in clinical practice are the organic nitrates. Due to its vasodilator effects, glyceryl trinitrate (GTN) has been used for decades in the regulation of arterial hypertension. GTN has also been proposed as an anticancer agent due to its ability to improve tumor perfusion, oxygenation, and therefore drug delivery [27]. One phase II clinical trial has studied the use of GTN in prostate cancer patients who have undergone definitive radiotherapy or radical prostatectomy. In this study, the administration of low doses of transdermal GTN resulted in prolongation of PSA doubling time and therefore less aggressive disease with less frequent or great time to biochemical recurrence [28]. NO was first observed to be influential on tumor progression when Hibbs et al. described that nitrite and nitrate, produced by murine macrophages, had a cytotoxic effect on murine L10 hepatocarcinoma tumor cells [29,30]. Although NO itself can be toxic, most of the cytotoxic effects of high doses of NO have been attributed to its oxidation products, such as peroxynitrite ($ONOO^-$) resulting from the reaction of NO with superoxide ($O_2^{\bullet -}$) [31]. As a result, excessive production of NO can lead to the production of the "reactive nitrogen species" (RNS) that are associated with cellular damage [32]. At the same time, NO doses in the micromolar range, in addition to being toxic, require a significant amount of oxygen to be stable. This produces a concomitant hypoxia that further contributes to NO toxicity [31].

There is further evidence that another NO donor, O_2-(2,4-dinitrophenyl) 1-[(4-ethoxycarbonyl)piperazin-1-yl]diazen-1-ium-1,2-diolate (JS-K), induces tumor growth arrest and necrosis in xenograft models of prostate cancer and prostate cancer cell lines [33–35]. JS-K has also been shown to induce autophagy-mediated cell death in breast cancer cells. However, this compound has shown limited toxicity in normal and nontransformed epithelial breast cells [36]. JS-K has shown antimetastatic effects through the activation of the tissue inhibitor of matrix metalloproteinase 2 and the downregulation of the mitogen-activated protein kinases (MAPK) signaling pathway in breast cancer cells [37].

Normal development of cellular metabolism leads to the production of "reactive oxygen species" (ROS). ROS are highly reactive molecules that can produce DNA alterations (mutations) by damaging the chemical structure of the DNA bases [38]. Under physiological conditions, low-to-moderate levels of ROS can induce cell proliferation and differentiation. On the other hand, ROS excess is scavenged by specific enzymes that protect the cell from the harmful effects of these superoxidant species [39,40]. In the cancer setting, the increased metabolism of a tumoral cell leads to increased ROS production. In the presence of NO, a reaction with ROS produces the RNS [41].

Some cancer-associated mutations have a role in protecting the cell against ROS and RNS. For example, breast cancer susceptibility protein (BRCA)-1 and BRCA are proteins that play an essential role in cell survival by allowing repair of DNA double-strand breaks by homologous recombination [42]. Germline mutations in the *BRCA* genes are associated with familial patterns of breast and ovarian cancer [43]. Importantly, BRCA1 prevents oxidative damage by diminishing ROS through the stabilization of the antioxidant enzyme Nrf2 [44,45]. NO has been shown to decrease BRCA1 expression via protein phosphatase 2 activation [46]. Based on this hypothesis, some authors have already suggested the possibility of using NO donors as sensitizers to ROS inducer therapeutic agents in BRCA mutant breast cancers [47]. Synthetic lethality is a mechanism that could render BRCA-deficient cancers more vulnerable [42,48]. The use of poly ADP-ribose polymerase (PARP) inhibitors has been shown to be effective in BRCA-mutated cancers when combined with DNA-damaging chemotherapy drugs, such as platinum salts or taxanes, due to the lack of homologous recombination in BRCA-deficient cells [49]. Many of these chemotherapeutic agents also produce ROS, and as a result, it has been suggested that these drugs be used in combination with NO donors to increase DNA damage by RNS generation.

NO AS A SENSITIZER

Drug resistance is one of the major causes of cancer relapse and therefore is associated with poor prognosis [27]. Initially, tumors may be sensitive to a specific chemotherapy drug. However, on relapse, the same tumors are often found to be resistant to this same drug. One mechanism by which drug resistance develops is due to the overexpression of survival and antiapoptotic pathways. In this way, the nuclear factor kappa-light-chain-enhancer of activated B cells (NF-κB) pathway is one of the more commonly hyperactivated pathways associated with drug resistance [50]. In fact, the inhibition of NF-κB by transglutaminase 2 depletion has been shown to reverse drug resistance in breast cancer cells [51]. Like many other pro-inflammatory stimuli, NO has a direct role in NF-κB modulation. In addition, NO has been proposed as a chemotherapy sensitizer due to its role in the inactivation of NF-κB signaling. Tumor necrosis factor (TNF)α can promote either cell apoptosis or cell survival depending on which signaling pathway is triggered [52]. TNFα signaling through NF-κB increases the expression of antiapoptotic genes that are often upregulated in tumor biology [53]. *S*-Nitroso-*N*-acetylpenicillamine increases NO levels and therefore sensitizes metastatic prostate cancer cell lines to TNFα-mediated cytotoxicity through TNFα-NF-κB signaling downregulation in vitro [54].

The TNF-related apoptosis-inducing ligand (TRAIL) is a homotrimeric type II transmembrane protein of the TNF superfamily that induces apoptosis through interaction with its receptors [55]. Once it is produced by the immune NK cells, TRAIL is a potent activator of apoptosis in transformed cells, whereas normal cells appear to be resistant [56]. Like most of the members of the TNF superfamily, TRAIL can be cleaved at the cell surface to form a soluble molecule, inducing apoptosis in different tissues or organs. Several clinical studies have suggested the use of recombinant human TRAIL or agonistic monoclonal antibodies against TRAIL receptors as a promising drug in cancer treatment [57,58]. However, it appears that the clinical therapeutic benefit of TRAIL is limited. Most cancer cells develop resistance to TRAIL monotherapy in early stages [59,60]. The employment of NO donors such as (*Z*)-1-[2-(2-aminoethyl)-*N*-(2-ammonio-ethyl)amino]diazen-1-ium-1, 2-diolate (DETANONOate) or saquinavir-NO has been shown to restore TRAIL sensitivity in prostate cancer cell lines and xenograft models, through inhibition of NF-κB and Ying-Yang 1 (YY1) [61–63].

The Fas receptor is another member of the receptor TNF receptor superfamily. Like the TRAIL protein, the activation of the Fas receptor by its ligand (FasL) in cytotoxic T lymphocytes plays a key role in the immune-mediated elimination of cancer cells [64]. However, many cancer cells are resistant to FasL-mediated apoptosis [65]. Some studies have demonstrated that NO donors can be used to resensitize tumor cell lines to FasL-mediated apoptosis. Prostate cancer cell lines treated with high doses of DETANONOate were shown to upregulate Fas expression via negative regulation of YY1 by S-nitrosylation. The inhibition of YY1 was subsequently shown to increase sensitivity of the prostate cancer cell lines to FasL-induced agonist monoclonal antibody therapy [66,67].

The tumor microenvironment has been proposed as a one of the major determinants of chemosensitivity [68]. Several studies suggest that hypoxia that is inherent to solid tumors decreases the efficacy of chemotherapy drugs [69,70]. Hypoxia in solid tumors is a direct consequence of both the increased proliferation rate and the anarchic cell architecture [71]. On one hand, excessive cell proliferation causes an increased mitochondrial function and increased oxygen consumption. On the other hand, the chaotic tumor architecture elicits an inadequate blood vessel distribution and structure. This produces intratumoral oxygen gradients that generate regions with low oxygen concentration or even anoxia [72,73]. This hypoxia seen in solid tumors has been widely associated with reduced efficacy of anticancer therapies. Both ionizing radiation and chemotherapeutic agents base their effectiveness, at least partly, on the production of increased levels of ROS that contribute to the cytotoxic mechanisms of several widely employed drugs, namely, doxorubicin, cisplatin, or 5-fluorouracil [74–76]. Hypoxia regulates intracellular NO levels in both a tissue and an isoform-dependent manner. In the tumoral context, hypoxia has been shown to inhibit iNOS gene expression in melanoma cells [8]. The differential influence of hypoxia on the different NOS isoforms can be explained by their different KMs for oxygen [77]. Accordingly, nNOS and iNOS activity would be limited in a hypoxic environment, while the effect on eNOS would be minimal. However, there is experimental evidence suggesting that suppression of intracellular NO production is an essential mechanism underlying tumor resistance to chemotherapy in a hypoxic environment [78]. Increased production of ROS and RNS is one of the mechanisms underlying both chemotherapy and radiotherapy efficiency; it results in an increased mutation rate that leads to cell-cycle arrest and cell death. Consequently, the combination of chemotherapy and radiotherapy with NO donors has been suggested as a plausible option to increase the success of standard anticancer therapies. Studies in breast and prostate cancer models have shown that low concentrations of NO donors like GTN or DETANONOate diminish hypoxia-induced resistance to doxorubicin and 5-fluorouracil by counteracting the diminished NO production inherent to hypoxia and increasing intracellular drug accumulation [68,78–80]. High doses of DETANONOate have also shown increased sensitivity to cisplatin-diamine-dichloro platinum in prostate cancer models, both in vitro and xenografts, by YY1 and Bcl-2/BclXL inhibition [81]. However, the role of NO donors in chemotherapy sensitization remains controversial. The NO donor propylamine propylamine nonoate (PPNO) has been shown in vitro to inhibit both the catalytic and relaxation activity of topoisomerase IIα and, therefore, contribute to etoposide resistance in MCF-7 estrogen-receptor-positive breast cancer cells. Controversially, the same NO donor had no effect over other topoisomerase IIα inhibitors like doxorubicin [82].

As previously described, hypoxic cells are also resistant to radiation therapy [83]. However, unlike with chemotherapy agents, studies assessing the role of NO as a radiation sensitizer in breast or prostate cancers are limited. As mentioned above, GTN improved the outcomes of prostate cancer patients undergoing surgery or radiotherapy for the treatment of localized disease [28]. Whether this was due to radiosensitization cannot be deduced at this time. Certainly NO donors such as the hypoxia-activated RRx-001 have been proposed as radiosensitizers [84]. Additionally, the NO conjugate NO-sulindac was shown to sensitize PC-3 prostate cancer cells to radiotherapy, which was accompanied by a reduction in HIF-1α levels [85]. It should however be noted that NO effects on radiosensitization are complex and can be dependent on both dose and cell status, as evidenced by the effect of p53 mutation status in lung cancer cell sensitivity to radiation [86].

NO CONJUGATES

For over a decade, studies have been assessing whether the chemical derivatives of NO-releasing products can be combined with other compounds to increase their efficiency and improve their pharmacodynamics. For example, combining these products with chemotherapeutic agents such as doxorubicin or PARP inhibitors or with liposomal vesicles has enhanced the cytotoxic potential of these drugs and has emerged as a promising future therapeutic strategy [87–89]. In breast cancer cells, a semisynthetic NO-releasing doxorubicin (Dox), called nitrooxy-Dox (NitDox), has been shown to be effective in doxorubicin-resistant MCF-7 cells. NitDox induces nitration of P-glycoprotein and multidrug resistance-related protein 1 that impairs doxorubicin efflux. NitDox avoids the cardiotoxicity profile usually associated with the use of doxorubicin. Interestingly however, there was no therapeutic effect observed with liposomal doxorubicin [90].

The nonsteroidal antiinflammatory drugs (NSAIDs) are universally used as analgesia and are characterized by specific inhibitors of cyclooxygenase-1 and cyclooxygenase-2. Several studies have identified NSAIDs as potential chemopreventative agents, although they may be highly toxic [91]. The combination of a NO donor with a NSAID (NO-NSAID) could boost the efficacy of NSAIDs and potentially minimize their side effects. Sulindac is one such NSAID drug. NO-releasing analogues of sulindac have shown enhanced cytotoxicity but less severe side effects compared with sulindac alone in prostate cancer lines in vitro [92]. The mechanism underlying this enhanced cytotoxicity remains unclear. Aspirin, also known as acetylsalicylic acid (ASA) is used in many clinical settings. The combination of ASA and NO donors has been widely studied in the treatment of cancer. O_2-(acetylsalicyloyloxymethyl)-1-(*N*-isopropylamino)-diazen-1-ium-1,2-diolate (IPA/NO-aspirin) and O_2-(acetylsalicyloyloxymethyl)-1-(*N,N*-diethylamino)-diazen-1-ium-1,2-diolate (DEA/NO-aspirin) produce increased oxidative species that results in increased DNA damage and GAPDH inhibition. As a direct result of this cellular damage, the tumoral cell undergoes apoptosis by caspase-3 activation. The combinations of both IPA/NO-aspirin and DEA/NO-aspirin have been shown to be significantly more effective compared with aspirin or nitric oxide donors alone, in both in vitro human breast adenocarcinoma cells and in vivo in nude mouse models [93].

Both NO-ibuprofen and NO-aspirin inhibit proliferation and induce apoptosis in the androgen-dependent (LNCaP) and androgen-independent (PC-3) prostate cancer epithelial cell lines [94]. Additionally, NO-sulindac displayed similar properties in prostate cancer cells, while sulindac alone failed to elicit any cytotoxic effect [95]. Further research showed that the cytotoxicity of NO-sulindac was partially attributed to the NO-releasing moiety and that the drug also inhibited the hypoxic response and invasiveness of the cells via PI3K/Akt signaling [96]. It is evident that the combination of NSAIDs with NO donors has identified a promising new approach in cancer therapeutics. Some studies even suggest that the NO-NSAID combination improves the potency in the order of 34,000-fold, when compared with the parent NSAID alone, in in vitro models of breast and prostate cancer cells [97,98]. Photodynamic therapy (PDT) has emerged as a promising therapeutic strategy against certain cancer types. PDT is based on the triggering of a photochemical reaction, resulting in the accumulation of ROS and NO. This, in turn, causes cell death in a specific tumor area [99]. PDT induces iNOS expression and therefore NO production in both the tumor and the surrounding microenvironment [100]. The combination of a nonsteroidal antiandrogen compound with a photosensitizer that releases NO in response to a light stimulus has shown increased efficiency compared with a photosensitizer alone in prostate cancer cell lines [101]. This can open a new area of research to significantly improve both the specificity and the efficacy of anticancer drugs.

NOS INHIBITION

As this chapter has outlined, NO exerts a dichotomous effect on the physiology of cancer cells, which varies depending on its concentration. As a result, in addition to NO donors, several NOS inhibitors have been explored both as chemopreventative and chemotherapeutic agents. NO is extensively associated with key molecular aspects, which leads to metastasis and a malignant phenotype in both in vitro and in vivo models [102,103]. Treatment with L-NG-nitroarginine methyl ester (L-NAME), a pan-NOS inhibitor, or 1400 W that is specific for iNOS, has been shown to significantly decrease cell proliferation and cause decreased migration in several breast cancer cell lines. In addition, it results in the impaired ability to establish bone and brain metastasis and tumor growth in MDA-MB-231 breast cancer cell xenograft models [17,20,104]. In addition to iNOS, other NOS isoforms have been implicated in cancer progression. Increased expression of eNOS and therefore increased intracellular NO contribute to the development of antiandrogen resistance in prostate cancer. The inhibition of eNOS with L-NAME has been shown to result in the resensitization of prostate cancer cells to bicalutamide treatment [105].

During the malignant process, cells undergo a series of transformations, which includes metabolism. Warburg's hypothesis suggests that, even in the presence of ample oxygen, during glycolysis, cancer cells stop using oxidative phosphorylation (OXPHOS) as their main source of ATP production. This is less effective in terms of ATP production but allows cancer cells ample aerobic substrates in order to sustain a high mitotic rate [106,107]. NO has also been shown to contribute to the metabolic transformation of cancer cells. In breast cancer cell lines, it has been demonstrated that endogenous NO production leads to the inhibition of cytochrome c oxidase and plays a substantial role in cell proliferation. Quenching NO production by using L-NAME as an iNOS inhibitor reduces the proliferation rate in these cancer cells by restoring a more oxidative metabolic phenotype [108].

CONCLUSIONS

The number of studies investigating the role of NO on cancer progression has exponentially increased in the last 30 years. Despite the remarkable progress that has been made understanding the role of NO on the different stages of cancer development and progression, it still remains as a highly controversial field with gaps in our knowledge. Different researchers across the globe continue to debate the role of this reactive molecule as a therapeutic target susceptible to being either inhibited or boosted. In this review, we summarize the studies that have investigated the different roles of NO in breast and prostate cancer progression. The overall conclusion is that there is definitive evidence that targeting NO for the treatment of breast and/or prostate cancer has definite merit, but further studies are needed to understand the specific role of NO in the different stages and environments of cancer progression.

The lack of therapeutic targets is one of the main reasons of the high mortality in some types of cancer, and the role of NO in cancer progression has seemed to emerge, in the last few years, as a plausible target in some of this cancer types. Despite further investigation is still needed, the employment of NO donors as a single therapy in the treatment of breast and prostate cancer does not seem to be a promising strategy in the clinical setting. Some cytostatic and apoptotic effects have been shown in in vitro and xenograft studies [34–36]. However, extrapolating these results to the clinical setting seems complicated because of the nonspecific effects of NO. Much more promising are the studies that propose NO as an adjuvant in cancer therapy. Different NO donors have increased the efficiency of induced synthetic lethality in some specific background like BRCA-deficient breast cancer [48,49]. NO has also been shown to resensitize breast and prostate cancer recurrences to first-choice drugs such as doxorubicin, platinum derivatives, or 5-FU and hypoxia-induced chemoresistance [76,79]. The employment of NO donors has also succeed as immune sensitizers, increasing the efficiency of immune-system-based therapies like those involving the activation of TRAIL or FasL signaling in the immune cells [61,66]. NO has increased the efficiency of other promising, but highly toxic drugs like NSAIDs and topoisomerase inhibitors in the treatment of different cancers. This has allowed investigating the employment of lower doses and therefore reducing the toxicity [90,93,98]. Triple-negative breast cancer is a heterogeneous disease with a poor prognosis because of the lack of effective therapies. iNOS overexpression has emerged as a poor prognosis factor in different patient cohorts all around the world [15,21], pointing to NO and the upregulated pathways as a consequence of the iNOS overexpression, like plausible therapies for these types of cancer [20,109].

In conclusion, despite the titanic effort that has been made and considerable findings achieved, understanding whether NOS overexpression and increased NO levels are involved in the progression or inhibition of cancer will require further investigation to understand the different pathways enhanced or deregulated in the different cancer settings. Understanding every single cancer setting and tumor microenvironments will make the difference in order to decide if the therapeutic approach will require NO donors or NO inhibitors.

Acknowledgments

This work was funded by a Breast Cancer Now Project Grant to Dr. Sharon Glynn (2013MayPR019). Dr. Pablo Garrido is funded by a Clarin-Marie Curie COFUND Postdoctoral Grants by the Principality of Asturias and the FICYT. Dr. Sharon Glynn is also funded by the Irish Cancer Society. Dr. Elaine Walsh is funded by a School of Medicine PhD Scholarship Grant from National University of Ireland, Galway.

Conflict of Interest: No potential conflicts of interest were disclosed.

References

[1] Kelm M. Nitric oxide metabolism and breakdown. Biochim Biophys Acta 1999;1411(2–3):273–89.

[2] Förstermann U, Sessa WC. Nitric oxide synthases: regulation and function. Eur Heart J 2012;33(7):829–37.

[3] Cianciulli A, Calvello R, Porro C, Trotta T, Salvatore R, Panaro MA. PI3k/Akt signalling pathway plays a crucial role in the anti-inflammatory effects of curcumin in LPS-activated microglia. Int Immunopharmacol 2016;36:282–90.

[4] Calabrese V, Mancuso C, Calvani M, Rizzarelli E, Butterfield DA, Stella AM. Nitric oxide in the central nervous system: neuroprotection versus neurotoxicity. Nat Rev Neurosci 2007;8(10):766–75.

[5] Borlaug BA, Melenovsky V, Koepp KE. Inhaled sodium nitrite improves rest and exercise hemodynamics in heart failure with preserved ejection fraction. Circ Res 2016;119(7):880–6.

[6] Monica FZ, Bian K, Murad F. The endothelium-dependent nitric oxide-cGMP pathway. Adv Pharmacol 2016;77:1–27.

[7] Totoson P, Maguin-Gate K, Nappey M, Wendling D, Demougeot C. Endothelial dysfunction in rheumatoid arthritis: mechanistic insights and correlation with circulating markers of systemic inflammation. PLoS One 2016;11(1):e0146744.

[8] Postovit LM, Adams MA, Lash GE, Heaton JP, Graham CH. Nitric oxide-mediated regulation of hypoxia-induced B16F10 melanoma metastasis. Int J Cancer 2004;108(1):47–53.

[9] Thomas DD, Heinecke JL, Ridnour LA, Cheng RY, Kesarwala AH, Switzer CH, et al. Signaling and stress: the redox landscape in NOS2 biology. Free Radic Biol Med 2015;87:204–25.

[10] Adams C, McCarthy HO, Coulter JA, Worthington J, Murphy C, Robson T, et al. Nitric oxide synthase gene therapy enhances the toxicity of cisplatin in cancer cells. J Gene Med 2009;11(2):160–8.

[11] Xu W, Liu LZ, Loizidou M, Ahmed M, Charles IG. The role of nitric oxide in cancer. Cell Res 2002;12 (5–6):311–20.

[12] Burke AJ, Sullivan FJ, Giles FJ, Glynn SA. The Yin and Yang of nitric oxide in cancer progression. Carcinogenesis 2013;34(3):503–12.

[13] Switzer CH, Glynn SA, Cheng RY, Ridnour LA, Green JE, Ambs S, et al. S-nitrosylation of EGFR and Src activates an oncogenic signaling network in human basal-like breast cancer. Mol Cancer Res 2012;10(9):1203–15.

[14] Ridnour LA, Barasch KM, Windhausen AN, Dorsey TH, Lizardo MM, Yfantis HG, et al. Nitric oxide synthase and breast cancer: role of TIMP-1 in NO-mediated Akt activation. PLoS One 2012;7(9):e44081.

[15] Glynn SA, Boersma BJ, Dorsey TH, Yi M, Yfantis HG, Ridnour LA, et al. Increased NOS2 predicts poor survival in estrogen receptor-negative breast cancer patients. J Clin Invest 2010;120(11):3843–54.

[16] Pervin S, Singh R, Hernandez E, Wu G, Chaudhuri G. Nitric oxide in physiologic concentrations targets the translational machinery to increase the proliferation of human breast cancer cells: involvement of mammalian target of rapamycin/eIF4E pathway. Cancer Res 2007;67(1):289–99.

[17] Heinecke JL, Ridnour LA, Cheng RY, Switzer CH, Lizardo MM, Khanna C, et al. Tumor microenvironment-based feed-forward regulation of NOS2 in breast cancer progression. Proc Natl Acad Sci U S A 2014;111 (17):6323–8.

[18] Maeda H, Fang J, Inutsuka T, Kitamoto Y. Vascular permeability enhancement in solid tumor: various factors, mechanisms involved and its implications. Int Immunopharmacol 2003;3(3):319–28.

[19] Ryan AE, Burke AJ, Giles FJ, Sullivan FJ, Glynn SA. Mechanisms of nitric oxide-dependent regulation of tumor invasion and metastasis. In: Bonavida B, editor. Nitric oxide and cancer: pathogenesis and therapy. Los Angeles: Springer; 2015. p. 49–63.

[20] Granados-Principal S, Liu Y, Guevara ML, Blanco E, Choi DS, Qian W, et al. Inhibition of iNOS as a novel effective targeted therapy against triple-negative breast cancer. Breast Cancer Res 2015;17:25.

[21] Jin Z, Wang W, Jiang N, Zhang L, Li Y, Xu X, et al. Clinical implications of iNOS levels in triple-negative breast cancer responding to neoadjuvant chemotherapy. PLoS One 2015;10(7):e0130286.

[22] Aaltoma SH, Lipponen PK, Kosma VM. Inducible nitric oxide synthase (iNOS) expression and its prognostic value in prostate cancer. Anticancer Res 2001;21(4B):3101–6.

[23] Cronauer MV, Ince Y, Engers R, Rinnab L, Weidemann W, Suschek CV, et al. Nitric oxide-mediated inhibition of androgen receptor activity: possible implications for prostate cancer progression. Oncogene 2007;26(13):1875–84.

[24] Qin Y, Dey A, Purayil HT, Daaka Y. Maintenance of androgen receptor inactivation by S-nitrosylation. Cancer Res 2013;73(22):6690–9.

[25] Pervin S, Singh R, Chaudhuri G. Nitric oxide-induced cytostasis and cell cycle arrest of a human breast cancer cell line (MDA-MB-231): potential role of cyclin D1. Proc Natl Acad Sci U S A 2001;98(6):3583–8.

[26] Pervin S, Singh R, Chaudhuri G. Nitric-oxide-induced Bax integration into the mitochondrial membrane commits MDA-MB-468 cells to apoptosis: essential role of Akt. Cancer Res 2003;63(17):5470–9.

[27] Yasuda H. Solid tumor physiology and hypoxia-induced chemo/radio-resistance: novel strategy for cancer therapy: nitric oxide donor as a therapeutic enhancer. Nitric Oxide Biol Chem 2008;19(2):205–16.

[28] Siemens DR, Heaton JP, Adams MA, Kawakami J, Graham CH. Phase II study of nitric oxide donor for men with increasing prostate-specific antigen level after surgery or radiotherapy for prostate cancer. Urology 2009;74(4):878–83.

[29] Hibbs Jr. JB, Taintor RR, Vavrin Z. Macrophage cytotoxicity: role for L-arginine deiminase and imino nitrogen oxidation to nitrite. Science 1987;235(4787):473–6.

[30] Hibbs Jr. JB, Taintor RR, Vavrin Z, Rachlin EM. Nitric oxide: a cytotoxic activated macrophage effector molecule. Biochem Biophys Res Commun 1988;157(1):87–94.

[31] Pacher P, Beckman JS, Liaudet L. Nitric oxide and peroxynitrite in health and disease. Physiol Rev 2007;87 (1):315–424.

[32] Willmot M, Gray L, Gibson C, Murphy S, Bath PM. A systematic review of nitric oxide donors and L-arginine in experimental stroke; effects on infarct size and cerebral blood flow. Nitric Oxide Biol Chem 2005;12 (3):141–9.

[33] Laschak M, Spindler KD, Schrader AJ, Hessenauer A, Streicher W, Schrader M, et al. JS-K, a glutathione/glutathione S-transferase-activated nitric oxide releasing prodrug inhibits androgen receptor and WNT-signaling in prostate cancer cells. BMC Cancer 2012;12:130.

[34] Shami PJ, Saavedra JE, Wang LY, Bonifant CL, Diwan BA, Singh SV, et al. JS-K, a glutathione/glutathione S-transferase-activated nitric oxide donor of the diazeniumdiolate class with potent antineoplastic activity. Mol Cancer Ther 2003;2(4):409–17.

[35] Kaur I, Terrazas M, Kosak KM, Kern SE, Boucher KM, Shami PJ. Cellular distribution studies of the nitric oxide-generating antineoplastic prodrug O(2) -(2,4-dinitrophenyl)1-((4-ethoxycarbonyl)piperazin-1-yl)diazen-1-ium-1,2-diolate formulated in pluronic P123 micelles. J Pharm Pharmacol 2013;65(9):1329–36.

[36] McMurtry V, Saavedra JE, Nieves-Alicea R, Simeone AM, Keefer LK, Tari AM. JS-K, a nitric oxide-releasing prodrug, induces breast cancer cell death while sparing normal mammary epithelial cells. Int J Oncol 2011;38(4):963–71.

[37] Simeone AM, McMurtry V, Nieves-Alicea R, Saavedra JE, Keefer LK, Johnson MM, et al. TIMP-2 mediates the anti-invasive effects of the nitric oxide-releasing prodrug JS-K in breast cancer cells. Breast Cancer Res 2008;10(3):R44.

[38] Cooke MS, Evans MD, Dizdaroglu M, Lunec J. Oxidative DNA damage: mechanisms, mutation, and disease. FASEB J 2003;17(10):1195–214.

[39] Giannoni E, Buricchi F, Raugei G, Ramponi G, Chiarugi P. Intracellular reactive oxygen species activate Src tyrosine kinase during cell adhesion and anchorage-dependent cell growth. Mol Cell Biol 2005;25(15):6391–403.

[40] Lee SR, Yang KS, Kwon J, Lee C, Jeong W, Rhee SG. Reversible inactivation of the tumor suppressor PTEN by H_2O_2. J Biol Chem 2002;277(23):20336–42.

[41] Hsieh HJ, Liu CA, Huang B, Tseng AH, Wang DL. Shear-induced endothelial mechanotransduction: the interplay between reactive oxygen species (ROS) and nitric oxide (NO) and the pathophysiological implications. J Biomed Sci 2014;21:3.

[42] Farmer H, McCabe N, Lord CJ, Tutt AN, Johnson DA, Richardson TB, et al. Targeting the DNA repair defect in BRCA mutant cells as a therapeutic strategy. Nature 2005;434(7035):917–21.

[43] Petrucelli N, Daly MB, Feldman GL. Hereditary breast and ovarian cancer due to mutations in BRCA1 and BRCA2. Genet Med 2010;12(5):245–59.

[44] Gorrini C, Harris IS, Mak TW. Modulation of oxidative stress as an anticancer strategy. Nat Rev Drug Discov 2013;12(12):931–47.

[45] Gorrini C, Gang BP, Bassi C, Wakeham A, Baniasadi SP, Hao Z, et al. Estrogen controls the survival of BRCA1-deficient cells via a PI3K-NRF2-regulated pathway. Proc Natl Acad Sci U S A 2014;111(12):4472–7.

[46] Yakovlev VA. Nitric oxide-dependent downregulation of BRCA1 expression promotes genetic instability. Cancer Res 2013;73(2):706–15.

[47] Somasundaram V, Nadhan R, KH S, Kumar Sengodan S, Srinivas P. Nitric oxide and reactive oxygen species: clues to target oxidative damage repair defective breast cancers. Crit Rev Oncol Hematol 2016;101:184–92.

[48] Bryant HE, Schultz N, Thomas HD, Parker KM, Flower D, Lopez E, et al. Specific killing of BRCA2-deficient tumours with inhibitors of poly(ADP-ribose) polymerase. Nature 2005;434(7035):913–7.

[49] Dent RA, Lindeman GJ, Clemons M, Wildiers H, Chan A, McCarthy NJ, et al. Phase I trial of the oral PARP inhibitor olaparib in combination with paclitaxel for first- or second-line treatment of patients with metastatic triple-negative breast cancer. Breast Cancer Res 2013;15(5):R88.

[50] Bentires-Alj M, Barbu V, Fillet M, Chariot A, Relic B, Jacobs N, et al. NF-kappaB transcription factor induces drug resistance through MDR1 expression in cancer cells. Oncogene 2003;22(1):90–7.

[51] Kim DS, Park SS, Nam BH, Kim IH, Kim SY. Reversal of drug resistance in breast cancer cells by transglutaminase 2 inhibition and nuclear factor-kappaB inactivation. Cancer Res 2006;66(22):10936–43.

[52] Bradley JR. TNF-mediated inflammatory disease. J Pathol 2008;214(2):149–60.

[53] Karin M, Lin A. NF-kappaB at the crossroads of life and death. Nat Immunol 2002;3(3):221–7.

[54] Garban HJ, Bonavida B. Nitric oxide disrupts H_2O_2-dependent activation of nuclear factor kappa B. Role in sensitization of human tumor cells to tumor necrosis factor-alpha -induced cytotoxicity. J Biol Chem 2001;276(12):8918–23.

[55] Hymowitz SG, Christinger HW, Fuh G, Ultsch M, O'Connell M, Kelley RF, et al. Triggering cell death: the crystal structure of Apo2L/TRAIL in a complex with death receptor 5. Mol Cell 1999;4(4):563–71.

[56] Lemke J, von Karstedt S, Zinngrebe J, Walczak H. Getting TRAIL back on track for cancer therapy. Cell Death Differ 2014;21(9):1350–64.

[57] Ashkenazi A, Pai RC, Fong S, Leung S, Lawrence DA, Marsters SA, et al. Safety and antitumor activity of recombinant soluble Apo2 ligand. J Clin Invest 1999;104(2):155–62.
[58] Walczak H, Miller RE, Ariail K, Gliniak B, Griffith TS, Kubin M, et al. Tumoricidal activity of tumor necrosis factor-related apoptosis-inducing ligand in vivo. Nat Med 1999;5(2):157–63.
[59] Huang Y, Yang X, Xu T, Kong Q, Zhang Y, Shen Y, et al. Overcoming resistance to TRAIL-induced apoptosis in solid tumor cells by simultaneously targeting death receptors, c-FLIP and IAPs. Int J Oncol 2016;49(1):153–63.
[60] Menke C, Bin L, Thorburn J, Behbakht K, Ford HL, Thorburn A. Distinct TRAIL resistance mechanisms can be overcome by proteasome inhibition but not generally by synergizing agents. Cancer Res 2011;71(5):1883–92.
[61] Huerta-Yepez S, Vega M, Escoto-Chavez SE, Murdock B, Sakai T, Baritaki S, et al. Nitric oxide sensitizes tumor cells to TRAIL-induced apoptosis via inhibition of the DR5 transcription repressor Yin Yang 1. Nitric Oxide Biol Chem 2009;20(1):39–52.
[62] Huerta-Yepez S, Vega M, Jazirehi A, Garban H, Hongo F, Cheng G, et al. Nitric oxide sensitizes prostate carcinoma cell lines to TRAIL-mediated apoptosis via inactivation of NF-kappa B and inhibition of Bcl-xl expression. Oncogene 2004;23(29):4993–5003.
[63] Donia M, Maksimovic-Ivanic D, Mijatovic S, Mojic M, Miljkovic D, Timotijevic G, et al. In vitro and in vivo anticancer action of saquinavir-NO, a novel nitric oxide-derivative of the protease inhibitor saquinavir, on hormone resistant prostate cancer cells. Cell Cycle 2011;10(3):492–9.
[64] Peter ME, Hadji A, Murmann AE, Brockway S, Putzbach W, Pattanayak A, et al. The role of CD95 and CD95 ligand in cancer. Cell Death Differ 2015;22(4):549–59.
[65] Algeciras-Schimnich A, Pietras EM, Barnhart BC, Legembre P, Vijayan S, Holbeck SL, et al. Two CD95 tumor classes with different sensitivities to antitumor drugs. Proc Natl Acad Sci U S A 2003;100(20):11445–50.
[66] Garban HJ, Bonavida B. Nitric oxide inhibits the transcription repressor Yin-Yang 1 binding activity at the silencer region of the Fas promoter: a pivotal role for nitric oxide in the up-regulation of Fas gene expression in human tumor cells. J Immunol 2001;167(1):75–81.
[67] Hongo F, Garban H, Huerta-Yepez S, Vega M, Jazirehi AR, Mizutani Y, et al. Inhibition of the transcription factor Yin Yang 1 activity by S-nitrosation. Biochem Biophys Res Commun 2005;336(2):692–701.
[68] Frederiksen LJ, Sullivan R, Maxwell LR, Macdonald-Goodfellow SK, Adams MA, Bennett BM, et al. Chemosensitization of cancer in vitro and in vivo by nitric oxide signaling. Clin Cancer Res 2007;13(7):2199–206.
[69] Milane L, Duan Z, Amiji M. Role of hypoxia and glycolysis in the development of multi-drug resistance in human tumor cells and the establishment of an orthotopic multi-drug resistant tumor model in nude mice using hypoxic pre-conditioning. Cancer Cell Int 2011;11:3.
[70] Lindsay D, Garvey CM, Mumenthaler SM, Foo J. Leveraging hypoxia-activated prodrugs to prevent drug resistance in solid tumors. PLoS Comput Biol 2016;12(8):e1005077.
[71] Harris AL. Hypoxia—a key regulatory factor in tumour growth. Nat Rev Cancer 2002;2(1):38–47.
[72] Jain RK. Normalization of tumor vasculature: an emerging concept in antiangiogenic therapy. Science 2005;307 (5706):58–62.
[73] Jain RK, Stylianopoulos T. Delivering nanomedicine to solid tumors. Nat Rev Clin Oncol 2010;7(11):653–64.
[74] Das DS, Ray A, Das A, Song Y, Tian Z, Oronsky B, et al. A novel hypoxia-selective epigenetic agent RRx-001 triggers apoptosis and overcomes drug resistance in multiple myeloma cells. Leukemia 2016;30(11):2187–97.
[75] Moeller BJ, Richardson RA, Dewhirst MW. Hypoxia and radiotherapy: opportunities for improved outcomes in cancer treatment. Cancer Metastasis Rev 2007;26(2):241–8.
[76] Chen F, Zhuang M, Zhong C, Peng J, Wang X, Li J, et al. Baicalein reverses hypoxia-induced 5-FU resistance in gastric cancer AGS cells through suppression of glycolysis and the PTEN/Akt/HIF-1alpha signaling pathway. Oncol Rep 2015;33(1):457–63.
[77] Ho JJ, Man HS, Marsden PA. Nitric oxide signaling in hypoxia. J Mol Med (Berl) 2012;90(3):217–31.
[78] Matthews NE, Adams MA, Maxwell LR, Gofton TE, Graham CH. Nitric oxide-mediated regulation of chemosensitivity in cancer cells. J Natl Cancer Inst 2001;93(24):1879–85.
[79] Frederiksen LJ, Siemens DR, Heaton JP, Maxwell LR, Adams MA, Graham CH. Hypoxia induced resistance to doxorubicin in prostate cancer cells is inhibited by low concentrations of glyceryl trinitrate. J Urol 2003;170 (3):1003–7.
[80] de Luca A, Moroni N, Serafino A, Primavera A, Pastore A, Pedersen JZ, et al. Treatment of doxorubicin-resistant MCF7/Dx cells with nitric oxide causes histone glutathionylation and reversal of drug resistance. Biochem J 2011;440(2):175–83.
[81] Huerta-Yepez S, Baritaki S, Baay-Guzman G, Hernandez-Luna MA, Hernandez-Cueto A, Vega MI, et al. Contribution of either YY1 or BclXL-induced inhibition by the NO-donor DETANONOate in the reversal of drug

resistance, both in vitro and in vivo. YY1 and BclXL are overexpressed in prostate cancer. Nitric Oxide Biol Chem 2013;29:17–24.

[82] Kumar A, Ehrenshaft M, Tokar EJ, Mason RP, Sinha BK. Nitric oxide inhibits topoisomerase II activity and induces resistance to topoisomerase II-poisons in human tumor cells. Biochim Biophys Acta 2016;1860(7):1519–27.

[83] Rockwell S, Dobrucki IT, Kim EY, Marrison ST, Vu VT. Hypoxia and radiation therapy: past history, ongoing research, and future promise. Curr Mol Med 2009;9(4):442–58.

[84] Scicinski J, Oronsky B, Ning S, Knox S, Peehl D, Kim MM, et al. NO to cancer: the complex and multifaceted role of nitric oxide and the epigenetic nitric oxide donor, RRx-001. Redox Biol 2015;6:1–8.

[85] Stewart GD, Nanda J, Katz E, Bowman KJ, Christie JG, Brown DJ, et al. DNA strand breaks and hypoxia response inhibition mediate the radiosensitisation effect of nitric oxide donors on prostate cancer under varying oxygen conditions. Biochem Pharmacol 2011;81(2):203–10.

[86] Su X, Takahashi A, Guo G, Mori E, Okamoto N, Ohnishi K, et al. Biphasic effects of nitric oxide radicals on radiation-induced lethality and chromosome aberrations in human lung cancer cells carrying different p53 gene status. Int J Radiat Oncol Biol Phys 2010;77(2):559–65.

[87] Suchyta DJ, Schoenfisch MH. Encapsulation of N-diazeniumdiolates within liposomes for enhanced nitric oxide donor stability and delivery. Mol Pharm 2015;12(10):3569–74.

[88] Seabra AB, de Lima R, Calderon M. Nitric oxide releasing nanomaterials for cancer treatment: current status and perspectives. Curr Top Med Chem 2015;15(4):298–308.

[89] Chegaev K, Riganti C, Lazzarato L, Rolando B, Guglielmo S, Campia I, et al. Nitric oxide donor doxorubicins accumulate into Doxorubicin-resistant human colon cancer cells inducing cytotoxicity. ACS Med Chem Lett 2011;2(7):494–7.

[90] Pedrini I, Gazzano E, Chegaev K, Rolando B, Marengo A, Kopecka J, et al. Liposomal nitrooxy-doxorubicin: one step over caelyx in drug-resistant human cancer cells. Mol Pharm 2014;11(9):3068–79.

[91] Ulrich CM, Bigler J, Potter JD. Non-steroidal anti-inflammatory drugs for cancer prevention: promise, perils and pharmacogenetics. Nat Rev Cancer 2006;6(2):130–40.

[92] Nortcliffe A, Ekstrom AG, Black JR, Ross JA, Habib FK, Botting NP, et al. Synthesis and biological evaluation of nitric oxide-donating analogues of sulindac for prostate cancer treatment. Bioorg Med Chem 2014;22(2):756–61.

[93] Basudhar D, Cheng RC, Bharadwaj G, Ridnour LA, Wink DA, Miranda KM. Chemotherapeutic potential of diazeniumdiolate-based aspirin prodrugs in breast cancer. Free Radic Biol Med 2015;83:101–14.

[94] Royle JS, Ross JA, Ansell I, Bollina P, Tulloch DN, Habib FK. Nitric oxide donating nonsteroidal anti-inflammatory drugs induce apoptosis in human prostate cancer cell systems and human prostatic stroma via caspase-3. J Urol 2004;172(1):338–44.

[95] Huguenin S, Fleury-Feith J, Kheuang L, Jaurand MC, Bolla M, Riffaud JP, et al. Nitrosulindac (NCX 1102): a new nitric oxide-donating non-steroidal anti-inflammatory drug (NO-NSAID), inhibits proliferation and induces apoptosis in human prostatic epithelial cell lines. Prostate 2004;61(2):132–41.

[96] Stewart GD, Nanda J, Brown DJ, Riddick AC, Ross JA, Habib FK. NO-sulindac inhibits the hypoxia response of PC-3 prostate cancer cells via the Akt signalling pathway. Int J Cancer 2009;124(1):223–32.

[97] Kodela R, Chattopadhyay M, Kashfi K. Synthesis and biological activity of NOSH-naproxen (AVT-219) and NOSH-sulindac (AVT-18A) as potent anti-inflammatory agents with chemotherapeutic potential, Med Chem Commun 2013;4(11). https://www.ncbi.nlm.nih.gov/pubmed/?term=24273639.

[98] Kashfi K, Ryan Y, Qiao LL, Williams JL, Chen J, Del Soldato P, et al. Nitric oxide-donating nonsteroidal anti-inflammatory drugs inhibit the growth of various cultured human cancer cells: evidence of a tissue type-independent effect. J Pharmacol Exp Ther 2002;303(3):1273–82.

[99] Agostinis P, Berg K, Cengel KA, Foster TH, Girotti AW, Gollnick SO, et al. Photodynamic therapy of cancer: an update. CA Cancer J Clin 2011;61(4):250–81.

[100] Rapozzi V, Varchi G, Della Pietra E, Ferroni C, Xodo LE. A photodynamic bifunctional conjugate for prostate cancer: an in vitro mechanistic study. Invest New Drugs 2016;34:1–9.

[101] Rapozzi V, Ragno D, Guerrini A, della Pietra E, Cesselli D, Castoria G. Androgen receptor targeted conjugate for bimodal photodynamic therapy of prostate cancer in vitro. Bioconjug Chem 2015;26(8):1662–71.

[102] Ranganathan S, Krishnan A, Sivasithambaram ND. Significance of twist and iNOS expression in human breast carcinoma. Mol Cell Biochem 2016;412(1–2):41–7.

[103] Jadeski LC, Hum KO, Chakraborty C, Lala PK. Nitric oxide promotes murine mammary tumour growth and metastasis by stimulating tumour cell migration, invasiveness and angiogenesis. Int J Cancer 2000;86(1):30–9.

[104] Iwasaki T, Higashiyama M, Kuriyama K, Sasaki A, Mukai M, Shinkai K, et al. NG-nitro-L-arginine methyl ester inhibits bone metastasis after modified intracardiac injection of human breast cancer cells in a nude mouse model. Jpn J Cancer Res 1997;88(9):861–6.
[105] Yu S, Jia L, Zhang Y, Wu D, Xu Z, Ng CF, et al. Increased expression of activated endothelial nitric oxide synthase contributes to antiandrogen resistance in prostate cancer cells by suppressing androgen receptor transactivation. Cancer Lett 2013;328(1):83–94.
[106] Hsu PP, Sabatini DM. Cancer cell metabolism: Warburg and beyond. Cell 2008;134(5):703–7.
[107] Warburg O. On the origin of cancer cells. Science 1956;123(3191):309–14.
[108] Sen S, Kawahara B, Chaudhuri G. Mitochondrial-associated nitric oxide synthase activity inhibits cytochrome c oxidase: implications for breast cancer. Free Radic Biol Med 2013;57:210–20.
[109] Switzer CH, Cheng RY, Ridnour LA, Glynn SA, Ambs S, Wink DA. Ets-1 is a transcriptional mediator of oncogenic nitric oxide signaling in estrogen receptor-negative breast cancer. Breast Cancer Res 2012;14(5):R125.

CHAPTER

5

Evaluation of the Impact of Nitric Oxide on Resistance to Platinum-Based Chemotherapeutics

Ahlam A. Ali, Cian M. McCrudden, Helen O. McCarthy

Queen's University Belfast, Belfast, United Kingdom

Abstract

One of the most persistent problems in the treatment and management of cancer is the development of resistance to therapy in the cancerous cells. The mechanisms of resistance can be intrinsic or can be acquired through exposure to chemotherapeutic agents. Accumulating clinical experience, supported by data from animal models, indicates that chemotherapy is most effective when given in combination to achieve additive or synergistic effects that reduce the potential development of drug resistance.

Platinum compounds are one of the most widely used classes of agents in cancer therapy; however, platinum drugs approved for clinical use are rarely given to patients as a monotherapy. A concept that is emerging is the utilization of a biological messenger to abrogate the development of resistance. One such agent is the gasotransmitter nitric oxide (NO), which has recently been implicated in several mechanisms of cytotoxicity by platinum agents. Release of NO can potentiate the cytotoxic effects of platinum chemotherapeutics, either by direct synergistic cytotoxic effects or by increase of blood supply and vascular permeability, thus increasing tumor perfusion and consequently the amount of chemotherapy reaching the tumor. Combined understanding of how platinum drugs work, and the mechanisms by which NO can abrogate resistance should allow for the development of new combination regimens. This chapter discusses the clinical importance of cisplatin and related platinum-based drugs for the treatment of various human cancers, highlighting the potential use of NO as a chemosensitizer. Special attention is paid to the molecular mechanisms underlying resistance, and the role of NO to reverse chemoresistance of tumor cells.

Abbreviations

ABC	ATP-binding cassette
AIF	apoptosis-inducing factor
AMS	amine-modified mesoporous silica
ASK-1	apoptosis signal-regulating kinase 1
b-Gal-NONOate	b-galactosyl-pyrrolidinyl diazeniumdiolate
Bax	apoptosis regulator protein

Nitric Oxide (Donor/Induced) in Chemosensitization, Volume 1
http://dx.doi.org/10.1016/B978-0-12-811020-1.00005-3

BBB	blood-brain barrier
Bclxl	B-cell lymphoma-extra large
CTR1	copper transporter proteins
DACH	diaminocyclohexane
DR5	death receptor 5
GLYN	glycidyl nitrate
GSH	glutathione
GSNO	*S*-nitrosoglutathione
GTN	nitroglycerin
HIF-1	hypoxia-inducible transcription factor
HMG	high-mobility group proteins
HMO	3,3-bis(hydroxymethyl)oxetane
HNSCC	head and neck squamous cell carcinoma
IC50	half maximal inhibitory concentration
IFN-γ	interferon gamma
iNOS	inducible nitric oxide synthase
ISMN	isosorbide-5-mononitrate
LD50	median lethal dose
MMR	mismatch repair protein
MRP	multidrug resistance protein
NER	nucleotide excision repair
NF-κB	nuclear factor kappa-light-chain-enhancer of activated B cells
NO	nitric oxide
NONOates	diazeniumdiolates
NOS	nitric oxide synthase
NPs	nanoparticles
NSCLC	nonsmall-cell lung cancer
OCT	organic cation transporters
OS	overall survival
p21	cyclin-dependent kinase inhibitor 1
p53	tumor suppressor gene
PARP	poly(ADP-ribosyl)ated proteins
PFS	progression-free survival
PSA	prostate-specific antigen
PTEN	phosphatase and tensin homologue
RNS	reactive nitrogen species
ROS	reactive oxygen species
SCC	squamous cell carcinoma
SNAP	*S*-nitroso-*N*-acetylpenicillamine
SNP	sodium nitroprusside
TRAIL	TNF-related apoptosis-inducing ligand
TTP	time to progression
YY1	Yin-Yang 1

PLATINUM COMPOUNDS; CLINICAL IMPLICATIONS

Platinum compounds are one of the most widely used classes of drugs in cancer therapy and have remained the backbone of systemic cancer therapies since 1980 [1]. Approximately half of all patients on anticancer chemotherapy regimens are treated with a platinum drug. Platinum drugs are employed in 32 of 78 treatment regimens in combination with a wide range of other drugs [2]. Cisplatin (*cis*-diamminedichloroplatinum(II)), the most potent platinum compound, is the first generation of platinum drugs approved by the FDA for the treatment of a wide spectrum of solid tumors [3]. Cisplatin is taken into cells by one of three possible mechanisms: passive diffusion, copper transporter proteins (e.g., CTR1), and/or organic cation transporters (OCTs) [4]. Once inside a cell, the chloride ligands of cisplatin are replaced by water molecules resulting in positively charged nucleophilic sites that react with intracellular macromolecules to form protein, RNA, and DNA adducts.

Treatment with cisplatin often leads to an initial therapeutic response associated with complete disease remission, partial response or disease stabilization. However, cisplatin therapy requires additional medication to alleviate the side effects that are frequently associated with use, which include severe kidney problems, reduced immunity to infections, allergic reactions, gastrointestinal disorders, and hearing loss especially in younger patients [5]. Despite these side effects and because of impressive potency, cisplatin is still used for the treatment of numerous human malignancies including bladder, head, and neck melanoma, lymphomas, and myelomas; nonsmall-cell lung cancer (NSCLC); small-cell lung cancer; and ovarian and testicular cancers (90% cure rate) [3]. The major hurdle to the success of cisplatin is the resistance that develops. Many tumors are intrinsically resistant to the platinum drug. Moreover, many sensitive tumors develop resistance gradually after initial response [5].

In order to reduce the dose-limiting toxicity of cisplatin, a second-generation of compounds was developed. One second-generation compound, carboplatin (*cis*-diammine (1,1-cyclobutanecarboxylate)platinum (II)), differs from cisplatin in that it has a bidentate dicarboxylate ligand in place of the two chloride ligands, which are the leaving groups in cisplatin. The mechanism of action is similar to that of cisplatin, but carboplatin exhibits lower reactivity and slower DNA binding kinetics than cisplatin. Depending on the type of cancer, carboplatin may be between 8- and 45-fold less potent than cisplatin [6]. With a reduced toxicity profile, carboplatin is suitable for more aggressive high-dose chemotherapy. Carboplatin is accepted worldwide and has almost replaced cisplatin in combination regimens with paclitaxel for the treatment of ovarian cancer [7]. Carboplatin has recently undergone additional phases II and III trials for the treatment of salivary gland cancer [8] and advanced Müllerian cancer [9]. Yet, carboplatin has limited efficacy against testicular germ cell cancers, squamous cell carcinoma (SCC) of the head and neck, and bladder cancer. As a consequence, cisplatin remains the drug of choice for treating those diseases [7].

As resistance still remains a problem with the first- and second-generation compounds, the third-generation platinum drug oxaliplatin was developed to overcome such resistance. Oxaliplatin is a platinum complex with a (1R,2R)-1,2 diaminocyclohexane (DACH) ligand and oxalate as leaving groups. The bidentate oxalate significantly reduces reactivity of oxaliplatin and thereby limits toxic side effects, such as peripheral sensory neuropathy. The DACH ligand is more lipophilic, which increases the passive uptake of oxaliplatin above that seen with cisplatin and carboplatin. In combination with 5-FU and folinic acid, oxaliplatin currently has widespread approval for the treatment of adjuvant and metastatic colorectal cancer that is intrinsically insensitive to cisplatin [6]. The increased cellular accumulation of oxaliplatin in colorectal cancer has been linked to the overexpression of organic cation transporters (OCT1 and OCT2) as no such effects were observed with either cisplatin or carboplatin [10]. Thus, to date, only three platinum-based drugs that have entered clinical trials have gained international marketing approval, while another three (nedaplatin, lobaplatin, and heptaplatin) have gained approval in certain markets (Table 1).

TABLE 1 Platinum Chemotherapeutics Clinically Approved for Human Use

Drug	Brand Names	Structure	Year of Approval	Market Status	Clinical Use	Dose-Limiting Toxicity
Cisplatin	Platinol Platidiam Platinex Platistin	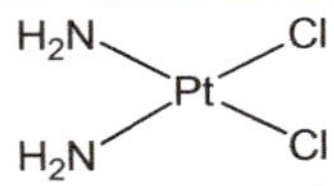	1979	Worldwide	Lung, ovarian, testicular, breast and brain cancer, sarcomas, lymphomas	Nephrotoxicity
Carboplatin	Paraplatin Paraplatine Carbomedac Carbosin	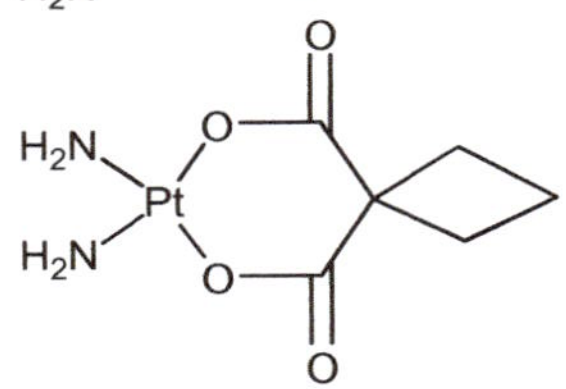	1989	Worldwide	Advanced ovarian carcinoma, head and neck cancer, and lung cancer	Myelosuppression
Oxaliplatin	Eloxatin Dacotin	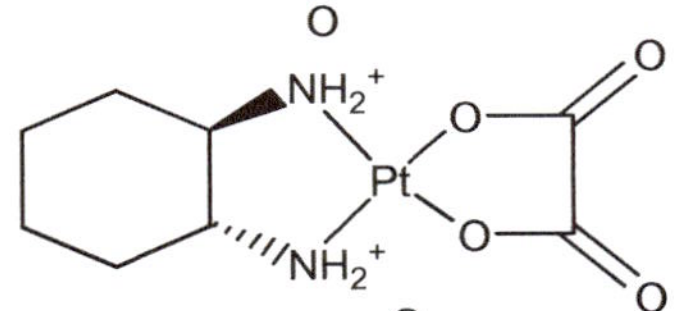	2002	Worldwide	Metastatic colorectal cancer, advanced gastric and ovarian cancer	Nephrotoxicity
Nedaplatin	Aqupla	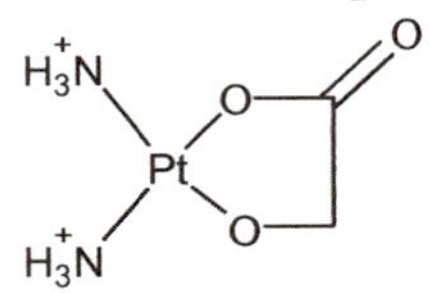	1996	Japan	Small and nonsmall-cell lung cancer, head and neck cancer, esophageal, bladder, ovary and cervix cancer	Myelosuppression
Lobaplatin	—	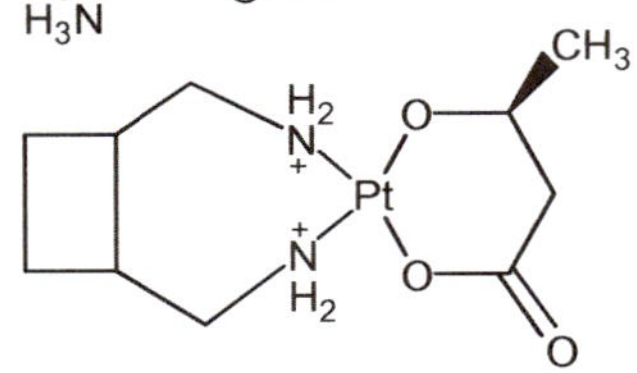	2004	China	Breast, testicular, small-cell lung, gastric, and ovarian cancer, chronic myeloid leukemia	Thrombocytopenia
Heptaplatin	Sunpla	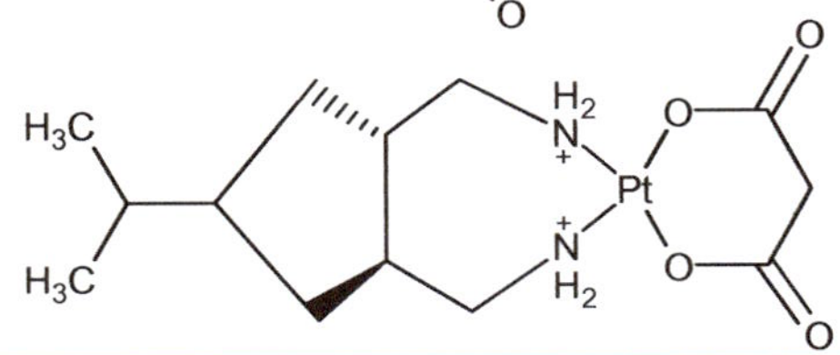	2005	Korea	Gastric cancer	Nephrotoxicity/ intraabdominal bleeding

Potentiating Platinum Compounds

Platinum compounds alkylate DNA purine bases, leading to the creation of guanine-guanine (GG) intra- and interstrand cross-links. This interferes with DNA repair mechanisms, causes DNA damage and subsequently induces apoptosis in cancer cells [11]. However, repeat treatment cycles often trigger the development of chemoresistance in a previous chemosensitive malignancy [12]. The emphasis is currently shifting toward new cancer treatment strategies that combine the utility of existing therapies with the potential of emerging novel modalities. Several lines of evidence suggest that nitric oxide (NO) can reverse the resistance of various cancers to chemotherapeutic drugs, thus enhancing the anticancer activity of such therapeutics [13–16].

MECHANISMS OF CHEMORESISTANCE

Recently, many research and clinical efforts have been focused on understanding the nature of resistance of tumor cells to either patient's immune system or chemo-, radio-, and immunotherapeutic approaches. Chemotherapeutic resistance could be due to the induction of drug transporters, DNA repair, changes in drug metabolism, gene amplification, or mutation of target proteins with alterations in survival/apoptotic pathways [17]. Usually, drug resistance is a cumulative effect of a combination of different resistance mechanisms (Fig. 1).

Mechanisms by Which NO May Be Exploited to Reverse/Combat Chemoresistance

NO is a relatively stable, free radical gas that rapidly diffuses into cells and cell membranes and reacts with molecular targets. NO is generated in tissues by the catalytic action of three main isoforms of nitric oxide synthase (NOS) on the amino acid, L-arginine [18]. NO plays an important physiological role in vasodilation [19], neurotransmission [20], and immune responses [21]. The role of NO in cancer biology has been the source of much debate. While NO may play a role in the development, growth and malignant progression of cancer [22,23], NO is also known to have a cytotoxic function, and plays a major role in the response of tumors to therapy [24]. The precise cellular response to NO depends on the concentration of NO achieved and on subtle variations in the composition of the intra- and extracellular milieu. In general, it is accepted that low physiological concentrations of NO foster a tumorigenic phenotype. Clinical and experimental studies support a positive relationship between tumor malignancy and NOS activity in certain tumors [25,26]. Higher levels of NO, in the μM range are correlated with induction of tumor cell cytotoxicity. The cytotoxicity evoked by NO is via conversion to oxidized intermediates that damage the DNA, exert lipid peroxidation on the membrane and inhibit certain proteins by S-nitrosylation [27].

A variety of methods have been used to elevate the intratumoral NO concentration to provoke sensitization to chemotherapies, which have included NO donor drugs and various NOS gene therapy strategies. The therapeutic effect can result from direct cytotoxicity and the reactive nitrogen and oxygen species associated with NO or the subtle modulation of

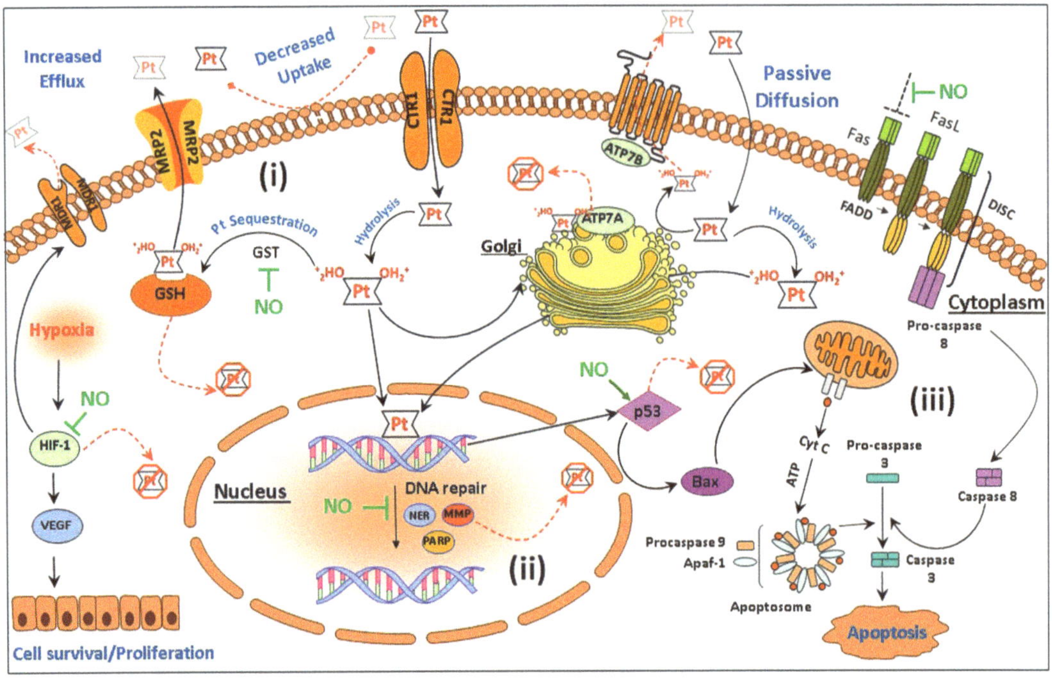

FIG. 1 Mechanism of action and postulated mechanisms of resistance of platinum drugs. After entering the cell via CTR1 receptors or passive diffusion, platinum drugs were activated via hydrolysis. The aquated platinum binds to DNA. Distortion of DNA activates apoptosis. Resistance to platinum includes the following: (i) reduced intracellular accumulation of platinum due to decreased uptake or increased efflux; glutathione binds to platinum to facilitate its efflux, ATP-binding cassette (ABC) transporters such as MDR1 and MRP1 also act as ATP-consuming drug efflux pumps that extrude platinum drug out of the cell, copper homeostasis proteins ATP7A and ATP7B are also involved in platinum efflux, and ATP7A sequesters platinum in vesicular structures preventing further distribution of the drug, while ATP7B is responsible for platinum efflux outside the cell; (ii) increased repair of platinum adducts; increased ability to replicate past platinum adducts; and (iii) defects in the apoptotic response pathway such as p53. NO could reverse resistance via inactivation of proteins such as glutathione or interfere with DNA repair mechanism or increase p53 content, leading to upregulation of proapoptotic proteins such as Bax and Fas. This in turn activates both the intrinsic (mitochondrial) and extrinsic (death receptor) apoptotic pathways.

apoptosis-related and resistance genes and the modification of drug accessibility, regulation of hypoxic conditions, and "normalization" of the tumor vasculature [28].

Drug Efflux Mechanism

Drug efflux transporters, generally highly expressed in drug-resistant tumor cells, are selectively involved in the efflux of small molecule drugs [29]. Overexpression of ATP-binding cassette (ABC) transporters has been shown to be associated with poor prognosis for response of tumors toward chemotherapy and survival of patients [30]. In addition, increased levels of reactive oxygen species (ROS) in hypoxic tumors provoke cellular adaptation through upregulation of the glutathione (GSH) redox system, which is also associated with drug resistance. GSH conjugation represents a major pathway for detoxification of platinum drugs; overexpression of glutathione-S-transferases allows tumor cells to gain a selective survival advantage over normal cells by enhancement of a GSH-platinum adduct for cellular efflux by the multidrug resistance protein (MRP2), a member of the ABC family [31]. NO has been shown to downregulate GSH, thereby increasing sensitivity of tumors to chemotherapy. For example, the NO donor NCX-4016 significantly reduced the surviving fraction of cisplatin resistance human ovarian cancer cells (HOCCs) ($70 \pm 10\%$) and caused a 50% reduction in the levels of cellular GSH [14]. Furthermore, NO-releasing diazeniumdiolates (NONOates), JS-K, and CB-3-100 increased cisplatin and arsenic cytotoxicity in rat liver carcinoma CAsE cells. The synergistic effect of NO was attributed to enhanced accumulation of arsenic and cisplatin within cells and through activation of C-Jun N-terminal kinases (JNK) and extracellular signal-regulated kinases (ERK) [32]. Treatment with NO donors such as DETA/NO was also found to correlate with tyrosine nitration of the multidrug resistance protein 3 transporters (MRP3) and a decrease of the ATPase activity of P-glycoprotein [33]. This could contribute to reduced efflux and accumulation of chemotherapeutics including platinum compounds. Therefore, NO could potentially reverse chemoresistance via modulation of the platinum efflux mechanism.

DNA Repair

The platinum atom binds covalently to DNA mainly at N^7 position of guanine, forming intra- and interstrand cross-links. The binding of platinum complexes to DNA induces a distortion of the structure of the double helix [34]. This DNA lesion accounts for cisplatin cytotoxicity and is recognized by many cellular proteins leading to DNA repair, replication bypass, or initiation of apoptosis. Among the protein families that are involved in the identification of platinum-DNA adducts are nonhistone chromosomal high-mobility group proteins 1 and 2 (HMG1 and HMG2), nucleotide excision repair (NER) proteins, poly(ADP-ribosyl)ated proteins (PARP), and mismatch repair (MMR) proteins. HMG proteins contribute to cisplatin cytotoxicity by shielding platinum-DNA adducts from repair, and HMGB1 expression is strongly correlated with both cisplatin and oxaliplatin cytotoxicity in vitro [35]. NER is the pathway employed by cancer cells for removal of platinum-DNA adducts and for DNA damage repair; indeed, the remarkable sensitivity of testicular germ cell cancer to cisplatin has been attributed to the reduced NER activity [36]. The MMR protein complex is involved in repairing cisplatin-induced DNA damage rather than repairing DNA adducts. Downregulation or mutation of MMR genes is consistently documented in the context of cisplatin resistance [35]. PARP, critically involved in base excision repair, is also receiving attention, as PARPs are highly expressed and constitutively hyperactivated in cisplatin-resistant cells [37]. Disruption of tumor suppressor p53 is one of the most predominant mechanisms known to induce resistance to chemotherapy. HMG1 and HMG2 facilitate the binding of p53 to DNA to induce the transactivation of several target genes involved in cell cycle progression (p21), DNA repair, and apoptosis (Bax) [37].

Interestingly, both NO and platinum compounds regulate the p53 signaling pathway. DNA damage induced by platinum drugs is known to upregulate p53 [38]. In addition,

NO-related reactive nitrogen species (RNS), such as peroxynitrite [39] or dinitrogen trioxide [40], can activate wild-type p53, which in turn regulates the apoptotic response to DNA damage. NO-mediated inhibition of DNA repair enzymes (e.g., PARP, DNA ligase, or Fapy-DNA glycosylase) contributes to sensitivity of tumors to chemotherapy [41]. These findings suggest that elevation of intracellular NO concentration to therapeutic superphysiological levels may increase response to anticancer drugs including cisplatin or oxaliplatin through the activation of p53 in cancer cells. The NO donors, spermine-NO, and nitroglycerin (GTN) activate and phosphorylate p53, increasing tumor sensitivity to cisplatin in vitro and in vivo [39], suggesting that NO plays a role in sensitizing the cells to platinum chemotherapy via interference with the DNA damaging mechanisms.

Role of Hypoxia

Tumor hypoxia and activation of hypoxia signaling pathways have consistently been identified as features that are strongly associated with aggressive malignancy and drug resistance. Exposure to a low-oxygen environment triggers a hypoxic response pathway centered on the regulated expression of the hypoxia-inducible transcription factor (HIF-1α). Other factors involved include the mitogen-activated protein kinase (MAPK) and phosphatidylinositide 3-kinases (PI3K) signal transduction pathways and transcription factors such as activator protein 1 (AP-1), NF-κB, and p53 [42], all of which are involved in the response of tumor cells to radio- and chemotherapy. Many chemotherapeutic drugs, including the platinum compounds, require oxygen in order to behave therapeutically [43]; a lack of normoxic concentrations leads to drug resistance in hypoxic tumors. Given the fact that NO production is reduced under hypoxic conditions, as the activity of NOS is oxygen-dependent [44], and that inhibition of NOS activity and NO signaling induces drug resistance [45], increasing NO levels by pharmacological interventions should improve the response of tumor cells to chemotherapy.

NO and NO-donating compounds sensitize tumor cells by inhibiting HIF-1α-mediated transcription in hypoxic cells, which in turn downregulates other downstream genes such as MDR1. NO is also known to modify and improve tumor perfusion by directly increasing blood flow, which in turn promotes better distribution of oxygen, and accumulation of elevated concentrations of chemotherapeutics [46]. Indeed, the main obstacle to chemotherapy in glioblastoma is the permeability of the blood-brain barrier (BBB). In an interesting study, the short-acting NO donor Proli/NO was used to disrupt the BBB to increase delivery of carboplatin to the brain in a rat C6 glioma model; simultaneous intravenous infusions of Proli/NO (1 μM) and carboplatin (20 mg/kg) led to long-term survival in 40% of rats harboring intracerebral C6 gliomas. The median length of survival for the untreated control group was 27 days (mean 26 days), whereas the median length of survival was 67 days (mean 155 days) for rats treated with intravenous infusions of carboplatin and Proli/NO. The results suggested that the NO donor Proli/NO increased the efficacy of carboplatin chemotherapy in rats via increased microvessel permeability in brain tumors [47]. A novel NO donor, glycidyl nitrate (GLYN), was also shown to significantly increase tumor blood flow but had no effect on the blood flow to normal tissues in mice bearing SCC VII tumors . The combination with cisplatin was significantly more efficacious than either cisplatin or GLYN alone with tumor growth delay time of 0.5 ± 0.5 days for GLYN alone, 1.1 ± 0.4 days for cisplatin alone, and

7.2 ± 3.1 days for 300 mg/kg GLYN plus cisplatin. This apparent specificity for tumor tissue may make GLYN particularly suitable for treating hypoxic tumors via the enhancement of chemotherapeutics [48].

In combination treatments, the survival of mice injected (i.p.) with L1210 leukemia cells was much higher when cisplatin (1 mg/kg) was combined with an NO donor (3,3-bis (hydroxymethyl)oxetane; HMO) [49]. The combination of NO donor NCX 4040 with cisplatin caused a 50% increase in survival in mice bearing ovarian cancer A2780 xenografts that were previously resistant to cisplatin [50]. In another study, the cytotoxicity of NCX 4040 and oxaliplatin was evaluated in xenograft models using the three human colon cancer cell lines (LoVo, LoVo-Dx, and WiDr). NCX 4040 was found to sensitize resistant WiDr tumors to oxaliplatin treatment. There was a 60% reduction in tumor growth compared with untreated (2.7%), NCX 4040 treated (5.1%), or oxaliplatin (5.7%) alone [51]. One possible mechanism underlying the synergistic interaction may be the oxidative stress whereby the NCX 4040-induced DNA damage further enhances the damage caused by oxaliplatin. Variation in drug sensitivity could be attributed to several factors related to the role of the microenvironment in the in vivo models. For example, it has been shown that high antioxidants levels are inversely correlated to chemotherapy-induced apoptosis in colon cancer by preventing ROS-induced damage [52]. Taken together, these studies suggest that NO delivery could abrogate hypoxia-induced chemoresistance.

Inhibition of Antiapoptotic Pathways

Several antiapoptotic pathways are involved in chemoresistance where NO could play a key role. Indeed, conditions favoring low NO levels conferred chemoresistance against cisplatin. Low doses of the NO donors *S*-nitroso-*N*-acetylpenicillamine (SNAP) and sodium nitroprusside (SNP) induced resistance against cisplatin-specific apoptosis in head and neck squamous cell carcinoma (HNSCC) cell lines. The chemoresistance was due to the enhanced expression of the antiapoptotic protein survivin. The chemoresistance was reversed using RNA interference to knock down survivin, which sensitized HNSCC cells to cisplatin-induced apoptosis. Nevertheless, higher NO concentrations induced by SNAP and SNP suppressed survivin levels in HNSCC, resulting in apoptosis [53]. Nitrosative stress induced by superphysiological levels of NO also suppressed the antiapoptotic protein Cdc25A. Nitrosative stress decreased the Cdc25A-bound fraction of apoptosis signal-regulating kinase 1 (ASK-1), which is involved in the control of cellular proliferation, differentiation, and survival, and sensitized cells to cisplatin-induced apoptosis, which also activates ASK-1 [54].

NO has been implicated as playing a role in numerous apoptotic pathways. For example, NO inhibits NF-κB activity, a major antiapoptotic pathway known to be dysregulated in many tumors [55]. Exposure of cancer cells to high levels of NO donors (greater than 500 mM DETA/NO) resulted in the downregulation of NF-κB, Snail, and Yin-Yang 1 (YY1) via S-nitrosylation. Inhibition of these apoptotic repressors resulted in the upregulation of Raf kinase inhibitor protein (RKIP) and phosphatase and tensin homologue (PTEN), which in turn inhibited several antiapoptotic gene products such as B-cell lymphoma-extra large (Bclxl) and X-linked inhibitor of apoptosis protein (XIAP), thus facilitating the activation of the apoptotic pathways with chemotherapy [56]. It was also shown that combination treatment of human prostate cancer cell lines with DETA/NO and cisplatin resulted in significant synergistic apoptosis. NO-mediated chemosensitization in prostate cancer cell lines revealed that both resistant factors YY1 and Bclxl were inhibited, which in turn sensitized tumor cells to cisplatin-induced apoptosis. The in vitro findings were confirmed in mice bearing PC-3 xenografts. Combination treatment with DETA/NO and cisplatin resulted in synergistic inhibition of tumor growth. Tumor tissues from nude mice treated with DETA/NO showed an inhibition of YY1 and upregulation of a TNF-related apoptosis-inducing ligand (TRAIL) receptor, death receptor 5 (DR5). YY1 negatively regulates DR5 transcription and expression, which is responsible for TRAIL-induced apoptosis [57]. Therefore, there is sufficient evidence to support the proposition that NO could play a role in sensitizing cells to platinum-based chemotherapy through enhancement of apoptotic pathways.

THERAPEUTIC APPLICATIONS OF NO AS A CHEMOSENSITIZER

Use of Donors for NO Delivery

NO has a short half-life (1–5 s) and can be rapidly inactivated by reacting with oxygen or hemoglobin. Numerous approaches have been developed for NO usage in biomedical applications [58]. A number of NO donors have been investigated as potential cancer therapeutics. Examples of these include organic nitrites (GTN), metal-nitrosyl complexes (SNP), *S*-nitrosothiols, SNAP, *S*-nitrosoglutathione (GSNO), sydnonimines 3-morpholinosydnonimine, and NONOates [59]. NO donors belonging to the class of NONOates have been the most extensively evaluated class of NO donors for the treatment of cancer and are promising as effective chemo- and radiosensitizing agents with long half-lives. By structural manipulations, NONOates can be designed for targeted delivery of NO and conjugated to other payloads to enhance the therapeutic potential [60]. Wang's group coupled PYRRO/NO to a chain of amino acids recognized to be a substrate for prostate-specific antigen (PSA). This conjugated NONOate only released NO in the presence of PSA, conferring a level of specificity to cancer cells [61]. Early evidence that NO might enhance the antitumor effects of certain cancer chemotherapy agents such as cisplatin was reported by Wink et al., who examined the effect of combining either NO gas or NO donor drugs DEA/NO or PAPA/NO with cisplatin in Chinese hamster V79 lung fibroblast cells. Combination treatment of cisplatin with all the three sources of NO caused massive enhancement of V79 cell killing compared with that from exposure to cisplatin alone [62]. Similar enhancement of cisplatin cytotoxicity was also observed in a HNSCC cell line. The effect was restricted to the longer acting NONOate DETA/NO, which resulted in an almost a threefold reduction in the LD50 of cisplatin in CCL23 cells, while the shorter acting agents, DPTA/NO and PAPA/NO had no sensitizing effect in vitro [63].

Metastatic SW620 colon cancer cells have undergone multiple gene product modifications resulting in resistance to cisplatin-induced apoptosis [64,65]. Metastatic SW620 cells have lower expression of apoptosis-inducing factor (AIF) than the primary SW480 cell from which they were derived. SW620 was sensitized by DETA/NO (500 and 1000 μM) for 6 h to cisplatin-induced apoptosis through upregulation of AIF. Sensitization by DETA/NO observed in vitro was confirmed in vivo in nude mice bearing SW620. Tumor tissues derived from the SW620 xenografts showed significant up regulation of AIF and increased apoptosis following combination treatment with DETA/NO and cisplatin [66]. Perrotta et al. (2007) demonstrated that NO donors could increase the efficacy of chemoimmunotherapies in vivo. The combined systemic treatment of cisplatin with intratumoral delivery of dendritic cells previously treated ex vivo with NO donors (DETA/NO or isosorbide dinitrate) were significantly more efficacious than cisplatin alone in tumor regression of the highly tumorigenic and poorly immunogenic B16 melanoma and yielded almost 90% animal survival [67].

The NO donor GTN, which has already been established for the management of coronary artery disease, has also been examined in the treatment of cancer. GTN injection twice a week (0.02 or 0.2 mg/kg/per injection) combined with cisplatin significantly inhibited tumor growth compared with cisplatin alone in a model of murine colon cancer (colon-26). The same study also showed that the combination of i.p. GTN injection with cisplatin significantly inhibited tumor growth compared with cisplatin alone in a Lewis lung carcinoma model inoculated into C57BL/6 mice [68]. Glycosylated NO donors such as b-galactosyl-pyrrolidinyl diazeniumdiolate (b-Gal-NONOate) were developed to increase tumor toxicity through site-specific intracellular release of NO. The LacZ gene is commonly over expressed in tumor cells but not in normal mammalian cells; b-Gal-NONOate is specifically hydrolysed in LacZ-expressing tumor cells to release large amounts of NO [69]. Cisplatin chemosensitization by b-Gal-NONOate was studied in C6/LacZ, 9L/LacZ, and HeLa/LacZ tumor cells, when sensitization to cisplatin by b-Gal-NONOate occurred in a dose-dependent manner. Furthermore, the amount of cisplatin required to evoke cytotoxicity was much lower when combined with b-Gal-NONOate [70].

Issues With NO Donor Delivery

Due to the complexity of NO biology, conflicting findings have also been obtained. The final effect of NO in cancer is dependent upon the cell type, the redox state of the reaction, and the final intracellular concentration and the duration of exposure to NO. The effect of NO donors is based on the induction of late apoptosis/necrosis of cells, which might be caused by depletion of cellular ATP level. Neither SPER/NO nor DETA/NO, alone or in combination with cisplatin, had any significant influence on PTEN, survivin, or ABCB1 gene expressions in all tested cell lines [71]. Likewise, in glioma cells, PABA/NO sensitized to temozolomide, but not carboplatin. Conversely, a low dose of PABA/NO (5 μmol) even reduced the antiproliferative effect of carboplatin [72]. Increases in platinum resistance could be attributed to the altered expression of regulatory proteins involved in apoptotic signaling pathways, or it is possible that the intracellular concentration of NO in the instances described above were not sufficiently high enough to buffer endogenous antioxidant levels and provoke ROS-/RNS-dependent cytotoxicity.

iNOS Gene Delivery

The mechanisms underlying the expression of inducible nitric oxide synthase (iNOS) and the molecular pathways affected by NO are receiving a lot of attention. Among all the methods used for high concentrations of NO production, iNOS gene transfer may be accepted as a superior way due to safety [73], excellent "bystander" effects, and a longer duration of expression [74]. iNOS gene delivery using a designer biomimetic vector and iNOS nanoparticles (NPs) were found to produce a maximum of 62% cell killing in ZR-75-1 breast cancer cells, yet transfection efficiency was only 16% [75].

Gene therapy with iNOS also offers the opportunity to enhance the antitumor effects of cisplatin. CMV/iNOS plasmid gene transfer in human prostate (DU145 and PC3) and colon (HT29) cancer cell lines produced increased amounts of NO, with increased sensitivity to cisplatin-induced cell death. In vivo transfer of CMV/iNOS by direct injection into established RIF-1 xenografts caused a significant delay in tumor growth, and sensitized xenografts to the cytotoxic properties of cisplatin [76]. In another study, combination treatment with cationic liposome-mediated iNOS gene therapy with low-dose cisplatin significantly enhanced cisplatin-mediated cell apoptosis, inhibited cell proliferation, metastasis, and invasion in human lung adenocarcinoma A549 cells in vitro and in vivo. The enhanced antitumor effects of low-dose cisplatin by iNOS gene therapy in lung adenocarcinoma were associated with the upregulation of p53 expression and the downregulation of matrix metalloproteinase-2 (MMP2) and mTOR protein expression [77]. iNOS activity can also be promoted by delivery of genes upstream of iNOS. For example, intratumoral liposomal IFN-γ gene therapy sensitized to cisplatin in a model of syngeneic ovarian carcinoma. This treatment suppressed tumor growth and increased the long-term survival of animals but was ineffective in iNOS (−/−) mice, indicating that NO was a direct mediator of chemosensitization [78].

Issues With NO Gene Delivery

NO induced by iNOS can either promote or inhibit tumor progression. The effect of NO depends on the level and duration of NOS expression and timing of NO delivery, the microenvironment, the genetic background, and the cell type. The primary determinant for the role of NO and subsequent RNS is dictated by the redox environment. Elevated NO interacts with ROS and ROS-detoxifying mechanisms such as the GSH redox cycle; the RNS produced with unbalanced redox status promotes DNA damage and cell cytotoxicity. On the other hand, low levels of NO are insufficient to overcome the detoxifying mechanisms, which can result in a protumorigenic effect. Further complexity is added with the inter- and intravariability in cellular redox levels; it becomes virtually impossible to predict the response to NO associated therapies [79]. For instance, most human melanoma tumor cells are known to constitutively express iNOS, which is responsible for driving proliferation and regulating resistance to apoptosis [80]. It was found that chemical quenching of NO using the NO scavenger c-PTIO resulted in G2 growth arrest followed by a gain of cisplatin-induced apoptosis in melanoma cells [81]. This finding was supported by another study that documented that inhibition of iNOS repressed the growth of human melanoma in vivo and synergized with cisplatin [82]. A recent study by Godoy et al. also found that modulation of intrinsic intracellular NO levels substantially increases cisplatin toxicity in melanoma cells in vitro [83]. Interestingly, chemoresistance to cisplatin-induced apoptosis in those studies was correlated with low levels of NO produced with constitutive iNOS. Low levels of NO in the studies above induced S-nitrosylation, which is known to regulate apoptosis in cancer cells. NO interferes with proapoptotic caspase (9 and 3) activity by S-nitrosylation of the active site and also by impairing the cleavage of proenzymes, which is required for their activation. Low nM levels of NO also promote stabilization of HIF-1α directly, S-nitrosylation of the oxygen-dependent domain, or indirectly, by inhibiting prolyl hydroxylase 2, the enzyme responsible for targeting the prosurvival transcription factor HIF-1α for proteasomal degradation. Increases in S-nitrosylation of caspases and HIF-1α are recognized features of the complex mechanisms of drug resistance [81–83].

IMPROVED NO DELIVERY; EMERGING ROLE OF NPs

Delivery of NO remains a challenge; therefore, strategies to increase the bioavailability and targeted delivery are desirable. Potential systemic side effects of NO donors in vivo such as hypotension and rapid scavenging by the hemoglobin present in red blood cells have limited the therapeutic application of NO donors. Low-molecular-weight NO donors disperse rapidly in biological medium, are easily cleared in the kidneys and liver, and are not easily manipulated for targeting in biomedical applications; therefore, NO donors can be incorporated into or chemically linked to the surface of nanomaterials. Polymeric NPs are an attractive approach to targeted release through chemical modifications to the surface of NPs, tailoring the site and duration of NO release [84]. Several papers describe the synthesis, characterization, and biomedical applications of NO-releasing nanomaterials, including metallic NPs, liposomes, dendrimers, silica NPs, polymeric NPs, carbon nanotubes, and quantum dots [85]. In 2010, the first investigation of NO-releasing NPs for potential use in cancer was performed. In this study, *N*-diazeniumdiolate-modified silica NPs were prepared with an average particle diameter of 125 nm. The toxicity of the particles was studied using ovarian carcinoma cells; with results showing that the NPs were more effective against the carcinoma cells than a small molecule NO donor (PYRRO/NO) [86].

Recently, combination treatment of NO-releasing NPs with cisplatin in the neuroblastoma BE(2)-C cell line was studied. The NO donor GSNO was conjugated into the hydrophobic domain of a block copolymer of oligo(ethylene glycol methyl ether) methacrylate and the functional monomer 2-vinyl-4,4-dimethyl-5-oxazolone. NP size was 40 nm, which increased to 50 nm upon incorporation of the GSNO. NPs formed substantially improved NO stability in aqueous media without affecting the efficacy of intracellular delivery. Pretreatment of BE(2)-C cells with NO-releasing NPs prior to cisplatin treatment provided a fivefold reduction in cisplatin IC50 compared with cells that did not received pretreatment. Results from the study indicate that the combination of nano-NO delivery and cisplatin chemotherapy displayed significant enhancement of antitumor activity with dose-sparing characteristics of the drug, which could reduce dose-limiting toxicities of platinum drugs. Furthermore, there was no normal cell toxicity in this combination therapy, partly due to the reduction in cisplatin dose [87]. In a more recent study, NO and cisplatin-releasing wrinkle-structured amine-modified mesoporous silica (AMS) NPs were developed for the treatment of NSCLC. The toxicity of NO- and cisplatin-loaded silica NPs (NO-Si-DETA-cisplatin-AMS) was significantly higher than that of silica NPs loaded with only cisplatin (Si-DETA-cisplatin-AMS) in NSCLC cell lines (H596 and A549). The potentiation of cisplatin by NO appeared to be tumor-specific, as the NO-loaded NPs were not more potent than the NO-lacking NPs in normal lung fibroblast (WI-38) and epithelial (BEAS-2B) cells [88].

TABLE 2 Platinum Drug Delivery Systems in Clinical Trials

	Drugs	Name	Nanoscale Dimensions (nm)	Indications	Clinical Trial Phase Status	References
Polymer	Cisplatin	NC-6004	28	Advanced solid tumors, lung and biliary cancer	Clinic trial phases II and III	[89]
	Oxaliplatin	NC-4016	40	Advanced solid tumors or lymphomas	Clinic trial phase I	[90]
	Oxaliplatin	ProLindac	110–114	Solid tumors	Clinic trial phase II	[91,92]
Liposomes	Cisplatin	Lipoplatin	90–130	Pancreatic/head and neck/breast cancer	Clinic trial phase III	[93]
	Cisplatin	Aroplatin	<200	Colorectal cancer	Clinic trial phase I/II	[94]
	Cisplatin	SPI-077	110	Solid tumors, ovarian cancer, head and neck cancer	Clinic trial phase II	[95]

In summary, recent publications in this field highlight that the combination of NO donors/nanomaterials in cancer cells represents an exciting and promising strategy to overcome the difficulties associated with the biomedical administration of NO. The advantages utilizing nanomaterials to deliver NO include preservation of the NO donor until reaching the target site, delivering high concentrations of NO, reductions in the systemic side effects, and toxicity to normal tissue [85]. Although significant progress has been achieved in the design of NO-releasing vehicles for biomedical applications, translation of this technology to the clinic remains slow. Similarly, although many factors that hinder the clinical utility of the platinum compounds (side effects, development of resistance, low bioavailability, and low water solubility) may be abrogated by incorporation of the platinum drugs into a range of NPs, only polymer conjugates and liposomal nanoformulations have progressed into clinical studies (Table 2). There is definitely a gap in the market for smarter delivery devices for the NO class of drugs.

CLINICAL STUDIES USING NO AS A CHEMOSENSITIZER TO PLATINUM COMPOUNDS

Undoubtedly, the most exciting development of NO/platinum therapy lies in the clinical setting. Although the initial Japanese trial reported by Yasuda et al. [96] in the field of chemosensitization of cancer cells by NO was encouraging, studies that followed showed that the use of NO donors with platinum-based chemotherapy did not seem to improve treatment efficacy.

In a randomized phase II trial, the efficacy and safety of GTN plus vinorelbine and cisplatin in 120 patients with previously untreated stage IIIB/IV NSCLC was investigated. The study revealed that GTN patch (25 mg/patient daily) combined with vinorelbine and cisplatin may improve overall response and time to progression (TTP) in patients with stage IIIB/IV NSCLC. They concluded that NO-donating drugs such as GTN may reduce chemoresistance via improvement of hypoxic conditions, reduced HIF-1 stabilization, and direct effect of NO on cancer cells and increase in p53 protein via an increase in drug delivery in tumor tissue [96]. However, using other NO donors such as isosorbide-5-mononitrate (ISMN) did not improve the treatment response to irinotecan/cisplatin in patients with advanced NSCLC. Upon closer examination ISMN is metabolized through an extramitochondrial low-affinity pathway that yields negligible quantities of ROS, which are the key players in the induction of apoptosis by NO [97]. In contrast, GTN undergoes mitochondrial biotransformation in a process that is associated with an uncoupling of the respiratory chain and the production of ROS enabling a greater apoptotic effect with NO.

Results from three more recent clinical trials published in 2014 also showed disappointing outcomes. One small-randomized phase II study from Germany was conducted in 68 patients with stage IIIB or IV NSCLC; patients received oral vinorelbine-cisplatin, plus patches containing 25 mg GTN or placebo. The primary outcome was response rate. This study was stopped prematurely due to slow accrual. Although the response rate was higher in the patients treated with GTN (35% vs. 18%), this difference was not significant and no difference in TTP or overall survival (OS) was observed [98]. Another study conducted by the Dutch Lung Physician Society (NVALT) (NCT01171170) indicated that adding GTN to first-line carboplatin-paclitaxel-bevacizumab did not improve progression-free survival (PFS) and OS in patients with stage IV nonsquamous NSCLC; response rate was 54% in control arm and only 38% in experimental arm. In contrast to the other trials using 25 mg/24 h GTN, patches delivering only 15 mg/24 h were used in this trial. There was also change in serum VEGF levels of patients treated with GTN in the study indicating that the levels of NO were not high enough to decrease VEGF through reduction of HIF-1α expression and to increase tumor blood flow and consequently drug delivery [99].

A wider rigorous multicenter randomized, phase III study conducted by the Australasian Lung Cancer Trials Group (ALCTG) was stopped after a planned interim analysis of 270 patients. This trial provided strong evidence that topical GTN is not beneficial when added to carboplatin with either gemcitabine or paclitaxel as first-line chemotherapy in NSCLC. PFS was 4.9 months in the patients receiving GTN (25 mg patches for 2 days before, the day of, and 2 days after) plus chemotherapy versus 4.7 months in the patients treated with GTN. Many factors may have contributed to the negative results of the study such as the chemotherapy regimen, pharmacogenomics of the cohort with advanced NSCLC, or the heterogeneous nature of metastatic NSCLC [100]. Without further evidence though, it is impossible to speculate why GTN patches produce such variable outcomes.

CONCLUSION

With the prominence of drug-resistant cancers, there is an urgent need to design therapeutic agents with the ability to sensitize or synergistically target cancerous cells over healthy cells to effectively reduce off-target effects. The effects of NO effects range from a direct cytotoxic response through RNS generation to the subtle modulation of apoptosis-related and resistance genes with the regulation of hypoxic conditions and "normalization" of tumor vasculature. Currently, the most critical obstacle in NO delivery is lack of an effective and selective release of NO at the tumors site. Many considerations should be taken in account while working with the current NO donors such as the in vivo half-life of the NO donor, which may be considerably affected by the biological microenvironment (pH, temperature, and the presence of ROS), in addition to the NO release profile from the NO donor in vivo, which may produce low concentrations of NO, thus promoting tumor growth. Special attention must also be given to the potential side effects of the systemic distribution of the NO donor such as vasodilation, and the toxicity by-products formed after the NO release. Combining NO donors with nanomaterials has been considered a promising strategy to overcome this issue. It should be noted that the cytotoxicity of NO-releasing nanomaterial would be mainly dependent on the cell type, the chemical nature of the nanomaterial, and the concentration of NO released from the nanomaterial. Smart NO delivery systems for clinical applications are aimed at efficiently releasing NO for controlled periods of time at the desired target site of application, with minimum side effects, in an economically feasible and safe manner. In addition, these NPs may be formulated to carry more than one chemotherapeutic, thus prompting an effective strategy to overcome cancer resistance.

Conflict of Interest

No potential conflicts of interest were disclosed.

References

[1] Wang G, Reed E, Li QQ. Molecular basis of cellular response to cisplatin chemotherapy in non-small cell lung cancer. Oncol Rep 2004;12:955–66 [Review].
[2] Wheate NJ, Walker S, Craig GE, Oun R. The status of platinum anticancer drugs in the clinic and in clinical trials. Dalton Trans 2010;39(35):8113–27.
[3] Galluzzi L, Senovilla L, Vitale I, Michels J, Martins I, Kepp O, et al. Molecular mechanisms of cisplatin resistance. Oncogene 2012;31(15):1869–83.
[4] Dilruba S, Kalayda GV. Platinum-based drugs: past, present and future. Cancer Chemother Pharmacol 2016;77(6):1103–24.
[5] Stewart DJ. Mechanisms of resistance to cisplatin and carboplatin. Crit Rev Oncol Hematol 2007;63(1):12–31.
[6] Perego P, Robert J. Oxaliplatin in the era of personalized medicine: from mechanistic studies to clinical efficacy. Cancer Chemother Pharmacol 2016;77(1):5–18.
[7] Ozols RF, Bundy BN, Greer BE, Fowler JM, Clarke-Pearson D, Mannel RS, et al. Phase III trial of carboplatin and paclitaxel compared with cisplatin and paclitaxel in patients with optimally resected stage III ovarian cancer: a Gynecologic Oncology Group study. J Clin Oncol 2003;21(17):3194–200.
[8] Laurie SA, Siu LL, Winquist E, Maksymiuk A, Harnett EL, Walsh W, et al. A phase 2 study of platinum and gemcitabine in patients with advanced salivary gland cancer. Cancer 2010;116(2):362–8.
[9] Penson RT, Dizon DS, Cannistra SA, Roche MR, Krasner CN, Berlin ST, et al. Phase II study of carboplatin, paclitaxel, and bevacizumab with maintenance bevacizumab as first-line chemotherapy for advanced mullerian tumors. J Clin Oncol 2010;28(1):154–9.
[10] Zhang S, Lovejoy KS, Shima JE, Lagpacan LL, Shu Y, Lapuk A, et al. Organic cation transporters are determinants of oxaliplatin cytotoxicity. Cancer Res 2006;66(17):8847–57.

[11] Dasari S, Tchounwou PB. Cisplatin in cancer therapy: molecular mechanisms of action. Eur J Pharmacol 2014;740:364–78.
[12] Kigawa J. New strategy for overcoming resistance to chemotherapy of ovarian cancer. Yonago Acta Med 2013;56(2):43–50.
[13] Bonavida B. Sensitization of cancer cells for chemo/immuno/radiotherapy. In: Teicher B, editor. Volume of cancer drug discovery and development series. Totowa: The Humana Press; 2008.
[14] Bratasz A, Weir NM, Parinandi NL, Zweier JL, Sridhar R, Ignarro LJ, et al. Reversal to cisplatin sensitivity in recurrent human ovarian cancer cells by NCX-4016, a nitro derivative of aspirin. Proc Natl Acad Sci U S A 2006;103(10):3914–9.
[15] Sullivan R, Graham CH. Chemosensitization of cancer by nitric oxide. Curr Pharm Des 2008;14(11):1113–23.
[16] Riganti C, Miraglia E, Viarisio D, Costamagna C, Pescarmona G, Ghigo D, et al. Nitric oxide reverts the resistance to doxorubicin in human colon cancer cells by inhibiting the drug efflux. Cancer Res 2005;65(2):516–25.
[17] Jabr-Milane LS, van Vlerken LE, Yadav S, Amiji MM. Multi-functional nanocarriers to overcome tumor drug resistance. Cancer Treat Rev 2008;34(7):592–602.
[18] Pacher P, Beckman JS, Liaudet L. Nitric oxide and peroxynitrite in health and disease. Physiol Rev 2007;87 (1):315–424.
[19] Palmer RM, Ferrige A, Moncada S. Nitric oxide release accounts for the biological activity of endothelium-derived relaxing factor. Nature 1987;327(6122):524–6.
[20] Moncada S, Bolaños JP. Nitric oxide, cell bioenergetics and neurodegeneration. J Neurochem 2006;97(6):1676–89.
[21] Liew F, Cox F. Nonspecific defence the role of nitric oxide. Parasitol Today 1991;7(3):17–21.
[22] Hirst D, Robson T. Targeting nitric oxide for cancer therapy. J Pharm Pharmacol 2007;59(1):3–13.
[23] Ying L, Hofseth LJ. An emerging role for endothelial nitric oxide synthase in chronic inflammation and cancer. Cancer Res 2007;67(4):1407–10.
[24] Wink DA, Vodovotz Y, Laval J, Laval F, Dewhirst MW, Mitchell JB. The multifaceted roles of nitric oxide in cancer. Carcinogenesis 1998;19(5):711–21.
[25] Thomsen LL, Miles DW. Role of nitric oxide in tumour progression: lessons from human tumours. Cancer Metastasis Rev 1998;17(1):107–18.
[26] Gallo O, Masini E, Morbidelli L, Franchi A, Fini-Storchi I, Vergari WA, et al. Role of nitric oxide in angiogenesis and tumor progression in head and neck cancer. J Natl Cancer Inst 1998;90(8):587–96.
[27] Bonavida B, Khineche S, Huerta-Yepez S, Garbán H. Therapeutic potential of nitric oxide in cancer. Drug Resist Updat 2006;9(3):157–73.
[28] Gottesman MM, Fojo T, Bates SE. Multidrug resistance in cancer: role of ATP–dependent transporters. Nat Rev Cancer 2002;2(1):48–58.
[29] Gillet J, Efferth T, Remacle J. Chemotherapy-induced resistance by ATP-binding cassette transporter genes. Biochim Biophys Acta Rev Cancer 2007;1775(2):237–62.
[30] Krishnamurthy P, Ross DD, Nakanishi T, Bailey-Dell K, Zhou S, Mercer KE, et al. The stem cell marker Bcrp/ABCG2 enhances hypoxic cell survival through interactions with heme. J Biol Chem 2004;279(23):24218–25.
[31] Gatti L, Beretta G, Cossa G, Zunino F, Perego P. ABC transporters as potential targets for modulation of drug resistance. Mini Rev Med Chem 2009;9(9):1102–12.
[32] Liu J, Li C, Qu W, Leslie E, Bonifant CL, Buzard GS, et al. Nitric oxide prodrugs and metallochemotherapeutics: JS-K and CB-3-100 enhance arsenic and cisplatin cytolethality by increasing cellular accumulation. Mol Cancer Ther 2004;3(6):709–14.
[33] Doublier S, Riganti C, Voena C, Costamagna C, Aldieri E, Pescarmona G, et al. RhoA silencing reverts the resistance to doxorubicin in human colon cancer cells. Mol Cancer Res 2008;6(10):1607–20.
[34] Takahara PM, Rosenzweig AC, Frederick CA, Lippard SJ. Crystal structure of double-stranded DNA containing the major adduct of the anticancer drug cisplatin. Nature 1995;377(6550):649–52.
[35] Huang JC, Zamble DB, Reardon JT, Lippard SJ, Sancar A. HMG-domain proteins specifically inhibit the repair of the major DNA adduct of the anticancer drug cisplatin by human excision nuclease. Proc Natl Acad Sci U S A 1994;91(22):10394–8.
[36] Welsh C, Day R, McGurk C, Masters JR, Wood RD, Köberle B. Reduced levels of XPA, ERCC1 and XPF DNA repair proteins in testis tumor cell lines. Int J Cancer 2004;110(3):352–61.
[37] Vaisman A, Varchenko M, Umar A, Kunkel TA, Risinger JI, Barrett JC, et al. The role of hMLH1, hMSH3, and hMSH6 defects in cisplatin and oxaliplatin resistance: correlation with replicative bypass of platinum-DNA adducts. Cancer Res 1998;58(16):3579–85.

[38] Lowe SW, Ruley HE, Jacks T, Housman DE. p53-dependent apoptosis modulates the cytotoxicity of anticancer agents. Cell 1993;74(6):957–67.
[39] Schneiderhan N, Budde A, Zhang Y, Brüne B. Nitric oxide induces phosphorylation of p53 and impairs nuclear export. Oncogene 2003;22(19):2857–68.
[40] Ali AA, Coulter JA, Ogle CH, Migaud MM, Hirst DG, Robson T, et al. The contribution of N_2O_3 to the cytotoxicity of the nitric oxide donor DETA/NO: an emerging role for S-nitrosylation. Biosci Rep 2013;33(2).
[41] Sidorkina O, Espey MG, Miranda KM, Wink DA, Laval J. Inhibition of poly (ADP-ribose) polymerase (PARP) by nitric oxide and reactive nitrogen oxide species. Free Radic Biol Med 2003;35(11):1431–8.
[42] Sullivan R, Graham CH. Hypoxia-driven selection of the metastatic phenotype. Cancer Metastasis Rev 2007;26 (2):319–31.
[43] Hwang IT, Chung YM, Kim JJ, Chung JS, Kim BS, Kim HJ, et al. Drug resistance to 5-FU linked to reactive oxygen species modulator 1. Biochem Biophys Res Commun 2007;359(2):304–10.
[44] McCormick CC, Li WP, Calero M. Oxygen tension limits nitric oxide synthesis by activated macrophages. Biochem J 2000;350(3):709–16.
[45] Frederiksen LJ, Siemens DR, Heaton JP, Maxwell LR, Adams MA, Graham CH. Hypoxia induced resistance to doxorubicin in prostate cancer cells is inhibited by low concentrations of glyceryl trinitrate. J Urol 2003;170 (3):1003–7.
[46] Vinogradov S, Wei X. Cancer stem cells and drug resistance: the potential of nanomedicine. Nanomedicine 2012;7(4):597–615.
[47] Weyerbrock A, Walbridge S, Pluta RM, Saavedra JE, Keefer LK, Oldfield EH. Selective opening of the blood-brain barrier by a nitric oxide donor and long-term survival in rats with C6 gliomas. J Neurosurg 2003;99 (4):728–37.
[48] Ning S, Bednarski M, Oronsky B, Scicinski J, Knox SJ. Novel nitric oxide generating compound glycidyl nitrate enhances the therapeutic efficacy of chemotherapy and radiotherapy. Biochem Biophys Res Commun 2014;447 (3):537–42.
[49] Konovalova N, Goncharova S, Volkova L, Rajewskaya T, Eremenko L, Korolev A. Nitric oxide donor increases the efficiency of cytostatic therapy and retards the development of drug resistance. Nitric Oxide 2003;8(1):59–64.
[50] Bratasz A, Selvendiran K, Wasowicz T, Bobko A, Khramtsov VV, Ignarro LJ, et al. NCX-4040, a nitric oxide-releasing aspirin, sensitizes drug-resistant human ovarian xenograft tumors to cisplatin by depletion of cellular thiols. J Transl Med 2008;6(1):1.
[51] Leonetti C, Scarsella M, Zupi G, Zoli W, Amadori D, Medri L, et al. Efficacy of a nitric oxide-releasing nonsteroidal anti-inflammatory drug and cytotoxic drugs in human colon cancer cell lines in vitro and xenografts. Mol Cancer Ther 2006;5(4):919–26.
[52] Fu Y, Yang G, Zhu F, Peng C, Li W, Li H, et al. Antioxidants decrease the apoptotic effect of 5-Fu in colon cancer by regulating Src-dependent caspase-7 phosphorylation. Cell Death Dis 2014;5(1):e983.
[53] Fetz V, Bier C, Habtemichael N, Schuon R, Schweitzer A, Kunkel M, et al. Inducible NO synthase confers chemoresistance in head and neck cancer by modulating survivin. Int J Cancer 2009;124(9):2033–41.
[54] Tomko Jr. RJ, Lazo JS. Multimodal control of Cdc25A by nitrosative stress. Cancer Res 2008;68(18):7457–65.
[55] Reynaert NL, Ckless K, Korn SH, Vos N, Guala AS, Wouters EF, et al. Nitric oxide represses inhibitory kappaB kinase through S-nitrosylation. Proc Natl Acad Sci U S A 2004;101(24):8945–50.
[56] Bonavida B, Garban H. Nitric oxide-mediated sensitization of resistant tumor cells to apoptosis by chemo-immunotherapeutics. Redox Biol 2015;6:486–94.
[57] Huerta-Yepez S, Baritaki S, Baay-Guzman G, Hernandez-Luna MA, Hernandez-Cueto A, Vega MI, et al. Contribution of either YY1 or Bcl XL-induced inhibition by the NO-donor DETANONOate in the reversal of drug resistance, both in vitro and in vivo. YY1 and Bcl XL are overexpressed in prostate cancer. Nitric Oxide 2013;29:17–24.
[58] Seabra AB, Justo GZ, Haddad PS. State of the art, challenges and perspectives in the design of nitric oxide-releasing polymeric nanomaterials for biomedical applications. Biotechnol Adv 2015;33(6):1370–9.
[59] Bonavida B, Baritaki S, Huerta-Yepez S, Vega MI, Chatterjee D, Yeung K. Novel therapeutic applications of nitric oxide donors in cancer: roles in chemo-and immunosensitization to apoptosis and inhibition of metastases. Nitric Oxide 2008;19(2):152–7.
[60] Keefer LK. Fifty years of diazeniumdiolate research. From laboratory curiosity to broad-spectrum biomedical advances. ACS Chem Biol 2011;6(11):1147–55.

[61] Tang X, Xian M, Trikha M, Honn KV, Wang PG. Synthesis of peptide-diazeniumdiolate conjugates: towards enzyme activated antitumor agents. Tetrahedron Lett 2001;42(14):2625–9.

[62] Wink DA, Cook JA, Christodoulou D, Krishna MC, Pacelli R, Kim S, et al. Nitric oxide and some nitric oxide donor compounds enhance the cytotoxicity of cisplatin. Nitric Oxide 1997;1(1):88–94.

[63] Azizzadeh B, Yip HT, Blackwell KE, Horvath S, Calcaterra TC, Buga GM, et al. Nitric oxide improves cisplatin cytotoxicity in head and neck squamous cell carcinoma. Laryngoscope 2001;111(11):1896–900.

[64] Huerta S, Harris DM, Jazirehi A, Bonavida B, Elashoff D, Livingston EH, et al. Gene expression profile of metastatic colon cancer cells resistant to cisplatin-induced apoptosis. Int J Oncol 2003;22(3):663–70.

[65] Huerta S, Heinzerling JH, Anguiano-Hernandez Y, Huerta-Yepez S, Lin J, Chen D, et al. Modification of gene products involved in resistance to apoptosis in metastatic colon cancer cells: roles of Fas, Apaf-1, NFκB, IAPs, Smac/DIABLO, and AIF. J Surg Res 2007;142(1):184–94.

[66] Huerta S, Baay-Guzman G, Gonzalez-Bonilla CR, Livingston EH, Huerta-Yepez S, Bonavida B. In vitro and in vivo sensitization of SW620 metastatic colon cancer cells to CDDP-induced apoptosis by the nitric oxide donor DETANONOate: Involvement of AIF. Nitric Oxide 2009;20(3):182–94.

[67] Perrotta C, Bizzozero L, Falcone S, Rovere-Querini P, Prinetti A, Schuchman EH, et al. Nitric oxide boosts chemoimmunotherapy via inhibition of acid sphingomyelinase in a mouse model of melanoma. Cancer Res 2007;67(16):7559–64.

[68] Yasuda H, Nakayama K, Asada M, Kaneta T, Sasaki T, Yoshida M, et al. Nitroglycerin may increase response to docetaxel and cisplatin in murine syngraft solid cancer models via reduction of HIF-1α, P-gp, and phosphorylated Akt and via increase in activated p53 in tumor tissues. Cancer Res 2008;68(9 Suppl.):2045.

[69] Chen C, Shi Y, Li S, Qi Q, Gu L, Song J, et al. A glycosylated nitric oxide donor, β-Gal-NONOate, and its site-specific antitumor activity. Arch Pharm 2006;339(7):366–71.

[70] Deng L, Zhang E, Chen C. Synergistic interaction of β-galactosyl-pyrrolidinyl diazeniumdiolate with cisplatin against three tumor cells. Arch Pharm Res 2013;36(5):619–25.

[71] Kielbik M, Szulc-Kielbik I, Nowak M, Sulowska Z, Klink M. Evaluation of nitric oxide donors impact on cisplatin resistance in various ovarian cancer cell lines. Toxicol In Vitro 2016;36:26–37.

[72] Kogias E, Osterberg N, Baumer B, Psarras N, Koentges C, Papazoglou A, et al. Growth-inhibitory and chemosensitizing effects of the glutathione-S-transferase-π-activated nitric oxide donor PABA/NO in malignant gliomas. Int J Cancer 2012;130(5):1184–94.

[73] Fitzpatrick B, Mehibel M, Cowen RL, Stratford IJ. iNOS as a therapeutic target for treatment of human tumors. Nitric Oxide 2008;19(2):217–24.

[74] Nagasawa H, Little JB. Induction of sister chromatid exchanges by extremely low doses of alpha-particles. Cancer Res 1992;52(22):6394–6.

[75] McCarthy HO, Zholobenko AV, Wang Y, Canine B, Robson T, Hirst DG, et al. Evaluation of a multi-functional nanocarrier for targeted breast cancer iNOS gene therapy. Int J Pharm 2011;405(1):196–202.

[76] Adams C, McCarthy HO, Coulter JA, Worthington J, Murphy C, Robson T, et al. Nitric oxide synthase gene therapy enhances the toxicity of cisplatin in cancer cells. J Gene Med 2009;11(2):160–8.

[77] Ye S, Yang W, Wang Y, Ou W, Ma Q, Yu C, et al. Cationic liposome-mediated nitric oxide synthase gene therapy enhances the antitumor effects of cisplatin in lung cancer. Int J Mol Med 2013;31(1):33–42.

[78] Son KK, Hall KJ. Nitric oxide-mediated tumor cell killing of cisplatin-based interferon-γ gene therapy in murine ovarian carcinoma. Cancer Gene Ther 2000;7(10):1324–8.

[79] Vannini F, Kashfi K, Nath N. The dual role of iNOS in cancer. Redox Biol 2015;6:334–43.

[80] Grimm EA, Ellerhorst J, Tang C, Ekmekcioglu S. Constitutive intracellular production of iNOS and NO in human melanoma: possible role in regulation of growth and resistance to apoptosis. Nitric Oxide 2008;19 (2):133–7.

[81] Tang CH, Grimm EA. Depletion of endogenous nitric oxide enhances cisplatin-induced apoptosis in a p53-dependent manner in melanoma cell lines. J Biol Chem 2004;279(1):288–98.

[82] Sikora AG, Gelbard A, Davies MA, Sano D, Ekmekcioglu S, Kwon J, et al. Targeted inhibition of inducible nitric oxide synthase inhibits growth of human melanoma in vivo and synergizes with chemotherapy. Clin Cancer Res 2010;16(6):1834–44.

[83] Godoy LC, Anderson CC, Chowdhury R, Trudel LJ, Wogan GN. Melanoma fights cisplatin with NO: S-nitrosation as a mechanism supporting drug-resistance. Cancer Res 2013;73(8 Suppl.):894.

[84] Seabra AB, Marcato PD, de Paula LB, Durán N. New strategy for controlled release of nitric oxide. J Nano Res 2012;20:61–7.

[85] Seabra AB, de Lima R, Calderón M. Nitric oxide releasing nanomaterials for cancer treatment: current status and perspectives. Curr Top Med Chem 2015;15(4):298–308.
[86] Stevens EV, Carpenter AW, Shin JH, Liu J, Der CJ, Schoenfisch MH. Nitric oxide-releasing silica nanoparticle inhibition of ovarian cancer cell growth. Mol Pharm 2010;7(3):775–85.
[87] Duong HT, Kamarudin ZM, Erlich RB, Li Y, Jones MW, Kavallaris M, et al. Intracellular nitric oxide delivery from stable NO-polymeric nanoparticle carriers. Chem Commun 2013;49(39):4190–2.
[88] Munaweera I, Shi Y, Koneru B, Patel A, Dang MH, Di Pasqua AJ, et al. Nitric oxide-and cisplatin-releasing silica nanoparticles for use against non-small cell lung cancer. J Inorg Biochem 2015;153:23–31.
[89] Plummer R, Wilson R, Calvert H, Boddy A, Griffin M, Sludden J, et al. A Phase I clinical study of cisplatin-incorporated polymeric micelles (NC-6004) in patients with solid tumours. Br J Cancer 2011;104(4):593–8.
[90] Murakami M, Cabral H, Matsumoto Y, Wu S, Kano MR, Yamori T, et al. Improving drug potency and efficacy by nanocarrier-mediated subcellular targeting. Sci Transl Med 2011;3(64):64ra2.
[91] Van der Schoot S, Nuijen B, Sood P, Thurmond I, Stewart D, Rice J, et al. Pharmaceutical development, quality control, stability and compatibility of a parenteral lyophilized formulation of the investigational polymer-conjugated platinum antineoplastic agent AP5346. Pharmazie 2006;61(10):835–44.
[92] Campone M, Rademaker-Lakhai JM, Bennouna J, Howell SB, Nowotnik DP, Beijnen JH, et al. Phase I and pharmacokinetic trial of AP5346, a DACH–platinum–polymer conjugate, administered weekly for three out of every 4 weeks to advanced solid tumor patients. Cancer Chemother Pharmacol 2007;60(4):523–33.
[93] Ding S, Bierbach U. Target-selective delivery and activation of platinum-based anticancer agents. Future Med Chem 2015;7(7):911–27.
[94] Parker JP, Ude Z, Marmion CJ. Exploiting developments in nanotechnology for the preferential delivery of platinum-based anti-cancer agents to tumours: targeting some of the hallmarks of cancer. Metallomics 2016;8(1):43–60.
[95] Seetharamu N, Kim E, Hochster H, Martin F, Muggia F. Phase II study of liposomal cisplatin (SPI-77) in platinum-sensitive recurrences of ovarian cancer. Anticancer Res 2010;30(2):541–5.
[96] Yasuda H, Yamaya M, Nakayama K, Sasaki T, Ebihara S, Kanda A, et al. Randomized phase II trial comparing nitroglycerin plus vinorelbine and cisplatin with vinorelbine and cisplatin alone in previously untreated stage IIIB/IV non-small-cell lung cancer. J Clin Oncol 2006;24(4):688–94.
[97] Han JY, Nam BH, Kim HY, Yoon SJ, Kim HT, Lee JS. A randomized phase II study of irinotecan plus cisplatin versus irinotecan plus capecitabine with or without isosorbide-5-mononitrate in advanced non-small-cell lung cancer. Ann Oncol 2012;23(11):2925–30.
[98] Reinmuth N, Meyer A, Hartwigsen D, Schaeper C, Huebner G, Skock-Lober R, et al. Randomized, double-blind phase II study to compare nitroglycerin plus oral vinorelbine plus cisplatin with oral vinorelbine plus cisplatin alone in patients with stage IIIB/IV non-small cell lung cancer (NSCLC). Lung Cancer 2014;83(3):363–8.
[99] Dingemans AM, Groen HJ, Herder GJ, Stigt JA, Smit EF, Bahce I, et al. A randomized phase II study comparing paclitaxel-carboplatin-bevacizumab with or without nitroglycerin patches in patients with stage IV nonsquamous nonsmall-cell lung cancer: NVALT12 (NCT01171170)dagger. Ann Oncol 2015;26(11):2286–93.
[100] Davidson A, Veillard AS, Tognela A, Chan MM, Hughes BG, Boyer M, et al. A phase III randomized trial of adding topical nitroglycerin to first-line chemotherapy for advanced nonsmall-cell lung cancer: the Australasian lung cancer trials group NITRO trial. Ann Oncol 2015;26(11):2280–6.

Further Reading

[1] Cheng Q, Liu Y. Multifunctional platinum-based nanoparticles for biomedical applications. Nanomedicine and nanobiotechnology. United States: Wiley Interdisciplinary Reviews; 2016.
[2] Boulikas T. Clinical overview on Lipoplatin™: a successful liposomal formulation of cisplatin. Expert Opin Investig Drugs 2009;18(8):1197–218.

CHAPTER

6

Nitric Oxide and Platinum-Derivative-Based Regimens for Cancer Treatment: From Preclinical Studies to Clinical Trials

Stéphanie Plenchette[*,†], *Catherine Paul*[*,†], *Ali Bettaieb*[*,†]

[*]University of Burgundy-Franche Comté, LIIC EA7269, F-21000 Dijon, France

[†]EPHE, PSL Research University, F-75014 Paris, France

Abstract

Chemoresistance to platinum-based antitumor agents remains a major hindrance faced by patients with a wide variety of solid tumors. New effective strategies are still needed to improve chemosensitization and overcome chemoresistance of tumors by platinum-based chemotherapies. Over the past decade, considerable knowledge on the antitumor effect of nitric oxide (NO) and its mechanisms of action has been gained. Here, we provide an overview of the basic mechanisms of resistance to platinum-based drugs and how NO can bypass this chemotherapy resistance. Preclinical and clinical studies focused on combination therapy using platinum chemotherapeutic drugs with NO donors have demonstrated antitumor benefits in different types of cancer, particularly non-small cell lung cancer. Altogether, this review aims to summarize the evidence that therapy regimens, including NO donors, are worth considering as new therapeutic options.

Abbreviations

Bcl-2	B-cell lymphoma-2
BRCA1/2	BReast CAncer 1/2
Ctr1/2	copper transporter 1/2
DETANONOate	diethylenetriamine NONOate
EGFR	epidermal growth factor receptor
GLYN	glycidyl nitrate
GSH	glutathione
GSNO	*S*-nitrosoglutathione
GST	glutathione-*S*-transferase

Nitric Oxide (Donor/Induced) in Chemosensitization, Volume 1
http://dx.doi.org/10.1016/B978-0-12-811020-1.00006-5

HIF-1α	hypoxia-inducible factor 1-alpha
iNOS	inducible nitric oxide synthase
ISMN	isosorbide-5-mononitrate
NER	nucleotide excision repair
NF-κB	nuclear factor-kappa B
NO	nitric oxide
NSCLC	nonsmall cell lung cancer
NTG	nitroglycerin
ORR	overall response rate
OS	overall survival
PARP	poly(ADP-ribose) polymerase
PFS	progression-free survival
pSTAT3	phosphorylated signal transducer and activator of transcription 3
PTM	posttranslational modification
SNAP	*S*-nitroso-*N*-acetylpenicillamine
SNP	sodium nitroprusside
STAT3	signal transducer and activator of transcription 3
TNF	tumor necrosis factor
TRAIL	tumor-necrosis-factor-related apoptosis-inducing ligand
VEGF	vascular endothelial growth factor
XPA	*Xeroderma pigmentosum* complementation group A

INTRODUCTION

Platinum-based chemotherapeutic drugs have displayed an active role for treating a broad spectrum of solid human cancers over the course of several decades. Cisplatin, carboplatin, and oxaliplatin are first-, second-, and third-generation platinum drugs, respectively, that have been developed over the years in the attempt to circumvent severe side effects and better efficiency of platinum-based drugs [1]. In the race to develop improved platinum-based drugs, a large number of new complexes have been designed, synthesized, and evaluated preclinically for anticancer activity. However, only a small number of worthed clinical trials, evaluations, and current studies are underway to further test these drugs [2]. Until now, cisplatin, carboplatin, and oxaliplatin remain the three platinum analogues approved for clinical use and are among the most widely used chemotherapeutic drugs for cancer therapy. However, their clinical efficiency is still often limited by either intrinsic or acquired tumor resistance leading to treatment failure and tumor recurrence. Several factors may contribute to resistance to platinum-based drugs. Although not fully understood, much effort has been put into the deciphering of the mechanism of resistance that provided valuable insights to pave the way toward new approaches to specifically target and circumvent tumor resistance.

Recent studies have shown that nitric oxide (NO) counteracts platinum-based drugs resistance cancer cells. NO is a short-lived, free radical gas that plays both tumoricidal and tumor-promoting effects. The biochemical mechanism of NO action largely relies on posttranslational modification (PTM) of proteins such as S-nitrosylation, nitration, or metal S-nitrosylation. If these PTM were first considered as atypical, they are now well studied and recognized as pivotal factors for the regulation of numerous protein functions and signaling pathways involved in virtually all biological processes. Among these PTM, S-nitrosylation represents the main MPT induced by NO on proteins that modulates their structures, functions, expressions, localizations, or even their interactions with other protein partners. Importantly, the dichotomous nature of NO in cancer and its impact on proteins likely depends on many factors including its timing, location, concentration, source of NO production, and other conditions such as the redox status of cells (i.e., concomitant production of superoxide anion that can react with NO to generate toxic peroxynitrite).

NO donors have been shown to sensitize cancer cells to various chemotherapeutic agents. In particular, numerous studies have proven the efficacy of NO in circumventing platinum-based chemotherapeutic drugs resistance.

Here, we focus on diverse molecular features of NO that overcome platinum resistance in tumors of different histological origins to restore and enhance sensitivity and killing of resistant malignant cells; the benefit afforded by NO donors in combination therapy regimens are summarized in recent preclinical in vivo findings and outcomes in clinical trials.

REVERSAL OF PLATINUM RESISTANCE: ROLE OF NO

Cisplatin, carboplatin, and oxaliplatin exert cytotoxic activity by altering multiple pathways. Although the cytotoxic activity predominantly relies on DNA platination (i.e., the formation of DNA adducts), inhibiting DNA replication and transcription leading to the induction of apoptosis, the actual mode of action is still incompletely understood [3,4]. A massive effort has been devoted to the understanding of the cellular mechanisms that underlie resistance of platinum-based anticancer drugs and particularly cisplatin resistance. Cancer cells develop resistance to cisplatin through several different processes classified into four main groups, namely, "pretarget resistance" (or alterations in drug uptake/accumulation), "on-target resistance" (or repair of platinum-DNA adducts), "posttarget resistance" (or cell death signaling pathways resistance), and "off-target resistance" (mechanisms of resistance that do not present a close link with cisplatin) [5]. Here, we briefly discussed how NO could circumvent cisplatin resistance along the first three processes listed above.

Cisplatin "Pretarget Resistance" and NO

Reduced intracellular drug accumulation can be due to a decreased rate of drug uptake or an increased rate of drug efflux. These two processes reflect major mechanisms underlying the development of resistance to platinum-based therapeutics. Plasma membrane transporters such as copper transporter 1 or 2 (Ctr1 or Ctr2), multidrug resistance-associated protein 2, or the two copper extruding P-type ATPases 7A and 7B are important modulators of cisplatin transport that have been found altered or deregulated in many malignant cancer cells [6–9]. Basically, Ctr1-deficient cells have demonstrated profound resistance to cisplatin [10]. Furthermore, clinically relevant results reported a correlation between the elevated expression of Ctr1 and the overall survival (OS) benefit in non-small cell lung cancer (NSCLC) patients treated with platinum-based chemotherapeutic agents (cisplatin or carboplatin) [11,12]. Thus, it is believed that the upregulation of Ctr1 expression or its increase effectiveness would represent an interesting strategy for improving cancer cell sensitivity to platinum-based chemotherapeutics. Interestingly, the *CTR1* gene has been demonstrated to be strongly induced upon exogenous NO treatment in yeast [13] suggesting that such a mechanism may be extrapolated to some cancer cells to restore drug accumulation and sensitivity.

Elevated glutathione (GSH) levels are also important components of chemoresistance in many cancer cells. The cisplatin effectiveness can be limited by the ability of GSH, catalyzed by glutathione-*S*-transferase (GST) or even other GSH-related enzymes, to sequester cisplatin or to promote its efflux, therefore reducing the amount of cisplatin-DNA adducts and subsequent cellular damage [14,15]. GSH is a crucial modulator of NO reactivity, as it acts as an efficient scavenger of NO (NO can react with the cysteine of GSH to form *S*-nitrosoglutathione (GSNO)) [16]. The modulation of GSH levels by the NO-donating derivative of aspirin, NCX-4016, has shown to overcome GSH-mediated cytoprotection in cisplatin-resistant ovarian cancer cells through a significant depletion of GSH and resulting in restored sensitivity to cisplatin treatment [17]. In addition, other NO donors have been specifically designed to act as substrates for GSTs to counteract their role in drug resistance. In this way, JS-K and CB-3-100 NO prodrugs have demonstrated synergism with cisplatin in resistant tumor cells by increasing the cisplatin intracellular accumulation [18].

Cisplatin "On-Target Resistance" and NO

DNA platination (reaction of activated platinum species with purines bases of DNA) induces significant DNA lesions leading to intrastrand or interstrand cross-links that largely account for platinum derivatives cytotoxicity effect [1]. DNA repair pathways, particularly nucleotide excision repair (NER) and homologous recombination (HR) pathways, play an active role in the repair of the DNA adducts elicited by cisplatin and, therefore, in the regulation of platinum-based drug resistance in cancer treatment [19]. Both sodium nitroprusside (SNP, NO donor) and 3-morpholinosydnonimine [SIN-1, peroxynitrite ($ONOO^-$) donor] have shown to inhibit cisplatin-DNA adducts excision in the NER pathway [20]. This effect suggests that NER repair proteins are likely to be modified by NO through S-nitrosylation of cysteine thiols or by peroxynitrite through nitration of tyrosine hydroxyls. Further studies are needed to clearly identify which protein in the NER pathway is targeted by NO and whether this effect could restore highly resistant cancer cells to cisplatin treatment. NO is known to affect the catalytic function of several key DNA repair enzymes, most notably, through S-nitrosylation of the active site of cysteine residues and resulting in the loss of enzyme activity including, for example, 0^6-alkylguanine-DNA-alkyltransferase and 8-oxoguanine glycosylase [21]. The *Xeroderma pigmentosum* complementation group A (XPA) is a pivotal protein involved in the early stage of the NER pathway that coordinates the assembly of key proteins participating in the recognition of DNA damage [22]. In acellular conditions, a study reported that XPA, as a cysteine-rich zinc finger protein, is susceptible to transnitrosylation mediated by GSNO [23] and strongly suggesting it may result in DNA repair inhibition. Further studies are needed to determine whether NO-mediated XPA modification occurs in cells and sensitizes cancer cells to platinum-derivative drugs. *BReast CAncer 1 and 2 (BRCA1* and *BCRA2)* are tumor-suppressor genes and important regulators of a multitude of key cellular processes including DNA repair. Loss-of-function mutations of both BRCA1 and BRCA2 in different types of cancer cells generally confer increased sensitivity to cisplatin over normal cell counterparts [24–26]. However, secondary mutations of

BRCA1 and *BRCA2* (that correct the frameshift caused by primary mutations) in ovarian tumors were found to rescue BRCA1 and BRCA2 protein functions and contribute to resistance to cisplatin [27,28]. Interestingly, the NO donors *S*-nitroso-*N*-acetylpenicillamine (SNAP) and diethylenetriamine NONOate (DETANONOate) have been shown to promote the inhibition of *BRCA1* expression, reducing the ability of cells to repair DNA through the HR pathway [29]. Importantly, poly(ADP-ribose) polymerase (PARP) constitutes a large protein of family in which PARP-1, an essential protein for initiating various forms of DNA repair, is frequently expressed and hyperactivated in many cisplatin-resistant cancer cells [30]. To circumvent this effect, PARP-1 inhibitors have shown to synergize with cisplatin in the killing of NSCLC cell lines [31]. Another important component of the inhibition of DNA repair is the S-nitrosylation of PARP-1 occurring in an inducible nitric oxide synthase (iNOS)-dependent manner [32]. Altogether, the regulation of DNA repair pathways by NO provides an attractive strategy to reverse cisplatin resistance.

Cisplatin "Posttarget Resistance" and NO

Alteration in either the cell cycle arrest or the cell death signaling pathways represents other ways and means by which cancer cells can escape the cytotoxicity of cisplatin. Defects in the mitochondrial apoptotic signaling pathway, in part, ascribe to cisplatin resistance via the inactivation of p53 function and alterations in B-cell lymphoma-2 (Bcl-2) family protein [33,34]. The loss of p53 function, via mutations, is extremely frequent in human cancers and is well-known to contribute to drug resistance [35]. Higher rates of clinical response to cisplatin in ovarian cancer patients have been reported for patients with wild-type p53 tumors in comparison with patients with mutant p53 tumors [36]. The role of p53 in DNA-damage-induced signaling pathway is well documented. Of note, p53-mediated signaling pathways may be either dependent or independent (involving p53 translocation to mitochondria as a cellular response to DNA damage) of transcriptional activation [37]. A report by Leung et al. has suggested an important role for iNOS in regulating resistance to cisplatin in a p53-dependent manner in ovarian cancer cells [38]. This finding was further supported by the observation that cisplatin (and not oxaliplatin) upregulated iNOS protein levels in colon carcinoma cells, which in turn mediates the S-nitrosylation of p53 and prevents its translocation to the mitochondria, therefore supporting cisplatin resistance [39]. Importantly, the study by Leung et al. pointed out, on the opposite, the ability of the NO donor SNAP in restoring the sensitivity of ovarian cisplatin-resistant cells to apoptosis via the upregulation of p53 during cisplatin treatment [38].

Elevated and reduced expressions of antiapoptotic and proapoptotic members of the Bcl-2 family, respectively, confer resistance to platinum-based treatment in various cellular contexts [5]. While cisplatin has been shown to mediate a decrease in Bcl-2 protein expression through the ubiquitin-proteasome pathway, the NO donors SNP and dipropylenetriamine NONOate seemed to contrive this effect through the ability of NO to *S*-nitrosylate Bcl-2 and inhibit its ubiquitination and subsequent degradation, thus conferring cisplatin resistance [40]. Treatment of cells with other stress inducers also resulted in the S-nitrosylation and subsequent stabilization of Bcl-2 in an endogenous NO-dependent manner [41]. Above all, a growing number of studies have demonstrated the chemosensitizing properties of NO donors.

The combination of NO donors with platinum-based drugs and/or other chemotherapeutic drugs has fully proven preclinical and clinical relevance (see next paragraphs II and III).

Modifications of a variety of other apoptotic factors including, but not limited to, caspases or inhibitor of apoptosis family of proteins, such as survivin, may also be linked to platinum agents resistance [5,42]. NO exerts a regulatory role over these apoptotic factors that are supported by the observation that their S-nitrosylation can confer a cell death advantage (for review, see Refs. [43,44]).

Interestingly, the activation of alternative cell death pathways (p53-independent) such as those initiated by death receptor agonists represents a promising strategy to overcome platinum-based drugs resistance. Several examples of proteins modified by S-nitrosylation regulate signaling pathways engaged by members of the TNF (Tumor Necrosis Factor) superfamily (Fas ligand, tumor-necrosis-factor-related apoptosis-inducing ligand (TRAIL) and TNFα) (for review see Ref. [43]). Consistently, recent findings have demonstrated that exposure of cisplatin-resistant cells to TRAIL restored sensitivity to cisplatin in a peroxynitrite-production-dependent manner [45], reflecting possibly an involvement of protein tyrosine nitration. Altogether, these data strongly support the notion that triggering death receptor signaling is a promising alternative as a treatment for cisplatin-resistant cancers.

The promotion of cell survival via signaling pathways such as phosphatidylinositol-3-kinase and protein kinase B; mitogen-activated protein kinase; and/or transcription factors including nuclear factor-kappa B (NF-κB), c-Fos, c-Jun, signal transducer and activator of transcription 3 (STAT3) (and even more) have been shown to be attributed to the development of platinum resistance [14]. Compelling studies have provided evidence that NO plays a critical role in regulating these survival pathways and the transcription factors mentioned above to overcome resistance and ultimately favors cell death [44,46].

SENSITIZING EFFECT OF NO DONORS TO PLATINUM DERIVATIVES ON TUMORS IN PRECLINICAL IN VIVO MODELS

Syngeneic Murine Tumor Models

Although several studies have used murine cancer cell lines to assess, in vitro, the sensitizing effect of NO in combination with platinum derivatives, very few studies that have analyzed this effect in vivo in syngeneic models of cancer. Indeed, Ning et al. showed that glycidyl nitrate (GLYN), an NO donor, when used with cisplatin, significantly increased the tumor growth inhibition effect of cisplatin in syngeneic murine squamous cell carcinoma models [47]. Mechanistically, the sensitizing effect of GLYN seems to be related to its ability to increase the blood perfusion rate within tumors without affecting the blood flow of the surrounding normal tissues. Since then, no other studies have analyzed the antitumor effect of this category of NO donors, alone or in combination with drugs, which limits the interest of such study.

Human Tumor Xenografts in Murine Models

Most preclinical studies involving NO donors to platinum derivatives are carried out in xenograft models in immunosuppressed mice. However, these models, lacking, at least, the T-cell immune response, do not evaluate the immunomodulatory effect of NO [48], and therefore do not measure the role that can play the immune system on the sensitizing effect of NO when associated with platinum derivatives.

In a previous study, the in vivo antitumor efficacy of NCX-4040 (5 mg/kg), a NO-releasing aspirin derivative, in combination with cisplatin (8 mg/kg) was studied in nude mice bearing cisplatin-resistant ovarian A2780 xenograft tumors [49]. This study showed that NCX-4040 enhanced the sensitivity of ovarian cancer cells to cisplatin-mediated tumor growth inhibition. This was correlated with the ability of NCX-4040 to deplete intracellular GSH and to reduce epidermal growth factor receptor (EGFR) and phosphorylated STAT3 (pSTAT3), two well-known factors involved in chemoresistance and cell proliferation in cancer, when combined with cisplatin. Another group also showed that NCX-4040 synergized with oxaliplatin to trigger tumor growth inhibition in vivo in tumor-bearing immunosuppressed mice xenografted with human colon WiDr, LoVo, and doxorubicin-resistant LoVo (LoVoDx) cell lines [50]. This effect was associated to the increase in the number of apoptotic cells in mice implanted with three different colon tumors treated with NCX-4040 and oxaliplatin. Huerta, Yepez et al. have also evaluated the antitumor efficacy of another NO donor, DETANONOate, in combination with cisplatin in colon cancer SW620 cell line [51] and in a drug-resistant prostate carcinoma cell line [52] xenografted in nude mice. In the first study, the combination of DETANONOate (0.4 mg/kg) with cisplatin (2 mg/kg) inhibited tumor growth of colon cancer cells xenografted in nude mice, and derived tumor tissues revealed significant upregulation of the apoptosis-inducing factor and an increase in apoptosis by the DETANONOate and cisplatin combination treatment [51]. In the second work, they showed that mice bearing human ovarian PC-3 tumor xenografts and treated with DETANONOate (0.4 mg/kg) and cisplatin (2 mg/kg) resulted in significant inhibition of tumor growth, while single agent alone did not [52]. Immunohistochemistry analysis of tumors from xenograft models revealed lower expression of both Yin-Yang 1 and B-cell lymphoma-extra large in mice treated with DETANONOate and cisplatin combination. Other studies have evaluated the antitumor activity of the NO donor GSNO associated with cisplatin in immunosuppressed mice bearing ovarian cancer [53] or head and neck squamous cell carcinoma [54]. Giri et al. showed that GSNO (1 mg/kg) potentiated cisplatin-mediated (4 mg/kg) tumor toxicity in nude mice bearing intraperitoneal xenografts of human A2780 ovarian carcinoma cell line. The sensitizing effect of GSNO is attributed to its ability to S-nitrosylated several proteins involved in cell survival and proliferation such as NF-κB, Akt, and EGFR; however, nothing has been investigated in vivo as yet. Kaliyaperumal et al. also showed that in luciferase labeled SCC14a cell xenograft in nude mice, the tumor growth was reduced by systemic treatment with GSNO (10 mg/kg) combined with cisplatin (2 mg/kg). Mechanistically, the ability of GSNO to sensitize cells to cisplatin correlates with GSNO-mediated decrease in the level of expression of pSTAT3, vascular endothelial growth factor (VEGF, angiogenic factor) and the antiapoptotic Bcl-2 protein in tumors.

However, attention must be paid to the antitumor responses to combined therapy. NO donors/platinum derivatives are significant but are partial. Moreover, no dose/response curves with the NO donors have been performed to optimize the in vivo experiments. Furthermore, the lack of understanding about the molecular mechanisms characterizing the sensitizing effect of these NO donors hampers their optimal use.

Altogether, the search for new therapeutic combinations involving NO donors and chemotherapeutic agents or targeted therapies should give more importance to the use of NO donors as a new strategy for chemosensitization.

NO DONORS AND CLINICAL TRIALS IN NSCLC

Most clinical trials involving NO donors are carried out in patients with NSCLC (Table 1). In a first pivotal phase-II trial, Yasuda et al. investigated the efficacy and safety of NTG (nitroglycerin) associated with vinorelbine and cisplatin in 120 patients with previously untreated stage IIIB/IV NSCLC [55]. In this study, it was demonstrated that the addition of NTG during chemotherapy may enhance the response rate (72% vs. 42%), elongate the time to disease progression (TTP; 327 vs. 185 days), and improve OS in these patients (median survival time; 413 vs. 289 days) without major adverse effects. The results obtained in this clinical trial were confirmed by a recent study evaluating the effect of adding NTG to oral vinorelbine and cisplatin in a Caucasian population (66 patients) with stage IIIB/IV NSCLC [56]. Despite the small sample size, this latter study showed an increase of the overall response rate (ORR) in the NTG compared with the placebo arms (35.3% vs. 18.8%).

The potentiating effect of NO donors on chemotherapies against NSCLC seems however to be conditioned by the type of chemotherapy on the one hand and by the nature of the NO donor used on the other hand. The former point is illustrated by the study of Davidson et al. that reported in an Australian randomized phase-III trial (270 patients) that the addition of NTG to carboplatin-based, doublet chemotherapy in NSCLC showed no benefit [58], either in terms of progression-free survival (PFS) or OS. In this study, chemotherapy was predominantly with carboplatin and gemcitabine (79%) or carboplatin and placlitaxel (18%), while only six patients (2%) received a vinorelbine plus cisplatin combination in a similar way as for the studies reported by Yasuda et al. and Reinmuth et al. In a randomized phase-II study in patients treated with either irinotecan plus cisplatin or irinotecan plus capecitabine, Han et al. demonstrated that the addition of isosorbide-5-mononitrate (ISMN, a NO donor) did not improve the outcome of these chemotherapies [60]. In this latter study, the lack of potentiating effect of the NO donor might be due not only to the different type of chemotherapy but also to the nature of this NO donor. Indeed, ISMN is denitrated by the liver more slowly than NTG; however, at different concentrations, NO can have opposite effects [61,62].

A key parameter that could explain the sensitizing effect of NTG on the association of cisplatin and vinorelbine is based on the NTG capacity to reduce the VEGF level via the reduction of hypoxia-inducible factor 1-alpha (HIF-1α). In the Yasuda's phase-II trial, plasma VEGF levels after NTG treatment (3 days) were significantly lower than before treatment, in patients receiving NTG, while no difference in VEGF levels was observed over time in patients receiving placebo [55]. Additionally, in a retrospective study, Yasuda et al. showed that NTG plus docetaxel and carboplatin in patients with NSCLC could improve the antitumor response, based on a decreased expression of HIF-1α and VEGF [63]. Since NTG has also been described to increase the response to radiotherapy, Arrieta et al. have performed a randomized phase-II clinical trial in NSCLC patients (35 patients) combining vinorelbine plus cisplatin chemotherapy plus radiotherapy with NTG [57]. In this study, NTG treatment was associated with longer PFS and OS, together with a decrease in VEGF levels. However, the small sample size of this study prevents statistical verification. Interestingly, the results of NVALT12, a

TABLE 1 Summary of Clinical Trials Studies of Therapeutic Regimens Combining Platinum Derivatives and NO Donors for NSCLC Patients

Clinical Trials (Patient Number)	NO Donor	Stage of NSCLC	Treatment	Results	Refs.
Phase II (120)	NTG	IIIB/IV	Four cycles every 3 weeks: Vinorelbine (25 mg/m^2, days 1 and 8) + cisplatin (80 mg/m^2, day 1) ± NTG patches (25 mg/patient from day −3 to +2 of cycle, arm A) ± placebo patches (5 days, arm B)	Arm A versus arm B: RR: 72% vs. 42% Median TTP: 327 vs. 185 days	[55]
Phase II (66)	NTG	IIIB/IV	3-Week cycle: Oral vinorelbine (60–80 mg/m^2, days 1 and 8) + cisplatin (80 mg/m^2) ± NTG patches (25 mg/patient daily, from day −3 to +2 of cycle, arm A) ± placebo patches (5 days, arm B)	Arm A versus arm B: ORR: 35.3% vs. 18.8% DCR: 61.8% vs. 53.1% TTP and OS similar	[56]
Phase II (35)	NTG	IIIA/IIIB	Two cycles CT + C-RT + 2 CT: Vinorelbine (25 mg/m^2, days 1 and 8) Cisplatin (70 mg/m^2, days 1 every 21 days) ± NTG (25 mg/patient daily, from day −1 to +4 after CT)	Proof of concept for adding NTG to chemotherapy and radiotherapy	[57]
Phase III (270)	NTG	III/IV	Six cycles every 3 weeks: Chemotherapy: carboplatin/ gemcitabine (79%), carboplatin/ paclitaxel (18%), vinorelbine/cisplatin (2.2%) ± NTG patches (25 mg/patient daily, from day −2 to +2, arm A) ± placebo patches (arm B)	No demonstrable effect of NTG on PFS, OS, OTR	[58]
Phase II (223)	NTG	IV nonsquamous	Four cycles every 3 weeks: Carboplatin (area under the curve 6) + paclitaxel (200 mg/m^2) + bevacizumab (15 mg/kg) ± NTG (15 mg, from day −2 to +2)	Adding NTG not improve PFS and OS	[59]
Phase II (146)	ISMN	IIIB/IV	3-week-cycle IP ± NTG or IX ± NTG: IP: irinotecan (80 mg/m^2, days 1 and 8)/cisplatin (60 mg/m^2, day 1) IX: irinotecan (90 mg/m^2, days 1 and 8)/oral capecitabine (1000 mg/m^2, twice daily on days 1–14) ± ISMN (20 mg twice daily, days −3 to +2)	Adding NTG not improve the treatment outcome	[60]

NTG, *nitroglycerin;* ISMN, *isosorbide-5-mononitrate;* RR, *response rate;* TTP, *time to progression;* ORR, *overall response rate;* DCR, *disease control rate;* OS, *overall survival;* CT, *chemotherapy;* C-RT, *chemoradiotherapy;* PFS, *progression-free survival;* OTR, *objective tumor response.*

randomized phase-II (223 patients) study in patients treated with a combination of paclitaxel, carboplatin, and bevacizumab—which dramatically decreases the VEGF levels by itself—showed that the addition of NTG loses its potential to improve PFS and OS in patients with stage IV nonsquamous NSCLC [59].

In summary, the ability of NTG to improve the therapeutic efficacy of chemotherapies in patients with NSCLC may be dependent on the type of therapy regimen. The combination of vinorelbine and cisplatin seems to be the most favorable chemotherapy for the use of NTG in NSCLC patients. However, other platinum-derived agents should be tested in combination with chemotherapies that may act on various cells of the tumor microenvironment like immune cells (myeloid-derived suppressor cells, regulatory T cells (Treg), etc.). In addition, the lung cancer model may not be the best type of cancer for the use of platinum-based chemotherapy and NTG, since unpublished laboratory results showed a strong antitumor effect of oxaliplatin and NTG in a preclinical model of colon cancer in mice.

CONCLUSION

Platinum derivatives are keystone chemotherapeutic drugs for treating a wide variety of solid cancers including lung, ovarian, colorectal, testicular, head, neck, and bladder cancer and other cancers. The occurrence of chemoresistance, a common feature of cancer therapy, inevitably leads to tumor relapse and poor patient's outcome. Improving the efficiency of platinum-based combination regimens is always a therapeutic challenge. In recent years, insights from preclinical and clinical studies for new combination treatment regimens including NO donors have unveiled promising therapeutic benefits. The studies summarized herein illustrate the complexity of NO responses and above all provide evidences that the overall efficacy of platinum derivatives might be improved with NO donors used as adjuvant. However, to improve therapeutic outcomes, additional studies are required to further investigate the potential therapeutic benefit of new combined therapy regimens with other conventional chemotherapeutic agents or targeted therapies (monoclonal antibodies, checkpoint inhibitors, etc.) in combination with NO donors. In this perspective, as an NO donor, NTG seems to present all the requirements for developing new combinations of treatments for cancer patients.

Acknowledgments

The authors gratefully acknowledge the financial support from "La Ligue Contre le Cancer," "Fondation ARC pour la Recherche sur le Cancer," and "Région Bourgogne."

Conflict of Interest: No potential conflicts of interest were disclosed.

References

[1] Dilruba S, Kalayda GV. Platinum-based drugs: past, present and future. Cancer Chemother Pharmacol 2016;77(6):1103–24.
[2] Wheate NJ, Walker S, Craig GE, Oun R. The status of platinum anticancer drugs in the clinic and in clinical trials. Dalton Trans 2010;39(35):8113–27.
[3] Wang D, Lippard SJ. Cellular processing of platinum anticancer drugs. Nat Rev Drug Discov 2005;4(4):307–20.
[4] Dasari S, Tchounwou PB. Cisplatin in cancer therapy: molecular mechanisms of action. Eur J Pharmacol 2014;740:364–78.
[5] Galluzzi L, Senovilla L, Vitale I, Michels J, Martins I, Kepp O, et al. Molecular mechanisms of cisplatin resistance. Oncogene 2012;31(15):1869–83.

[6] Blair 1 BG, Larson CA, Safaei R, Howell SB. Copper transporter 2 regulates the cellular accumulation and cytotoxicity of cisplatin and carboplatin. Clin Cancer Res 2009;15(13):4312–21.
[7] Chen HH, Chen WC, Liang ZD, Tsai WB, Long Y, Aiba I, et al. Targeting drug transport mechanisms for improving platinum-based cancer chemotherapy. Expert Opin Ther Targets 2015;19(10):1307–17.
[8] Holzer AK, Manorek GH, Howell SB. Contribution of the major copper influx transporter CTR1 to the cellular accumulation of cisplatin, carboplatin, and oxaliplatin. Mol Pharmacol 2006;70(4):1390–4.
[9] Korita PV, Wakai T, Shirai Y, Matsuda Y, Sakata J, Takamura M, et al. Multidrug resistance-associated protein 2 determines the efficacy of cisplatin in patients with hepatocellular carcinoma. Oncol Rep 2010;23 (4):965–72.
[10] Ishida S, Lee J, Thiele DJ, Herskowitz I. Uptake of the anticancer drug cisplatin mediated by the copper transporter Ctr1 in yeast and mammals. Proc Natl Acad Sci U S A 2002;99(22):14298–302.
[11] Chen HH, Yan JJ, Chen WC, Kuo MT, Lai YH, Lai WW, et al. Predictive and prognostic value of human copper transporter 1 (hCtr1) in patients with stage III non-small-cell lung cancer receiving first-line platinum-based doublet chemotherapy. Lung Cancer 2012;75(2):228–34.
[12] Kim ES, Tang X, Peterson DR, Kilari D, Chow CW, Fujimoto J, et al. Copper transporter CTR1 expression and tissue platinum concentration in non-small cell lung cancer. Lung Cancer 2014;85(1):88–93.
[13] Nasuno R, Aitoku M, Manago Y, Nishimura A, Sasano Y, Takagi H. Nitric oxide-mediated antioxidative mechanism in yeast through the activation of the transcription factor Mac1. PLoS One 2014;9(11):e113788.
[14] Stewart DJ. Mechanisms of resistance to cisplatin and carboplatin. Crit Rev Oncol Hematol 2007;63(1):12–31.
[15] Chen HH, Kuo MT. Role of glutathione in the regulation of cisplatin resistance in cancer chemotherapy. Met-Based Drugs 2010;2010.
[16] Aquilano K, Baldelli S, Ciriolo MR. Glutathione: new roles in redox signaling for an old antioxidant. Front Pharmacol 2014;5:196.
[17] Bratasz A, Weir NM, Parinandi NL, Zweier JL, Sridhar R, Ignarro LJ, et al. Reversal to cisplatin sensitivity in recurrent human ovarian cancer cells by NCX-4016, a nitro derivative of aspirin. Proc Natl Acad Sci U S A 2006;103(10):3914–9.
[18] Liu J, Li C, Qu W, Leslie E, Bonifant CL, Buzard GS, et al. Nitric oxide prodrugs and metallochemotherapeutics: JS-K and CB-3-100 enhance arsenic and cisplatin cytolethality by increasing cellular accumulation. Mol Cancer Ther 2004;3(6):709–14.
[19] Martin LP, Hamilton TC, Schilder RJ. Platinum resistance: the role of DNA repair pathways. Clin Cancer Res 2008;14(5):1291–5.
[20] Chien YH, Bau DT, Jan KY. Nitric oxide inhibits DNA-adduct excision in nucleotide excision repair. Free Radic Biol Med 2004;36(8):1011–7.
[21] Tang CH, Wei W, Liu L. Regulation of DNA repair by S-nitrosylation. Biochim Biophys Acta 2012;1820(6):730–5.
[22] Fadda E. Role of the XPA protein in the NER pathway: a perspective on the function of structural disorder in macromolecular assembly. Comput Struct Biotechnol J 2015;14:78–85.
[23] Smirnova J, Zhukova L, Witkiewicz-Kucharczyk A, Kopera E, Oledzki J, Wysłouch-Cieszyńska A, et al. Reaction of the XPA zinc finger with *S*-nitrosoglutathione. Chem Res Toxicol 2008;21(2):386–92.
[24] Lohse I, Borgida A, Cao P, Cheung M, Pintilie M, Bianco T, et al. BRCA1 and BRCA2 mutations sensitize to chemotherapy in patient-derived pancreatic cancer xenografts. Br J Cancer 2015;113(3):425–32.
[25] Pothuri B. BRCA1- and BRCA2-related mutations: therapeutic implications in ovarian cancer. Ann Oncol 2013;24(Suppl. 8):viii22–7.
[26] Turner NC, Tutt AN. Platinum chemotherapy for BRCA1-related breast cancer: do we need more evidence? Breast Cancer Res 2012;14(6):115.
[27] Sakai W, Swisher EM, Karlan BY, Agarwal MK, Higgins J, Friedman C, et al. Secondary mutations as a mechanism of cisplatin resistance in BRCA2-mutated cancers. Nature 2008;451(7182):1116–20.
[28] Swisher EM, Sakai W, Karlan BY, Wurz K, Urban N, Taniguchi T. Secondary BRCA1 mutations in BRCA1-mutated ovarian carcinomas with platinum resistance. Cancer Res 2008;68(8):2581–6.
[29] Yakovlev VA. Nitric oxide-dependent downregulation of BRCA1 expression promotes genetic instability. Cancer Res 2013;73(2):706–15.
[30] Michels J, Vitale I, Galluzzi L, Adam J, Olaussen KA, Kepp O, et al. Cisplatin resistance associated with PARP hyperactivation. Cancer Res 2013;73(7):2271–80.
[31] Michels J, Vitale I, Senovilla L, Enot DP, Garcia P, Lissa D, et al. Synergistic interaction between cisplatin and PARP inhibitors in non-small cell lung cancer. Cell Cycle 2013;12(6):877–83.

[32] Zhou X, Cooper KL, Huestis J, Xu H, Burchiel SW, Hudson LG, et al. S-nitrosation on zinc finger motif of PARP-1 as a mechanism of DNA repair inhibition by arsenite. Oncotarget 2016;7(49):80482–92.
[33] Fraser M, Bai T, Tsang BK. Akt promotes cisplatin resistance in human ovarian cancer cells through inhibition of p53 phosphorylation and nuclear function. Int J Cancer 2008;122(3):534–46.
[34] Cetintas VB, Kucukaslan AS, Kosova B, Tetik A, Selvi N, Cok G, et al. Cisplatin resistance induced by decreased apoptotic activity in non-small-cell lung cancer cell lines. Cell Biol Int 2012;36(3):261–5.
[35] Reles A, Wen WH, Schmider A, Gee C, Runnebaum IB, Kilian U, et al. Correlation of p53 mutations with resistance to platinum-based chemotherapy and shortened survival in ovarian cancer. Clin Cancer Res 2001;7 (10):2984–97.
[36] Gadducci A, Cosio S, Muraca S, Genazzani AR. Molecular mechanisms of apoptosis and chemosensitivity to platinum and paclitaxel in ovarian cancer: biological data and clinical implications. Eur J Gynaecol Oncol 2002;23(5):390–6.
[37] Erster S, Mihara M, Kim RH, Petrenko O, Moll UM. In vivo mitochondrial p53 translocation triggers a rapid first wave of cell death in response to DNA damage that can precede p53 target gene activation. Mol Cell Biol 2004;24 (15):6728–41.
[38] Leung EL, Fraser M, Fiscus RR, Tsang BK. Cisplatin alters nitric oxide synthase levels in human ovarian cancer cells: involvement in p53 regulation and cisplatin resistance. Br J Cancer 2008;98(11):1803–9.
[39] Hernlund E, Kutuk O, Basaga H, Linder S, Panaretakis T, Shoshan M. Cisplatin-induced nitrosylation of p53 prevents its mitochondrial translocation. Free Radic Biol Med 2009;46(12):1607–13.
[40] Chanvorachote P, Nimmannit U, Stehlik C, Wang L, Jiang BH, Ongpipatanakul B, et al. Nitric oxide regulates cell sensitivity to cisplatin-induced apoptosis through S-nitrosylation and inhibition of Bcl-2 ubiquitination. Cancer Res 2006;66(12):6353–60.
[41] Azad N, Vallyathan V, Wang L, Tantishaiyakul V, Stehlik C, Leonard SS, et al. S-nitrosylation of Bcl-2 inhibits its ubiquitin-proteasomal degradation. A novel antiapoptotic mechanism that suppresses apoptosis. J Biol Chem 2006;281(45):34124–34.
[42] Engels K, Knauer SK, Loibl S, Fetz V, Harter P, Schweitzer A, et al. NO signaling confers cytoprotectivity through the survivin network in ovarian carcinomas. Cancer Res 2008;68(13):5159–66.
[43] Plenchette S, Romagny S, Laurens V, Bettaieb A. S-Nitrosylation in TNF superfamily signaling pathway: implication in cancer. Redox Biol 2015;6:507–15.
[44] Plenchette S, Romagny S, Laurens V, Bettaieb A. NO and cancer: itinerary of a double agent. Med Sci 2016;32 (6–7):625–33.
[45] Seah S, Low IC, Hirpara JL, Sachaphibulkij K, Kroemer G, Brenner C, et al. Activation of surrogate death receptor signaling triggers peroxynitrite-dependent execution of cisplatin-resistant cancer cells. Cell Death Dis 2015;6: e1926.
[46] Bettaieb A, et al. S-Nitrosylation in cancer cells: to prevent or to cause? In: Bonavida B, editor. Nitric oxide and cancer: pathogenesis and therapy. Switzerland: Springer International Publishing; 2015. p. 97–105.
[47] Ning S, Bednarski M, Oronsky B, Scicinski J, Knox SJ. Novel nitric oxide generating compound glycidyl nitrate enhances the therapeutic efficacy of chemotherapy and radiotherapy. Biochem Biophys Res Commun 2014;447 (3):537–42.
[48] Predonzani A, Calì B, Agnellini AH, Molon B. Spotlights on immunological effects of reactive nitrogen species: when inflammation says nitric oxide. World J Exp Med 2015;5(2):64–76.
[49] Bratasz A, Selvendiran K, Wasowicz T, Bobko A, Khramtsov VV, Ignarro LJ, et al. NCX-4040, a nitric oxide-releasing aspirin, sensitizes drug-resistant human ovarian xenograft tumors to cisplatin by depletion of cellular thiols. J Transl Med 2008;6:9.
[50] Tesei A, Zoli W, Fabbri F, Leonetti C, Rosetti M, Bolla M, et al. NCX 4040, an NO-donating acetylsalicylic acid derivative: efficacy and mechanisms of action in cancer cells. Nitric Oxide 2008;19(2):225–36.
[51] Huerta S, Baay-Guzman G, Gonzalez-Bonilla CR, Livingston EH, Huerta-Yepez S, Bonavida B. In vitro and in vivo sensitization of SW620 metastatic colon cancer cells to CDDP-induced apoptosis by the nitric oxide donor DETANONOate: involvement of AIF. Nitric Oxide 2009;20(3):182–94.
[52] Huerta-Yepez S, Baritaki S, Baay-Guzman G, Hernandez-Luna MA, Hernandez-Cueto A, Vega MI, et al. Contribution of either YY1 or BclXL-induced inhibition by the NO-donor DETANONOate in the reversal of drug resistance, both in vitro and in vivo. YY1 and BclXL are overexpressed in prostate cancer. Nitric Oxide 2013;29:17–24.

[53] Giri S, Rattan R, Deshpande M, Maguire JL, Johnson Z, Graham RP, et al. Preclinical therapeutic potential of a nitrosylating agent in the treatment of ovarian cancer. PLoS One 2014;9(6):e97897.
[54] Kaliyaperumal K, Sharma AK, McDonald DG, Dhindsa JS, Yount C, Singh AK, et al. *S*-Nitrosoglutathione-mediated STAT3 regulation in efficacy of radiotherapy and cisplatin therapy in head and neck squamous cell carcinoma. Redox Biol 2015;6:41–50.
[55] Yasuda H, Yamaya M, Nakayama K, Sasaki T, Ebihara S, Kanda A, et al. Randomized phase II trial comparing nitroglycerin plus vinorelbine and cisplatin with vinorelbine and cisplatin alone in previously untreated stage IIIB/IV non-small-cell lung cancer. J Clin Oncol 2006;24(4):688–94.
[56] Reinmuth N, Meyer A, Hartwigsen D, Schaeper C, Huebner G, Skock-Lober R, et al. Randomized, double-blind phase II study to compare nitroglycerin plus oral vinorelbine plus cisplatin with oral vinorelbine plus cisplatin alone in patients with stage IIIB/IV non-small cell lung cancer (NSCLC). Lung Cancer 2014;83(3):363–8.
[57] Arrieta O, Blake M, de la Mata-Moya MD, Corona F, Turcott J, Orta D, et al. Phase II study. Concurrent chemotherapy and radiotherapy with nitroglycerin in locally advanced non-small cell lung cancer. Radiother Oncol 2014;111(2):311–5.
[58] Davidson A, Veillard AS, Tognela A, Chan MM, Hughes BG, Boyer M, et al. A phase III randomized trial of adding topical nitroglycerin to first-line chemotherapy for advanced nonsmall-cell lung cancer: the Australasian lung cancer trials group NITRO trial. Ann Oncol 2015;26(11):2280–6.
[59] Dingemans AM, Groen HJ, Herder GJ, Stigt JA, Smit EF, Bahce I, et al. A randomized phase II study comparing paclitaxel-carboplatin-bevacizumab with or without nitroglycerin patches in patients with stage IV non-squamous nonsmall-cell lung cancer: NVALT12 (NCT01171170). Ann Oncol 2015;26(11):2286–93.
[60] Han JY, Nam BH, Kim HY, Yoon SJ, Kim HT, Lee JS. A randomized phase II study of irinotecan plus cisplatin versus irinotecan plus capecitabine with or without isosorbide-5-mononitrate in advanced non-small-cell lung cancer. Ann Oncol 2012;23(11):2925–30.
[61] Frederiksen LJ, Sullivan R, Maxwell LR, Macdonald-Goodfellow SK, Adams MA, Bennett BM, et al. Chemosensitization of cancer in vitro and in vivo by nitric oxide signaling. Clin Cancer Res 2007;13(7):2199–206.
[62] Mocellin S, Bronte V, Nitti D. Nitric oxide, a double edged sword in cancer biology: searching for therapeutic opportunities. Med Res Rev 2007;27(3):317–52.
[63] Yasuda H, Nakayama K, Watanabe M, Suzuki S, Fuji H, Okinaga S, et al. Nitroglycerin treatment may enhance chemosensitivity to docetaxel and carboplatin in patients with lung adenocarcinoma. Clin Cancer Res 2006;12 (22):6748–57.

CHAPTER

7

NO-H_2S-Releasing Chimeras as a Multifaceted Approach to Cancer Therapy

Khosrow Kashfi, Mahnoush Esmaili

City University of New York School of Medicine, New York, NY, United States

Abstract

Resistance to chemotherapy whether it is intrinsic or acquired is potentially a major problem that often times leads to poor patient outcomes. Therefore, designing novel anticancer drugs with improved efficacy, lower side effects, and potential synergy with existing chemotherapeutic drugs is of utmost importance. Nitric oxide (NO) and hydrogen sulfide (H_2S) are two gasotransmitters that regulate a number of key biological functions. Both appear to have dichotomous actions in cancer biology, yet they may prove to hold major key roles in unraveling the challenges that are encountered in cancer treatment. In this chapter, we have presented the rationale for designing anticancer compounds that incorporate these gasotransmitters, briefly discussing the various classes of NO-releasing agents that are in various stages of development. We have then focused on the more recent H_2S-releasing pharmaceuticals and also the so-called NOSH compounds that release both NO and H_2S.

Abbreviations

ADT-OH	5-(4-hydroxyphenyl)-3H-1,2-dithiole-3-thione
cGMP	cyclic guanosine monophosphate
COX-1	cyclooxygenase-1
COX-2	cyclooxygenase-2
H_2S	hydrogen sulfide
iNOS	inducible nitric oxide synthase
MPO	myeloperoxidase
NO	nitric oxide
NOS	nitric oxide synthase
NSAIDs	nonsteroidal antiinflammatory drugs
PDE5A	phosphodiesterase 5A
RSNO	*S*-nitrosothiols

Nitric Oxide (Donor/Induced) in Chemosensitization, Volume 1
http://dx.doi.org/10.1016/B978-0-12-811020-1.00007-7

RSSH persulfides
sGC soluble guanylate cyclase
TBZ 4-hydroxy benzothiazamide
VDCC voltage-dependent Ca^{2+} channels

INTRODUCTION

Nitric oxide (NO) a gaseous free radical and hydrogen sulfide (H_2S) a gas that bears the pungent smell of rotten eggs are two gasotransmitters that have multiple roles in normal physiology. For example, NO regulates vascular relaxation [1], controls inflammation by inhibiting NF-κB activation [2,3], and suppresses expression of pro-inflammatory mediators in mast cells, macrophages, and vascular smooth muscles [4–6]. NO regulates blood flow [1,7]; modulates platelet; and induces leukocyte activation, adhesion, and aggregation [8]. The important role of NO in human biology was recognized in 1998 when the Nobel Prize in Physiology and Medicine was awarded to Robert F. Furchgott, Louis J. Ignarro, and Ferid Murad for establishing NO as messenger molecule. In retrospect, the acceptance of H_2S in human physiology was less arduous, and in the past 20 years, there has been an exponential growth of publications in various aspects of its use and effects. A landmark study was that of Abe and Kimura in 1996, which documented the role of H_2S as a neuromodulator [9]. H_2S has also been shown to have vasorelaxant effects [10] and be protective in an animal model of myocardial ischemia-reperfusion injury [11] for a critical review on the protective effects of H_2S on the cardiovascular system. [12]. Similarly to NO, H_2S also inhibits leukocytes adherence to blood vessel walls [13] and appears to be intimately involved in reducing and promoting the resolution of inflammation [14]. A quick inspection of the literature regarding the various aspects of NO and H_2S in health and disease leads one to conclude that there appears to be an uncanny resemblance between the general actions of these two gasotransmitters.

NO is synthesized from L-arginine by the enzyme nitric oxide synthase (NOS) [15,16] of which there are three isoforms (Fig. 1A). Neuronal (nNOS or NOS1) and endothelial (eNOS or NOS3) are constitutive (cNOS) and are calcium-dependent forms of the enzyme that are regulated by negative feedback [17], phosphorylation [18], and interaction with regulatory molecules [19]. cNOS produces low levels of NO over a short period of time and nanomolar concentrations for seconds or minutes, to regulate neural and vascular function, respectively [20,21]. The third isoform (iNOS or NOS2) is calcium-independent; is regulated transcriptionally; and is induced by inflammatory cytokines, oxidative stress, hypoxia, and various endotoxins [22,23]. Inducible nitric oxide synthase (iNOS) produces micromolar levels of NO that may be for hours or days [24–26] and is involved in immune surveillance.

Biosynthesis of H_2S (Fig. 1B) requires tissue-specific enzymes that generate L-cysteine, a metabolite of dietary methionine. Three such enzymes have been identified: cystathionine β-synthase (CBS) and cystathionine γ-lyase (CSE, aka CGL) that require pyridoxal 5′-phosphate (PLP) as a cofactor and use L-cysteine as substrate [27,28] and 3-mercaptopyruvate sulfurtransferase (3-MST) that produces H_2S using 3-mercaptopyruvate, a metabolite of L-cysteine [29]. CBS, CSE, and MST are constitutive enzymes and have different expression levels in different tissues and organs [30], although there is some recent evidence suggesting that CSE may be inducible [31]. Expression of CSE is modulated in diverse conditions such as inflammation mediated by tumor

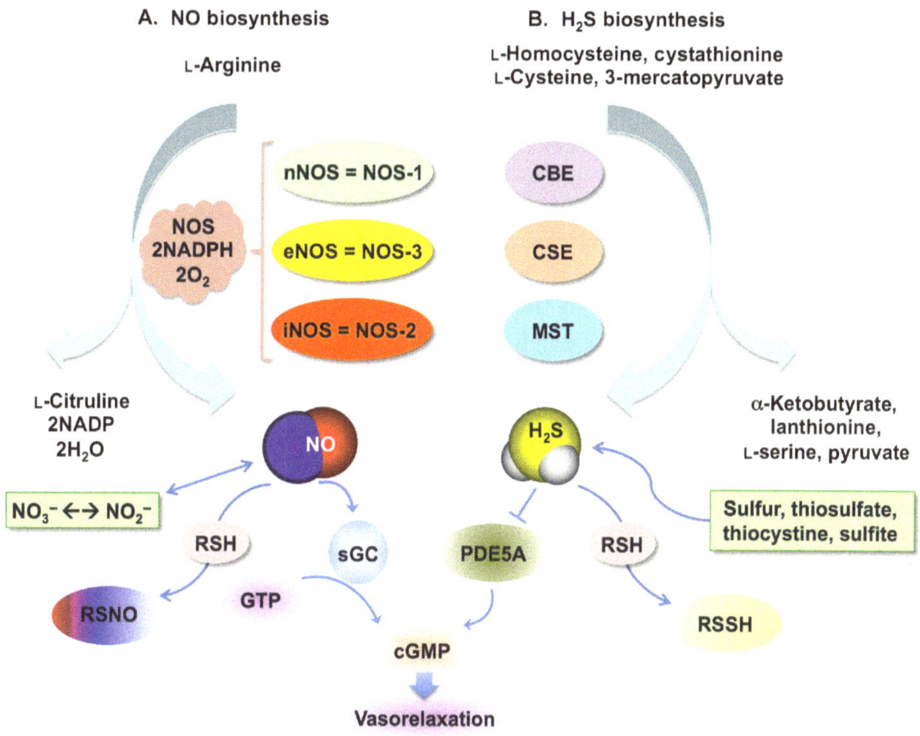

FIG. 1 Biosynthesis of nitric oxide and hydrogen sulfide. NO is produced by three nitric oxide synthase (NOS) isoforms: neuronal, endothelial, and inducible (nNOS, eNOS, and iNOS) that catalyze the oxidation of L-arginine to L-citrulline. NO is also produced through the reduction of nitrite/nitrate under low-oxygen conditions. H_2S is generated from the oxidation of the substrates L-homocysteine, cystathionine, L-cysteine, and 3-mercaptopyruvate through the enzymes cystathionine β-synthase (CBS) and cystathionine γ-lyase (CSE) and 3-mercaptopyruvate sulfurtransferase (3-MST). α-Ketobutyrate, lanthionine, L-serine, and pyruvate are the secondary products formed. Alternatively, the production of H_2S occurs nonenzymatically from various storage forms of sulfur like thiosulfate, thiocysteine, and sulfite. NO reacts with the active site of soluble guanylate cyclase (sGC) and produces cyclic GMP, leading to vasorelaxation, whereas H_2S raises cGMP levels through the inhibition of phosphodiesterase 5A (PDE5A), an enzyme that catabolizes it. NO can affect cellular proteins by producing peroxynitrite, which in turn can interact with cysteine residues to form *S*-nitrosothiols (RSNO). H_2S can also interact with the sulfhydryl group of cysteines and proteins to form persulfides (RSSH).

necrosis factor-α (TNF-α) and lipopolysaccharides (LPS), metabolites such as glucocorticoids and glucose, and dietary restriction and endoplasmic reticulum (ER) stress [31–35]. H_2S can also be generated endogenously through the nonenzymatic reduction of elemental sulfur in the blood [36] using reducing equivalents supplied through the glycolytic pathway [37] or from various storage forms of sulfur such as thiosulfate, thiocysteine, and sulfite [38].

Most of NO and H_2S signaling leading to vasorelaxation appears to be through the second messenger, cyclic guanosine monophosphate (cGMP). NO directly activates the enzyme soluble guanylate cyclase (sGC) that catalyzes the conversion of GTP to cGMP, while H_2S raises cGMP levels through inhibition of phosphodiesterase 5A (PDE5A), an enzyme that catabolizes it (Fig. 1) [39]. NO may also react with molecular oxygen or superoxide anion

($^{\cdot}O_2^-$) to form other biologically active nitrogen oxides [40–42], which in turn can react with cellular targets. For example, peroxynitrite ($ONOO^-$) that is formed when NO reacts with $^{\cdot}O_2^-$ can interact with cysteine residues to form *S*-nitrosothiols (RSNO), a process that has become known as S-nitrosylation, recent detailed reviews in this general area are presented in Refs. [42,43]. Analogous to NO, H_2S can also interact with sulfhydryl group of cysteines and proteins to form persulfides (RSSH) by a process that has been termed sulfhydration [44]. Although the term sulfhydration is "commonly" used in the literature, it does not describe what is actually happening at the chemical level. Correctly, H_2S reacts with oxygen to produce sulfane sulfur, which in turn is introduced into the proteins' SH group by a well-documented mechanism involving a thiosulfoxide intermediate; the correct term for this reaction would be "sulfuration" [45–47]. In this chapter, the correct terminology will be used. In general, nitrosylation and sulfuration exert opposite effects, with nitrosylation that occurs first typically reducing the reactivity of the cysteine thiols and sulfuration enhancing them, thus rendering them more nucleophilic [31]. For example, the activity of glyceraldehyde 3-phosphate dehydrogenase (GAPDH), a glycolytic enzyme, is inhibited by S-nitrosylation at Cys-150 [48], whereas the sulfuration of the enzyme increases its activity manyfold [44]. However, there are also instances where sulfuration precedes nitrosylation on the same cysteine residue. For example, the p65 subunit of NF-κB is sulfurated within the first 2 h of TNF-α treatment and then is nitrosylated [32]. On the whole, sulfuration appears to be substantially more prevalent than nitrosylation. Typically, nitrosylation affects only about 1%–5% of most proteins, whereas around 10%–25% of endogenous GAPDH, β-tubulin, and actin are basally sulfurated [44].

Many cellular effects of H_2S are mediated by its interaction with membrane ion channels. Notably, the ATP-sensitive potassium (K_{ATP}) channel is targeted/opened by H_2S, leading to vasorelaxation in vascular smooth muscle [10] and enhancing cardiovascular function [49]. Studies using INS-1E cells, which are an insulin-secreting cell line, led to the conclusion that the interaction between H_2S, glucose, and the K_{ATP} channel may be an important and novel mechanism for the control of insulin secretion from pancreatic beta-cells [50]. Using primary human airway smooth muscles and GYY4137, a slow H_2S-releasing compounds, it was shown that H_2S decreased stiffness, an effect that was mediated by K_{ATP} channel, thus suggesting that this class of gasotransmitter may have a benefit against obstructive lung disease [51]. Actions of H_2S in the hypothalamus decrease blood pressure and heart rate in freely moving rats, an effect that was mediated by a K_{ATP}-channel-dependent mechanism [52]. Further, using a colorectal distension model in conscious rats or a chemically induced acute colitis model, it was shown that H_2S inhibited nociception in both models and that K_{ATP} channels and NO mediated this effect. Thus, it may be concluded that H_2S-releasing drugs might be beneficial in treating painful intestinal disorders [53]. Further, by using various H_2S donors, it was also shown that H_2S elicited an acceleration of gastric emptying and relaxation of pyloric sphincter muscle mediated by K_{ATP} channel and TRPV1 (capsaicin) receptor activations [54].

Apart from the K_{ATP} channel, voltage-dependent calcium channels are also important targets of H_2S, albeit there appears to be tissue specificity [55]. In isolated mouse pancreatic β-cells [56] and rat cardiomyocytes [57], H_2S inhibited L-type voltage-dependent Ca^{2+} channels (VDCC) currents, whereas in neurons the activities of Cav3.2T-type VDCCs and Cav1.2L-type VDCCs were enhanced by endogenous and/or exogenous H_2S [58,59]. As eluded to above, H_2S can directly affect its target proteins and indirectly modulate second messengers by inhibiting various phosphodiesterases. Thus, cGMP and cAMP levels can be increased [39,60,61]. H_2S can also increase intracellular calcium levels in vascular endothelial cells by stimulating calcium influx [62] or by releasing calcium from various intracellular stores. These may, in turn, activate many calcium-dependent signaling pathways [55].

INFLAMMATION, CANCER AND NSAIDs

Acute inflammation is an adaptive host defense mechanism against infection or injury and is self-limiting; however, chronic inflammation may lead to various ailments including cancer [63]. For example, inflammatory diseases increase the risk of developing bladder, cervical, gastric, intestinal, esophageal, ovarian, prostate, and thyroid cancers [64]. Since inflammation has become a common theme in cancer pathogenesis, it is not surprising then that nonsteroidal antiinflammatory drugs (NSAIDs) have become the prototypical chemopreventive agents against many forms of cancer [65], for example, colon [66–68], breast [69–71], pancreas [72], prostate [73], esophageal [74,75], head and neck [76], ovarian [77,78], skin [79,80], and bladder cancers [81,82]. Although NSAIDs are chemopreventive, their long-term use is problematic essentially due to their significant and at times life-threatening side effects. All NSAIDs eventually cause some degree of GI erosions, which eventually may lead to ulcers, with most having cardiovascular and renal side effects [43,65]. Also, in terms of efficacy, NSAIDs have serious limitations. For example, for colon cancer, the one most thoroughly studied, NSAIDs can prevent at best 50% of the cases [83]. Given the proved usefulness of NSAIDs on the one hand and their undesirable side effects on the other has prompted the search to a "better and safer" NSAID. To that end a number of approaches have been made, two of which have made use of NO and H_2S.

THE ROLE OF NO AND H_2S IN INFLAMMATION AND TUMOR BIOLOGY

NO has a dichotomous role in cancer biology, with some reports suggesting that NO possesses antitumor and antiinflammatory properties, while others implicate NO in tumor promotion and inflammation. For example, the production of endogenous NO is associated with apoptosis in cancer cells [84] characterized by upregulation of the tumor suppressor p53, changes in the expression of pro- and antiapoptotic Bcl-2 family members, activation of caspases, chromatin condensation, and DNA fragmentation [85]. On the other hand, a large body of experimental and clinical data suggests a promoting role of NO in tumor progression and metastasis. Expression of iNOS has been reported in malignancies of the breast [86–88], prostate [89], lung [90], brain [91], and colon [92,93]. It appears that low concentrations of NO of less than 100 nM prevent certain cell types from apoptosis and thereby favor tumorigenesis and progression [94,95]. A sustained low efflux of NO derived from eNOS stimulates the proliferation of the endothelial cells [96]. However, high concentrations of NO, more than 500 nM, may be proapoptotic, producing cytotoxicity and antitumor activity [97,98]. High NO levels that produced cytotoxic effects on tumor cells are derived from macrophages, neutrophils, endothelial cells, hepatocytes, cardiac myocytes, and chondrocytes [99–102]. Thus, NO displays a dual nature with respect to carcinogenesis, which may be attributed to low and high fluxes of NO [103–108]. What has also become evident is that the microenvironment of the tumor is a contributing factor that may ultimately steer the cellular outcomes [108].

Much like NO, the role of H_2S in inflammation, immunity, and cancer biology is also controversial, with some studies suggesting antiinflammatory, anticancer effects, whereas others suggest a contribution to immune-mediated tissue injury and tumor promotion. For example, CSE and myeloperoxidase (MPO) activity were increased in the carrageenan-induced rat hindpaw model of inflammation. This edema was reduced by pretreatment with DL-propargylglycine, a CSE inhibitor [109]. Endotoxin administration to mice increased in CSE gene expression in the liver and kidney and increased H_2S levels in the plasma, liver, and kidney. There was also histological evidence of lung, liver, and kidney tissue inflammatory damage. Administration of sodium hydrosulfide (a H_2S donor) resulted in histological signs of lung inflammation, increased lung and liver MPO activity, and increased plasma TNF-α levels [110]. Elevated H_2S levels have also been observed in the plasma of septic-shock patients [110]. Of note is when the NO-releasing flurbiprofen was administered to LPS-treated rats, there was a dose-dependent inhibition of liver H_2S synthesis and CSE mRNA levels [111,112]. Taken together, these observations suggest a proinflammatory role for H_2S. Proinflammatory actions of H_2S have also been reported in an animal model of pancreatitis [113,114].

Antiinflammatory effects of H_2S have also been reported. For example, the administration of carrageenan into a rat air pouch resulted in leukocytes and neutrophils infiltration. Pretreatment with various H_2S donors (NaHS, *N*-acetylcysteine, and Lawesson's reagent) reduced the number of leukocytes in a dose-dependent manner [13]. This reduction was comparable with that seen by using diclofenac, the NOS inhibitor, L-NAME, or dexamethasone. Pretreatment with β-cyanoalanine, a CSE inhibitor reversed the inhibitory effects of *N*-acetylcysteine. Pretreatment with diclofenac, NaHS, or Na_2S similarly reduced carrageenan-induced rat paw edema, while β-cyanoalanine showed a significant increase in paw swelling in response to carrageenan [13]. Through peroxynitrite scavenging, H_2S may inhibit tissue oxidative damage [115,116]. H_2S has been shown to induce apoptosis in neutrophils [117], which could contribute to the resolution of inflammation [118]. In LPS-stimulated microglia and astrocytes, H_2S donors inhibited TNF-α secretion, an antiinflammatory effect, due to inhibition of iNOS and p38-MAPK signaling [119]. In the neuroblastoma cell line SH-SY5Y, H_2S inhibited rotenone-induced cell apoptosis via the regulation of the p38- and JNK-MAPK pathways [120]. Thus, it appears that, just like NO, physiological concentrations of H_2S produce antiinflammatory effects, whereas higher concentrations may be pro-inflammatory [121,122].

NO-RELEASING COMPOUNDS AS CANCER THERAPEUTICS

Different approaches have been used in utilizing NO as an anticancer target. These have included using NO as radiotherapy and chemotherapy sensitizer, NO-releasing drugs, and NOS inhibitors [123]. Because tumors generally do not have a well-developed vasculature, administered drugs may not reach their intended target. NO donors can enhance tumor blood flow and oxygen supply, thereby resensitizing it to chemotherapeutic agents or radiation therapy [124–126]. NOS inhibitors such as L-NAME, L-NIL, and 1400 W have been evaluated in several preclinical in vivo animal models for their antitumorigenic properties by blocking iNOS- and eNOS-derived NO production [42,127,128]. There are also many synthetic "designer" NO-releasing compounds that are at various stages of development; some are preclinical, some are in different phases of clinical trials, and some have been abandoned for various reasons. While NOS inhibitors and NO donors have opposing effects, they can both serve as anticancer agents, as it has often been said, "the devil is in the details." Below, we have very briefly indeed discussed some of these promising NO-releasing compounds that have utility against various cancers and have provided some recent in-depth reviews of the subject matter.

Nitric oxide-releasing NSAIDs (NO-NSAIDs) were developed to reduce the potential side effects associated with traditional NSAIDs [129,130]. The rationale for their development was largely based on the observations that in the gastrointestinal (GI) system NO can enhance the local mucosal defense mechanisms, thus increasing blood flow, mucus, and bicarbonate secretions [7], reviewed in Refs. [43,65]. NO-NSAIDs are traditional NSAIDs linked to a NO-releasing moiety via a chemical spacer, which can be either aliphatic or aromatic (Fig. 2A). All NO-NSAIDs tested were more potent in inhibiting the growth of various human cancer cell lines compared with their traditional counterparts [131]. Aromatic spacers generated more potent NSAID derivatives than aliphatic ones [132,133], and positional isomerism

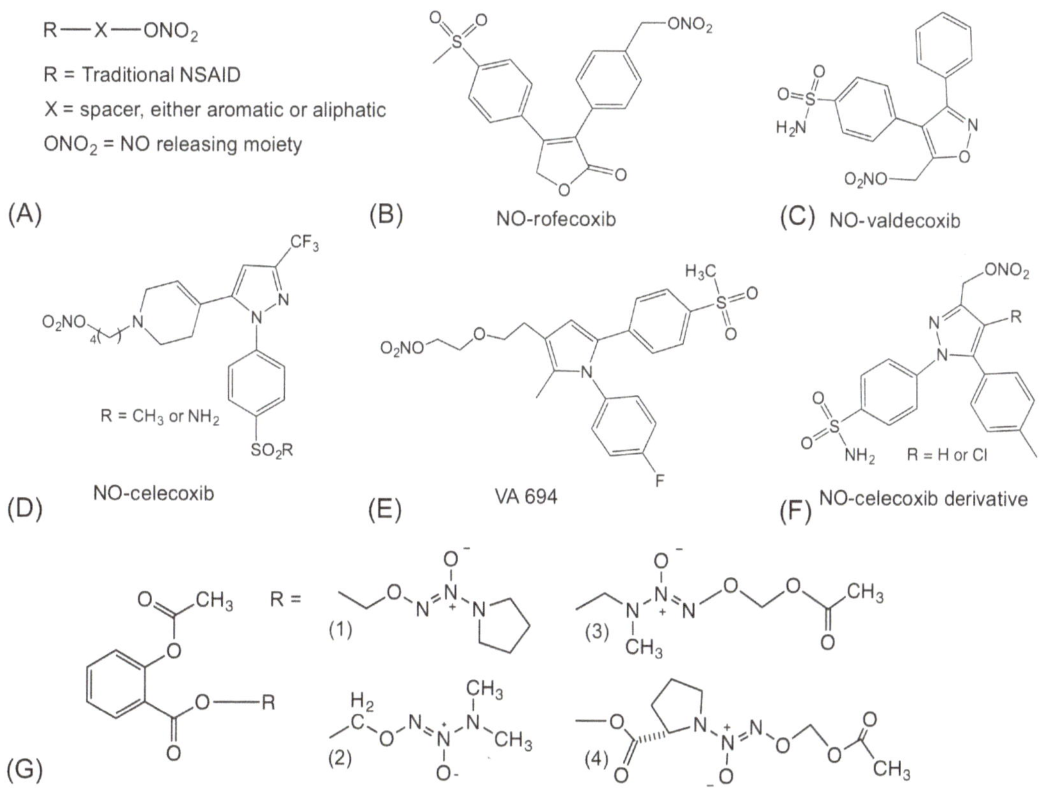

FIG. 2 The chemical structures of various NO-donating agents.

greatly influenced all cell kinetic parameters that influence cellular mass. It should be noted that, although structure-activity studies with NO-aspirin had indicated that NO was pivotal for its anticancer effects [134], it also became apparent that, in cases where the spacer was aromatic, the biological actions were due to the formation of a putative quinone methide (QM) intermediate [135–137].

Using a rat model of colonic adenocarcinoma, aspirin treatment for 4 weeks reduced the number of aberrant crypt foci by 64%, while *meta*-NO-aspirin (NCX 4016) produced an 85% reduction [138]. In *Min* ($APC^{Min/+}$) mice, treatment with *para*-NO-aspirin (NCX 4040) for 3 weeks reduced tumor growth by 55%, but it did not affect cell proliferation in the small intestinal mucosa [139]. Given that tumors had already formed when *para*-NO-aspirin was administered, it may be concluded that this compound displayed chemotherapeutic and chemopreventive properties. In another *Min* mouse study, *para*-NO-aspirin was more efficacious than *meta*-NO-aspirin in reducing the number of intestinal tumors [132]. In F344 rats treated with the carcinogen and azoxymethane, to induce colon cancers, NO-indomethacin and *meta*-NO-aspirin significantly suppressed both tumor incidence and multiplicity, with NO-indomethacin being more effective [140]. Using colon cancer xenografts, the simultaneous treatment with oxaliplatin and *para*-NO-aspirin or oxaliplatin followed by *para*-NO-aspirin showed comparable additive effects. However, treatment with *para*-NO-aspirin followed by oxaliplatin showed synergistic interactions possibly by sensitizing the cancer cells to the cytotoxic effects of the antitumor drug [141]. In a Syrian golden hamster model of pancreatic cancer, *meta*-NO-aspirin reduced the incidence and multiplicity of pancreatic cancer by 88.9% and 94%, respectively, while aspirin had no effect [142]. More recently, NO-naproxen at human-equivalent doses (HED) was shown to be highly effective in reducing urinary bladder cancer using the hydroxybutyl(butyl) nitrosamine (OH-BBN) model, whereas NO-sulindac was inactive. [82].

A second generation of NO-releasing aspirins followed the nitrate esters where furoxan derivatives were used as NO donors [143], reviewed in Refs. [43,65]. Unlike the nitrate esters that required enzymatic metabolism for NO release [144–146], the furoxan-NSAID hybrids released NO in the presence of plasma, GSH, or albumin, which is through thiol mechanisms [147].

Molecular targets of NO-aspirin in cancer have been thoroughly reviewed. Briefly, NO-aspirin alters cell kinetics; that is, it inhibits cell proliferation, it enhances apoptosis, and it also blocks the transitions of the cells through the cell cycle S [131–133,148,149]. Other targets affected in various models of cancer include, NF-κB [150–152], Wnt/β-catenin [133,140,149,153–156], iNOS [42,86–88,90,92,93,127,128,157–160], peroxisome proliferator-activated receptor δ (PPARδ) [161], mitogen-activated protein kinase (MAPK) [162,163], xenobiotic-metabolizing enzymes [142,164], oxidative stress [153,165], modulation of pro-inflammatory cytokines [166,167], and S-nitrosylation [168–170].

NO-RELEASING SELECTIVE CYCLOOXYGENASE-2 INHIBITORS (COXIBS)

Drugs such as celecoxib, rofecoxib, and valdecoxib have been successful in limiting the upper GI ulceration associated with traditional NSAIDs, reviewed by Solomon [171]. However, these drugs can reduce the beneficial vasodilatory effects of prostacyclin without affecting the levels of thromboxane A_2, a potent platelet aggregator. This eicosanoid imbalance can lead to an increased incidence of CV events [172]. In fact, several large-scale clinical trials have shown that long-term use of coxibs and even traditional NSAIDs is associated with an increased risk of adverse myocardial events, reviewed by Refs. [65,134]. NO could induce peripheral vasodilation to circumvent the elevation in blood pressure exhibited by coxibs; thus, coxibs that release NO should exhibit a safer CV profile. Indeed, recent data are in agreement with this thinking. NO-rofecoxib [173], NO-valdecoxib [174], NO-celecoxib [175], VA 694 [176], and a (pyrazolyl)benzenesulfonamide derivative of celecoxib [177] are shown in Fig. 2B–F, respectively. Recently, this field has expanded rapidly as evidenced by the number of NO-coxibs reported [178–185]. Reports are also emerging showing the anticancer properties of NO-coxibs [185].

NONO-NSAIDs

These class of compounds are based on linking a *N*-diazen-1-ium-1,2 diolate functional group to a classical NSAID (Fig. 2G) yielding compounds that have the potential of generating two equivalents of NO [186]. The first agent reported had a O^2-unsubstituted *N*-diazen-1-ium-1,2-diolate (NONOate) attached via a one-carbon methylene spacer to the carboxylic acid group of a traditional NSAID (aspirin, ibuprofen, and indomethacin) [187]. A second generation of O^2-acetoxymethyl-protected (PROLI/NO) releasing NONO-NSAIDs was developed where a diazeniumdiolate ion obtained from an amine such as L-proline was used, the *N*-nitroso derivative of which is nontoxic [188].

NONO-NSAIDs are an attractive class of compounds; however, there are limited data on their chemopreventive potential. In a recent study, evaluating the potential use of NONO-NSAIDs as antimetastasis agents and the effects of NONO-aspirin and NONO-naproxen were compared with those exerted by their respective parent NSAIDs on their avidities on human melanoma M624 cells [189]. Both NONO-NSAIDs, but not their corresponding parent NSAIDs, reduced M624 adhesion on vascular cellular adhesion molecule-1 (VCAM-1) and fibronectin under fluid flow conditions and static conditions, respectively. The NONOate moiety on the NSAIDs was shown to be critical for their function in reducing the avidity of melanoma cells. These findings suggest that NONO-NSAIDs could potentially be a new class of antimetastasis drugs for cancer treatment.

DIAZENIUMDIOLATE BASED NO-RELEASING COMPOUNDS AND HNO-NSAIDs

Diazeniumdiolates or NONOates can decompose to form nitroxyl (HNO) and/or NO [190,191] (Fig. 3A). HNO has emerged as an important pharmacological agent with beneficial effects in overcoming heart failure [192], preconditioning against myocardial infarction [193], and treating alcohol abuse [194]. The first anticancer activity of HNO that was generated using Angeli's salt ($Na_2N_2O_3$) was reported in 2008 [195] where it was shown that it suppressed the proliferation of both estrogen-receptor-positive (MCF-7) and estrogen-receptor-negative (MDA-MB-231) human breast cancer cell lines under both hypoxic and

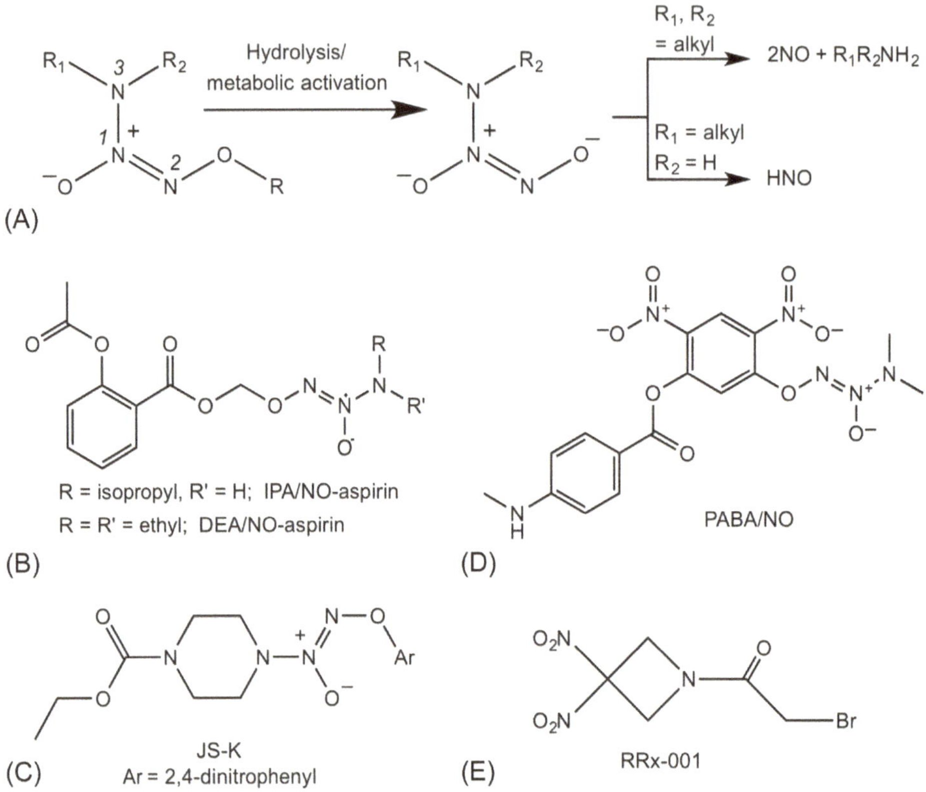

FIG. 3 (A) Activation of diazeniumdiolate prodrugs to release NO or HNO. (B) Structures of IPA/NO-aspirin and DEA/NO-aspirin, (C) JS-K, (D) PABA/NO, and (E) RRx-001.

normoxic conditions. In a xenograft mouse model of ER-negative breast cancer, HNO treatment caused significant reduction in tumor growth, which was accompanied by increases in apoptosis and suppression of angiogenesis [195]. Recently, two new NONO-NSAIDs were prepared by derivatizing both a primary and secondary amine diazeniumdiolate with aspirin to produce O^2-(acetylsalicyloyloxymethyl)-1-(*N*-isopropylamino)- diazen-1-ium-1,2-diolate (IPA/NO-aspirin) (Fig. 3B) and O^2-(acetylsalicyloyloxymethyl)-1-(*N*,*N*-diethylamino)-diazen-1-ium-1,2-diolate (DEA/NO-aspirin) [196]. Both of these agents had enhanced GI safety profiles, showed strong antiinflammatory properties, and exhibited significantly enhanced cytotoxicity compared with either aspirin or the parent diazeniumdiolate toward nonsmall-cell lung carcinoma cells (A549), but they were not appreciably toxic toward endothelial cells (HUVECs) suggesting cancer-specific sensitivity. These prodrugs also inhibited cyclooxygenase-2 (COX-2) and glyceraldehyde 3-phosphate dehydrogenase activities (GAPDH). Both IPA/NO-aspirin and DEA/NO-aspirin were also shown to reduce the growth of breast cancer cell lines MDA-MB-231, MDA-MB-468 (both ER-negative), and MCF-7 (ER-positive), more effectively than the parent compounds while not being appreciably cytotoxic in a related nontumorigenic MCF-10A cell line [179]. Notably, the cytotoxicity of these agents was about the same in ER-negative and the ER-positive cell lines suggesting an effect independent of the estrogen status of the cells.

JS-K AND PABA/NO

O^2-(2,4-dinitrophenyl)1-[(4-thoxycarbonyl)piperazin-1-yl]diazen-1-ium-1,2-diolate (JS-K, Fig. 3C) and O^2-[2,4-dinitro-5-(*N*-methyl-*N*-4-carboxyphenylamino) phenyl] 1-(*N*,*N*-dimethylamino)diazen-1-ium-1,2-diolate (PABA/NO, Fig. 3D) are members of the diazeniumdiolate class of NO prodrugs that were designed to be activated for anticancer effects by glutathione-*S*-transferase (GST)-induced release of NO reviewed in Ref. [197]. The rationale for this was based on the observation that GST (specifically GST-π), a key phase-II detoxification enzyme, is frequently overexpressed in cancer tissue [198,199]. JS-K [200,201] and PABA/NO [202,203] have shown promise as anticancer agents. U87 glioma cells exposed to PABA/NO showed a strong dose-dependent growth-inhibitory effect in vitro, and a strong synergistic effect was observed after concomitant treatment with temozolomide (TMZ), but not with carboplatin (CPT) [204]. However, in an animal model, PABA/NO administered by routes did not lead to an extension of survival time despite decreases in tumor proliferation and increases in apoptosis [204].

JS-K has shown potent antileukemic activity [200], induced apoptosis in human multiple myeloma cell lines [205], and was effective against leukemia, renal, prostate, and brain cancer cells [206,207], as well as both ER-positive and ER-negative breast cancer cell lines without affecting normal mammary epithelial cells [208]. In further evaluating the metabolic actions of JS-K, it was determined that the activating step in the metabolism of JS-K in the cell was the dearylation of the diazeniumdiolate by GSH resulting in the release of two equivalents of NO [209]. This caused rapid depletion of GSH, resulting in alterations in the redox potential of the cellular environment, which led to the activation of the MAPK stress signaling pathways and induction of apoptosis.

JS-K diminished WNT signaling in the androgen-receptor-positive 22Rv1 cells [198]. In human U87 glioma cells and primary glioblastoma cells, JS-K showed a concentration-dependent cytotoxicity [210]. In 18 different nonsmall-cell lung cancer (NSCLC) cell lines, JS-K was most effective against a subset of NSCLC cell lines where the endogenous levels of reactive oxygen and nitrogen species (ROS/RNS) were high [211]. Treatment of the lung adenocarcinoma cells with JS-K resulted in oxidative/nitrosative stress in cells with high basal levels of ROS/RNS, which, combined with the arylating properties of the compound, was reflected in glutathione depletion and alteration in the cellular redox potential, mitochondrial membrane permeabilization, and cytochrome *c* release.

β-Catenin has a central role in the Wnt signaling pathway (Chung et al., 2002) and is highly expressed in acute lymphoblastic leukemia (ALL) tumor lines of hematopoietic origin and primary leukemia cells but is undetectable in normal peripheral blood T cells. Among the leukemia cell lines, β-catenin is expressed at high levels in Jurkat T cells [212]. In this cell line, JS-K strongly inhibited its growth, reduced proliferation, and caused G_2/M cell-cycle arrest, which led to increased apoptosis [213]. JS-K reduced the transcriptional activity of β-catenin and also reduced the nuclear β-catenin protein levels but had no appreciable effect on levels in the cytosol. The underlying mechanism for these observations was shown to be through S-nitrosylation of nuclear β-catenin [213,214]. In order to improve the selectivity for cancer cells, diazeniumdiolates-oleanolic hybrids have been synthesized that caused apoptosis in HepG2 cells, G2/M cell-cycle arrest, activating the MAPK pathway, and enhancing the intracellular production of ROS [215].

RRx-001 AN AEROSPACE COMPOUND FOR CANCER TREATMENT

RRx-001 (Fig. 3E) also known as 1-bromoacetyl-3,3-dinitroazetidine (ABDNAZ) is a novel aerospace-derived compound under active investigation as a chemo-, immuno-, and radiosensitizer in phase-II clinical trials [216]. This compound contains a unique high energetic organic nitro functional group, a gem dinitroazetidine that has not been used to date for medical and pharmaceutical applications. In an aerospace setting, compounds containing this energetic functionality, such as 1,3,3 trinitroazetidine, are designed to fragment explosively to propel rockets [217]. Modification of this structure by removing one of the nitro groups and substituting it with a bromoacetate group resulted in RRx-001, a nonexplosive that may be used to treat cancer [218]. RRx-001 effectively inhibited the growth of 11 human cancer cell lines of varying tissue origins [125]. The cell lines consisted of oral squamous cell carcinoma (22B), pancreatic carcinoma (PANC-1), melanoma (M21), glioblastoma (U87 and SNB75), colon carcinoma (RKO), colorectal adenocarcinoma (HT-29), breast adenocarcinoma (MCF-7 and ER-positive), renal cell carcinoma (A498), brain neuroblastoma (IMR32), nonsmall-cell lung carcinoma (A549), and a murine squamous cell carcinoma (SCC VII). The cytotoxic activity of RRx-001 in these cell lines was enhanced under hypoxic conditions, which is an observable phenomena when a solid tumor exceeds a critical diameter of 1–2 mm^3 [126]. Using HL60 cells (human promyelocytic leukemia cells) that have a doubling time of about 48 h and HT-29 cells, RRx-001 induced intracellular ROS and RNS generation in a dose- and time-dependent manner [125]. Pretreatment of the cells with the GSH precursor and *N*-acetylcysteine (NAC) inhibited RRx-001-induced apoptosis, whereas pretreatment with BSO (buthionine-(*S*,*R*)-sulfoximine and GSH synthesis inhibitor) increased RRx-001-induced apoptosis. It should be noted that DNA damage (γH2AX expression) was increased in HT-29 and SCC cells as a result of RRx-001 exposure in a concentration- and time-dependent manners. These imply that RRx-001 has utility as a chemotherapeutic agent but not as a chemopreventive one.

RRx-001 differs from other NO-donating compounds in that the molecule induces local, endogenous, and biphasic production or release of NO, rather than fragmenting to release NO systemically. This activity is closely linked to the metabolism of RRx-001. Metabolic and disposition studies on RRx-001 have shown that, on infusion, the compound rapidly, irreversibly, and selectively binds to hemoglobin at a key NO binding site [219] and with glutathione [220,221] in directly increasing oxidative stress [222]. While the RRx-001 glutathione adduct is rapidly excreted, RRx-001-bound hemoglobin remains in the circulation for the duration of the lifetime of the red blood cell [223]. Results from a phase-I clinical trial of RRX-001 in heavily pretreated patients with advanced/refractory solid tumors were notable for the lack of systemic toxicity at all doses tested and evidence of anticancer efficacy, with 15 of 25 patients enrolled in the study, exhibiting partial response or stable disease [224]. Furthermore, after progression on RRx-001 treatment, four patients were rechallenged with drugs to which they had previously acquired resistance, and all four exhibited resensitization. Preliminary results from ongoing phase-II clinical trials have recapitulated the favorable safety and tolerability profile from phase I and provided initial confirmation of antitumor priming to subsequent therapy post-RRx-001 [225–228].

H_2S-RELEASING OR -DONATING COMPOUNDS AS CANCER THERAPEUTICS

Interest in the biological actions of H_2S has grown very rapidly, and there has been a lot of focus on H_2S-donating or H_2S-releasing compounds and on the way in which these compounds deliver this volatile gas to target tissues. Although the terms H_2S-donating and H_2S-releasing are used synonymously, they convey two different phenomena. H_2S-releasing compounds can be classified as sulfide salts such as NaHS, Na_2S, and CaS, which form HS^- and H_2S immediately upon solvation in physiological buffers. Although these compounds are extensively used in studying the biological actions of H_2S, they are of limited therapeutic value since their effective residence time in tissues is relatively short. H_2S is rapidly lost from solution by volatilization in laboratory conditions [229] or transferred across respiratory membranes [230]. H_2S-donating compounds are either naturally occurring compounds (mainly in foods) or synthetic hybrids. There are a large number of "foods" that may release H_2S when metabolized. For example, garlic and some of its metabolites (allicin, diallyl disulfide (DADS), and diallyl trisulfide (DATS)) have been shown to release H_2S in vivo [230,231]. Sulforaphane the isothiocyanate compound from broccoli and allyl isothiocyanate derived from wasabi, mustard, and horseradish have health benefits that are quite similar to those that are attributed to H_2S or other H_2S-donating compounds, although it has not yet been determined if any of these are indeed metabolized to produce H_2S.

The therapeutic potential of H_2S-donating compounds has been described in the cardiovascular system [232,233], urology [234], neuroinflammation [235], and ophthalmology [236]. These general areas have been thoroughly reviewed elsewhere [30,55,237]. In this chapter, we have focused on the role of H_2S-donating compounds that are in different stages of development as agents that may have utility against cancer, although, where appropriate, we have also summarized some novel compounds that have other indications. We have also discussed the potential applications of NOSH-NSAIDs that are dual NO-H_2S-donating hybrids.

SYNTHETIC H_2S-DONORS

GYY4137 and its Structural Analogs

In the field of H_2S studies, the bulk of the scientific literature makes use of high concentrations of crude sulfide salts such as sodium hydrosulfide (NaSH) or sodium sulfide (Na_2S) where they have been used as H_2S-releasing compounds. While these salts are convenient for use in the laboratory and generate solutions containing known quantities of H_2S, they are highly problematic [238]. For example, NaSH and Na_2S are simple salts and not H_2S

donors. They undergo instantaneous and pH-dependent dissociation by generating, not releasing, H_2S. This is in sharp contrast to enzymatic synthesis of H_2S, which is not instantaneous but a slow and a sustained process [239–241]. This kinetic difference means that these salts are not suitable for studying physiological processes and outcomes [233]. Further, in order to see biological responses, it is often necessary to use high concentrations of these salts, >100 μM and often 1–10 mM presumably because the H_2S generated from these salts dissipates in seconds [242,243]. Hence, there has been a need for the development of slow H_2S-donor molecules, which can overcome these obvious problems and can mimic endogenously produced H_2S levels.

The first water-soluble slow H_2S-donating molecule to be described was morpholin-4-ium 4 methoxyphenyl(morpholino) phosphinodithioate (GYY4137) (Fig. 4A and Table 1). GYY4137 liberated H_2S when dissolved in acidic (pH 3.0) phosphate buffer; however, when dissolved in buffer at pH 7.4 or 8.5, relatively much smaller amounts of H_2S were detected [242]. When administered to rats, either intraperitoneally (ip) or intravenously (iv), significant increases in plasma H_2S were observed for up to 3 h. The anticancer effects of GYY4137 were evaluated by determining its cell-growth-inhibitory effects using seven different human cancer cell lines. These represented cervical carcinoma (HeLa), colorectal carcinoma (HCT-116), hepatocellular carcinoma (HepG2), osteosarcoma (U2OS), breast adenocarcinoma (MCF-7), acute promyelocytic leukemia cells (HL-60), and myelomonocytic leukemia (MV4-11) [244]. Five-day treatment with GYY4137 (400 μM) significantly reduced proliferation of MCF-7, MV4-11, and HL-60 cells. At 800 μM, GYY4137 killed greater than 90% of all these cancer cells (MCF-7, MV4-11, HL-60, HeLa, HCT-116, HepG2, and U2OS), but did not affect survival of noncancer human diploid lung fibroblasts (IMR90 and WI-38). The apparent IC_{50} for GYY4137 in MCF-7, HL60, and MV4-11 cells was around 350 μM. GYY4137 promoted apoptosis as seen by the presence of cleaved PARP and cleaved caspase 9. In MCF-7 cells, it also caused cell-cycle arrest in the G_2/M phase; however, it did

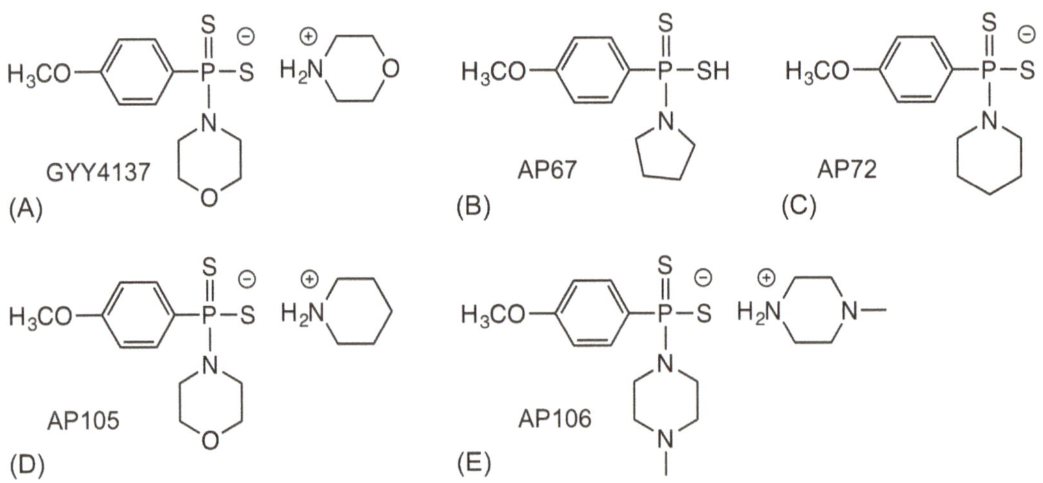

FIG. 4 (A) GYY4137 a slow H_2S-generating compound, (B–E) modified phosphorodithioate core resulting in agents, which generate H_2S at different rates.

TABLE 1 H_2S-Based Therapeutics in Development

Company	Drug Lead	Clinical Indication and Derivative	Status
Avicenna Pharma	NBS-1120	Cancer, antiinflammatory, pain—aspirin	Preclinical
	AVT-219	Osteoarthritis—naproxen	Preclinical
Antibe Therapeutics	ATB-346	Osteoarthritis—naproxen	Phase II
	ATB-352	Acute pain—ketoprofen	Preclinical
	ATB-340	Thrombosis—aspirin	Preclinical
	ATB-429	Ulcerative colitis—mesalamine	Preclinical
gIcare Pharma	GIC-1001	Colonic pain—trimebutine salt	Phase II[a]
SulfaGENIX	SG-1001	Oxidative stress, heart failure—polyvalent sulfur	Phase II[b]
University of Singapore	GYY4137	Cancer, inflammation, hypertension	Unknown
University to Exeter	AP39	Inflammation, oxidative stress	Preclinical

[a] *ClinicalTrials.gov identifiers, NCT01926444 and NCT02276768.*
[b] *ClinicalTrials.gov identifier, NCT01989208.*

not affect the noncancerous IMR90 cells. In a xenograft model using SCID mice, daily ip injection of GYY4137 (at 100–300 mg/kg) also dose-dependently decreased the growth of HL-60 and MV4-11 tumors [244]. Further, using HepG2 and Bel7402 cells, it has been reported that GYY4137 suppressed STAT3 activation by reducing p-STAT3 (Y705) levels [245]. In vivo, GYY4137 significantly inhibited tumor growth in the subcutaneous HepG2 xenograft model by inhibiting STAT3 activation and its target gene expression. Using MCF-7 and human breast epithelial MCF10A cells and normal human lung fibroblast WI38 and HepG2 cells, it has been shown that GYY4137 significantly increased glycolysis, leading to the overproduction of lactate and inducing intracellular acidification in cancer cell lines. H_2S also decreased anion exchanger and sodium/proton exchanger activities. It has been suggested that these combinations of increased metabolic acid production and defective pH regulation resulted in an uncontrolled intracellular acidification, leading to cancer cell death [246].

It should be noted that commercial preparations of GYY4137 are sold as a dichloromethane complex, and it has been reported that dichloromethane is metabolized to carbon monoxide (CO) [247,248], an important gasotransmitter. Since CO is reported to have similar biological effects as H_2S [249–251], it is possible that some of the observed effects of GYY4137 could also be due, at least in part, to a contribution by CO [233].

An inspection of the published reports where GYY4137 has been used as the H_2S-donating compound clearly indicates that the doses needed to see any biological/pharmacological effects are rather high. In order to improve the H_2S generation from GYY4137, modifications to the phosphorodithioate core have been made resulting in agents that generate H_2S at different rates (AP67, AP72, AP105, and AP106 (Fig. 4B–E), respectively). Using cultured human synoviocytes and rat arterial smooth muscle cells and ex vivo using precontracted mouse aorta, it has been shown that the cellular and tissue effects of the H_2S donors were dependent upon the rates at which the compounds generated H_2S. For example, H_2S generation from equimolar concentrations of each compound was in the following order

AP67 > AP72 > AP105 > AP106 > GYY4137, which correlated with the observed biological responses [233]. Any anticancer activity of these agents is yet to be reported.

ANETHOLE TRITHIONE (ADT-OH) DERIVATIVES

5-(4-Hydroxyphenyl)-3*H*-1,2-dithiole-3-thione (ADT-OH), the main metabolite of anethole dithiolethione (ADT) (Fig. 5A and B), has been used extensively as a H_2S donor with many therapeutic agents including a good number of NSAIDs, reviewed in Refs. [252,253]. The rationale for developing hybrid NSAIDs that release H_2S was based on the observation that, in animal models of chronic ulcer, an increase in the expression of CSE and CBS enzymes and H_2S levels was observed, suggesting a defensive response [254]. Administration of structurally diverse H_2S donors enhanced ulcer healing, since these agents did not affect gastric pH or inhibited acid secretion, and K_{ATP} channel agonists or antagonists had no effect on ulcer healing. Since H_2S is a vasodilator, it was postulated that this may have contributed to the healing process observed in this study [254]. A number of H_2S-donating NSAIDs have shown

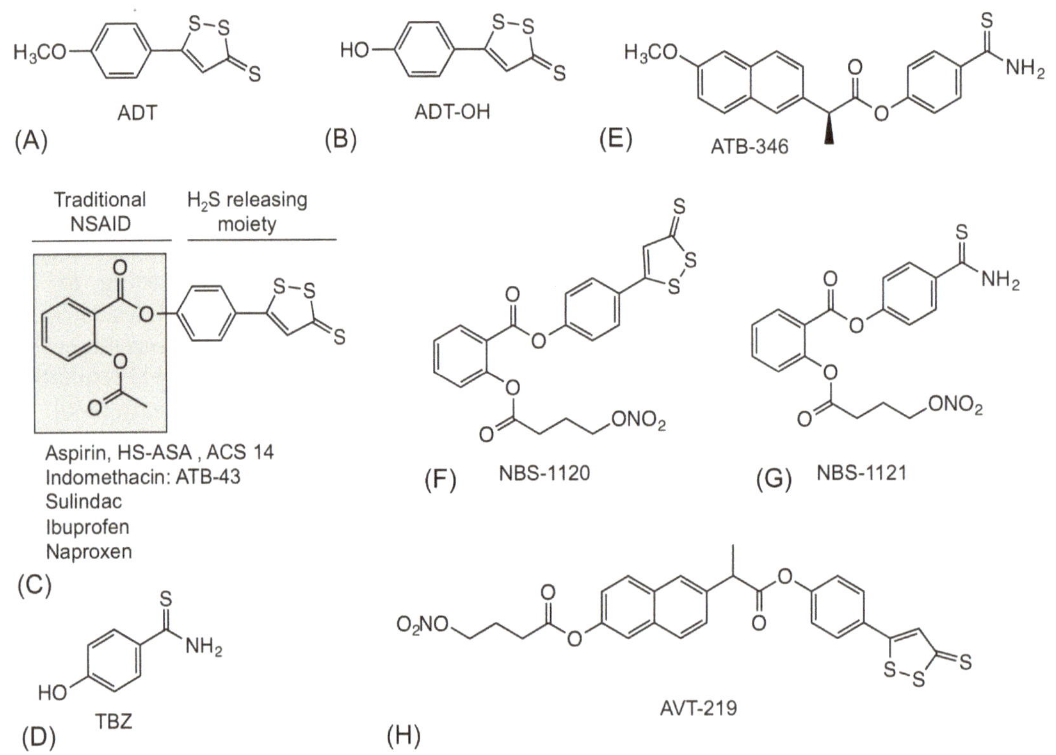

FIG. 5 Synthetic H_2S donating compounds. (A) ADT (anethole dithiolethione); (B) ADT-OH, (5-(4-hydroxyphenyl)-3*H*-1,2-dithiole-3-thione); (C) various H_2S-donating NSAIDs; (D) TBZ (4-hydroxy benzothiazamide); (E) ATB-346, H_2S-donating naproxen; (F and G) NBS-1120 and NBS-1121, NOSH-aspirin, dual NO- and H_2S-donating hybrid of aspirin; (H) AVT-219, NOSH-naproxen, dual NO- and H_2S-donating hybrid of naproxen.

considerable protective effects in the stomach compared with their traditional counterparts. For example, H_2S-diclofenac had significant GI-sparing effects compared with diclofenac [254–256] and inhibited both cyclooxygenase (COX)-1 and COX-2 enzymatic activities to the same extent as its parent counterpart [256]. It also reduced the plasma levels of the pro-inflammatory cytokines, IL-1β and TNF-α, while it elevated the antiinflammatory, IL-10 [255]. Similarly, H_2S-indomethacin [122], H_2S-mesalamine [256], and H_2S-naproxen [257] were also shown to significantly cause less damage to the stomach compared with their respective parent compound. In a recent study, the H_2S-donating aspirin (ACS14) was shown to spare the stomach of rats from the harmful side effects of aspirin when given 30 min before aspirin administration [258]. ACS14 also abrogated the aspirin-induced upregulation of COX-2 expression, but had no effect on the reduced PGE_2 levels; it also attenuated aspirin-suppressed superoxide dismutase-1 expression and GSH activity [258]. These data suggest that ACS14 enhances mucosal integrity by inhibiting oxidative stress in the gastric tissue.

H_2S-Donating NSAIDs as Chemopreventive Agents

H_2S-donating NSAIDs (HS-aspirin, HS-sulindac, HS-ibuprofen, and HS-naproxen (Fig. 5C)) were shown to have utility as anticancer agents [259–264]. Human safety data on some H_2S-donating NSAIDs are beginning to emerge although not yet published (pharmaceutical press releases), underscoring the great potential of these agents against many diseases including cancer.

H_2S-donating NSAIDs in general inhibit cell proliferation, enhance cell death by inducing apoptosis, and block transitions through the cell cycle, predominantly those between G_0/G_1 [259,260]. The net result being a significant inhibition of cell growth, which presumably accounts for the ultimate inhibition of carcinogenesis that is observed *in vivo*. HS-aspirin, HS-sulindac, HS-naproxen, and HS-ibuprofen inhibited growth of eleven different cancer cell lines of six different histological subtypes [260]. The cell lines were that of colon (HT-29, COX-1 and COX-2 positive; HCT 15, COX-null; and SW480, COX-1 positive and low levels of endogenous COX-2), breast (MCF7 (ER(+)), SKBR3 (ER(−)), and MDA-MB-231 (ER-, PR- , which does not over express Her2)), T-cell leukemia (Jurkat), pancreatic (BxPC3, both COX-1 and COX-2 positive; MIAPaCa-2, COX-null), prostate (LNCaP), and lung (A549). Since colon HT-29 and pancreatic BxPC3 cells express both COX-1 and COX-2 whereas colon HCT15 and pancreatic MIA PaCa2 cells are COX-null [131,265], it thus appears that the observed cell growth inhibition is independent of COX inhibition. The cell-growth-inhibitory effect of HS-NSAIDs is at least, in part, dependent on the release of H_2S since preincubation of MDA-MB-231 breast cancer cell with NaF, a carboxylesterase enzyme inhibitor, significantly reversed HS-ASA-mediated cell growth inhibition [259]. It is important to note that HS-NSAIDs preferentially affect cancer cells compared with normal cells [259].

Another H_2S-donating moiety that has been employed with NSAIDs to produce HS-NSAIDs hybrids is 4-hydroxy benzothiazamide (TBZ, Fig. 5D). This moiety has been conjugated with naproxen (ATB-346, Fig. 5E), and this hybrid has been shown to have a favorable GI safety profile and antiinflammatory properties that were similar to its parent compound, naproxen [257]. In an animal model of synovitis, ATB-346 reduced edema and pain score suggesting possible use in the treatment of patients with arthritis [266]. In evaluating the effects of ATB-346 against colon cancer, weekly administration of azoxymethane over a 4-week period resulted in the formation of ~50 aberrant crypt foci in the colon. Twice-daily treatment with naproxen at high doses significantly reduced the number of aberrant crypt foci. However, a significantly greater effect was observed with ATB-346, and it was also effective at much lower doses, where naproxen was ineffective. TBZ, the H_2S-releasing moiety of ATB-346, did not significantly affect the number of aberrant crypt foci, suggesting that both the inhibition of cyclooxygenase activity and H_2S were necessary for the enhanced chemopreventative effect. ATB-346 suppressed colonic prostaglandin synthesis and whole blood thromboxane synthesis as effectively as naproxen, but did not induce any gastrointestinal injury [262]. ATB-346 was also more effective at preventing intestinal polyp formation than naproxen in the APC^{Min+} mice. In this model, ATB-346, but not naproxen, significantly decreased the expression of intestinal cancer-associated signaling molecules, cMyc and β-catenin [264]. In another study, the proliferation of human melanoma cells was reduced, and apoptosis was increased by ATB-346, an effect that was modulated by the inhibition of NF-κB activation and the inhibition of the Akt signaling pathway. Moreover, daily oral dosing of ATB-346 (43 μmol/kg) significantly reduced melanoma development in vivo [263]. ATB-346 is the lead drug of Antibe Therapeutics (Table 1) and is being developed for the treatment of osteoarthritis and has completed phase-I clinical trials in healthy volunteers and is now under phase-II clinical evaluation.

MOLECULAR TARGETS AFFECTED BY H_2S-DONATING NSAIDs IN CANCER

Apart from the effects on cell kinetics (proliferation, cell-cycle arrest, and apoptosis), which ultimately determine cellular mass, HS-NSAIDs also modulate a number of molecular targets that are important in cancer. Some of these are briefly discussed below.

The NF-κB Signaling Pathway

The transcription factor NF-κB has a central role in cancer biology perpetuating the conversion of cells from normal to cancerous. NF-κB can suppress apoptosis, thus promoting tumorigenesis [267]. Most chemopreventive agents (i) suppress the activation of the NF-κB pathway through the inhibition of the NF-κB signaling pathway and (ii) sensitize the tumors to chemotherapeutic agents through the abrogation of NF-κB activation.

GYY4137 reduced LPS-induced NF-κB activation in mouse macrophages [243]. Also, when rats were injected with LPS, GYY4137 decreased hepatic LPS-evoked rise in NF-κB activation [268]. H_2S-donating diclofenac was shown to reduce inflammation in rats following LPS administration, and this was attributed to the inhibition of liver NF-κB activation [255].

NF-κB is constitutively activated in estrogen receptor (ER)-negative breast cancer cell lines and primary tumors [269–271], and this led to resistance to apoptosis [272,273] and promotion of an invasive and metastatic phenotype [274]. Using MDA-MB 231 cells that are ER(−), treatment with HS-ASA resulted in a concentration-dependent inhibition of DNA-binding activity of NF-κB (p65). Levels of nuclear p65, which is the functionally active subunit of NF-κB, were also reduced by HS-ASA in a concentration-dependent manner. One of the most critical steps in NF-κB activation is its dissociation from IκB, which is mediated through the phosphorylation and subsequent proteolytic degradation of this inhibitory subunit [275,276]. HS-ASA inhibited IκBα phosphorylation, while total levels were unaltered, thus inhibiting the translocation of p65 to the nucleus through the blockade of IκBα phosphorylation. The phosphorylation of the IκBs is known to be mediated by Iκappa kinases (IκKα and IκKβ), which are activated by phosphorylation at specific serine residues in the activation loop of IκKβ [275]. HS-ASA decreased the phosphorylation of both IκKα and IκKβ without affecting the total IκK levels. Therefore, HS-ASA prevents the activation of IκKs, inhibits phosphorylation-mediated degradation of IκBα, and prevents the translocation of NF-κB-p65 into the nucleus [259]. Recently, using ER(−) breast cancer cells, it has been reported that dithiolethiones inhibit NF-κB transcriptional activity not through H_2S release but rather via a covalent reaction with the NF-κB p50 and p65 subunits [277]. In a xenograft mouse model of colon cancer, HS-naproxen caused significant inhibition of tumor growth, which was associated with mark reductions in NF-κB expression [278].

Reactive Oxygen Species

High reactive oxygen species (ROS) levels can disrupt redox homeostasis leading to cellular damage and apoptosis [279,280]. Many anticancer agents induce ROS as part of their mechanism of action. When the equilibrium between ROS formation and endogenous antioxidant defense mechanisms is disturbed, a state of oxidative stress is generated within the cell, which ultimately may lead to cell death. Using molecular probes that detect intracellular levels of peroxides and superoxide anions ($O_2^{\cdot}$), it was shown that HS-ASA dose-dependently induced ROS levels in ER(−) breast cancer cells [259].

Thioredoxin/Thioredoxin Reductase-1 (Trx/TrxR)

Thioredoxin (Trx) is a ubiquitous redox-active thiol protein associated with the enzyme thioredoxin reductase-1 (TrxR) that maintains it in the reduced state [281]. Trx can suppress apoptosis [282], and increases in Trx and TrxR are found in several human primary cancers that are associated with aggressive tumor growth including breast carcinoma [283]. Also, there exist some relationships between TrxR inactivation and growth inhibition, cell-cycle arrest, and apoptosis in colorectal cancer cell lines [284]. In ER(−) breast and colon cancer cells, it was shown that TrxR activity was strongly inhibited by HS-ASA and HS-naproxen, respectively [259,278]. Moreover, it was also shown that HS-ASA caused a moderate inhibition of NF-κB-activity through the inhibition of IKK activation [259]. Thus, targeting the Trx/TrxR system, which is required for the maintenance of essential cysteine residues in the active thiol forms in certain proteins including the NF-κB protein [285], is an attractive target for cancer drug development [253].

Wnt/β-Catenin Signaling Pathway

The Wnt signaling cascade has a major role in human colon carcinogenesis [286], and its deregulation has also been implicated in T-cell acute lymphoblastic leukemia (ALL) and in B-cell chronic lymphocytic leukemia (CLL) [212,287]. The β-catenin protein is intimately involved in the Wnt signaling pathway that regulates cell-cell adhesion and may promote leukemia cell proliferation [212]. β-Catenin is expressed in T-ALL cells, tumor lines of hematopoietic origin, and primary leukemia cells but is undetectable in normal peripheral blood T cells. Using Jurkat T, it has been shown that HS-ASA dose-dependently reduced β-catenin protein levels whereas it had no effect on total β-catenin expression [288]. In synchronized cells, HS-ASA dose-dependently reduced cyclin D1 protein expression [288]. In some B- and T-cell leukemias, ASA and its NO-donating derivatives are reported to activate caspases [155,289]. Similarly, HS-ASA also increased caspases-3 enzyme activity and protein expression in a concentration-dependent manner [288].

Xenobiotic Metabolizing Enzymes

The balance between phase-I carcinogen-activating and phase-II detoxifying xenobiotic-metabolizing enzymes may determine an individual's risk for cancer. HS-ASA induced the activities of GST and NAD(P) H:quinone oxidoreductase (NQO1), two phase-II enzymes, in HT-29 human colorectal cancer cells and mouse hepatoma Hepa-1c1c7 cells [261]. Treatment of both cell lines with HS-ASA resulted in increases in NQO1, GST, and UDP-glucuronyltransferase (UGT1, another phase-II enzyme) protein levels and increases in the phase-I enzymes CYP1A1 and CYP2E1 [261]. The transcription factor Nrf2, which is involved in the induction of many phase-II enzymes, showed enhanced expression in the HS-ASA-treated cells. In vivo studies showed that HS-ASA induced hepatic GST activity by 3.4-fold and increased the hepatic protein levels of phase-II enzymes, GST, NQO1, and UGT but had no effect on the protein levels of CYP1A1 and CYP2E1 [261]. These finding strongly suggest that in vivo HS-ASA is a monofunctional inducer of phase-II metabolizing enzymes.

DUAL NO- AND H_2S-DONATING NSAIDs

Based on the chemistry of NO and H_2S and the structural components of NO-NSAIDs and HS-NSAIDs, recently, it was postulated that an NSAID possessing moieties that could donate both of these gasotransmitters might be more potent and effective than either one alone [290]. This led to the synthesis and characterization of a new class of antiinflammatory and anticancer compounds, the so-called NOSH-NSAIDs, of which six have been reported [214,290,291]. The structural components of the three of these compounds are shown in Fig. 5G–H. In NBS-1120 (Table 1) and NBS-1121, aspirin was used as a scaffold to which NO- and H_2S-releasing moieties were coupled with one of the 1, 2 positions. Nitrate (–ONO_2) was used for NO release, and this was attached to the aspirin through an aliphatic spacer, while ADT-OH and TBZ were used as the H_2S-donating moieties in NBS-1120 and NBS-1121, respectively (Fig. 5F and G). For NOSH-naproxen (AVT-219, Fig. 5H and Table 1), nitrate and ADT-OH were coupled to the parent drug as the NO- and H2S-donating moieties [291].

In terms of GI safety, NBS-1120 [292] and AVT-219 [293] and NOSH-sulindac (AVT-18A) [294] were essentially devoid of any GI toxicity. All three compounds exhibited comparable pharmacological properties to their respective parent compound, that is, they exhibited strong antiinflammatory, antipyretic, analgesic, and antiplatelet activities but, unlike their native counterparts, were GI safe. Furthermore, lipid peroxidation induced by aspirin, naproxen, and sulindac was higher than that exerted by NBS-1120, AVT-219, and AVT-18A. SOD activity was significantly inhibited by aspirin, naproxen, and sulindac but increased by NOSH-NSAIDs.

In a head-to-head study, the antinociceptive activity of NBS-1120 was compared with aspirin in different models of inflammatory pain [295]. It was found that NBS-1120 inhibited acetic acid-induced writhing response and carrageenan (Cg)-induced inflammatory hyperalgesia in a dose-dependent (5–150 μmol/kg) manner, which was superior to the effect of the same doses of aspirin. NBS-1120 exhibited greater antinociceptive effect that also lasted longer compared with aspirin upon complete Freund's adjuvant (CFA)-induced inflammatory hyperalgesia. Mechanistically, NBS-1120, but not aspirin, was able to reduce the production/release of interleukin-1 beta (IL-1β) during Cg-induced paw inflammation. Furthermore, NBS-1120, but not aspirin, reduced prostaglandin E2-induced hyperalgesia, which was prevented by treatment with an ATP-sensitive potassium channel (K_{ATP}) blocker (glibenclamide). Noteworthy, the antinociceptive effect of NBS-1120 was not associated with motor impairment. Thus, NBS-1120 seems to present greater potency than aspirin in reducing inflammatory pain in several models. The enhanced effects appear to be due to its ability to reduce the production of pronociceptive cytokines such as IL-1β and directly block hyperalgesia caused by a directly acting hyperalgesic mediator in a mechanism dependent on the modulation of K_{ATP} channels.

Comparing the effects of NBS-1120 to that of HS-aspirin and NO-aspirin in vitro against toxins released by activated microglia and astrocytes, it was found that NBS-1120 had significantly greater antiinflammatory and neuroprotective effects [296]. When activated by LPS/IFNγ, human microglia and THP-1 cells release materials that are toxic to differentiated SH-SY5Y cells. These phenomena also occur with IFNγ-stimulated human astroglia and U373 cells. When the cells were treated with the HS-aspirin or NO-aspirin, there was a significant enhancement of neuroprotection. However, NBS-1120 had significantly more potent protective properties. The effect was both concentration and incubation time-dependent. Such treatment not only reduced the release of the TNFα and IL-6 but also attenuated the activation of the p38-MAPK and NF-κB proteins. Based on these observations, it was suggested that NBS-1120 might be a new candidate for treating neurodegenerative disorders that have a prominent neuroinflammatory component such as Alzheimer's disease and Parkinson's disease [296].

The cell-growth-inhibitory properties of NBS-1120, NBS-1121, AVT-219, and AVT-18A were evaluated in eleven different human cancer cell lines of six different histological subtypes [292–294]. The cell lines were that of colon (HT-29, COX-1 and COX-2 positive; HCT 15, COX-null; and SW480, COX-1 positive and low levels of endogenous COX-2), breast (MCF7 (ER(+)) and MDA-MB-231 and SKBR3 (ER(−))), T-cell leukemia (Jurkat), pancreatic (BxPC3, both COX-1 and COX-2 positive; MIAPaCa-2, COX-null), prostate (LNCaP), and lung (A549). All four NOSH compounds were extremely effective in inhibiting the growth of these cell lines; however, NBS-1120 was the most potent, with IC_{50}s for cell growth inhibition ranging from 48 to 280 nM at 24 h [290,293,294].

Treatment of athymic nude bearing a human colon cancer xenograft with NBS-1120 at a dose of 100 mg/kg caused an 85% reduction in tumor growth after 18 days of treatment

[214]. Further, in a xenograft mouse model of colon cancer chemoprevention, aspirin at a dose of 50 mg/kg and NBS-1120 at different doses (25–100 mg/kg) inhibited tumor growth and tumor mass with NBS-1120 being approximately fivefold more potent than aspirin [292]. Of note was the observation that the stomachs of the aspirin-treated mice exhibited erosions consistent with preulcer development [297]. Thus, as a chemopreventive agent, NBS-1120 appears to be superior to aspirin both in terms of efficacy and safety. In another study comparing the chemotherapeutic efficacy of AVT-219 with that of naproxen at equimolar doses (0.18 mmol kg^{-1}) in a xenograft mouse model of colon cancer, it was reported that all the mice treated with naproxen died within 2 weeks from what appeared to be sever GI bleeding and damage to the spleen and liver. However, the AVT-219-treated mice did not show any overt signs of toxicity, and after 30 days of treatment, the AVT-219-treated mice showed a considerable reduction in tumor growth compared with untreated mice, which was as a result of increased apoptosis and reduced cell proliferation [293].

The underlying mechanism for the enhanced potency of NOSH-NSAIDs is not apparent, and there is no information available about the kinetics of NO and H_2S release and their potential interactions. However, it is known that both contribute toward the potency of the intact molecules. This is based on the observations where it was shown that the biological activity of aspirin and SNAP (*S*-nitroso-*N*-acetyl-penicillamine, which releases NO) and ADT-OH, which releases H_2S, was not the same as the biological activity of the intact NBS-1120 molecule [214]. Thus, the sum of parts did not equal the whole. However, recent reports indicate that NO can react with H_2S to produce HSNO that is a highly reactive intermediate [298,299]. Furthermore, NO and H_2S signaling pathways appear to be intimately intertwined with mutual potentiation of biological responses [300].

OTHER H_2S-DONATING AGENTS OF THERAPEUTIC INTEREST

An appreciation of the physiological and pathological importance of H_2S has fueled impetus to develop novel therapeutics that aim to deliver H_2S or to suppress its endogenous production (Table 1).

H_2S appears to have a significant role in the mitochondria, including the antioxidant effects, the modulation of mitochondrial cell death pathways, and the regulation of cellular bioenergetics [301–303]. Multiple studies have demonstrated that H_2S donors can maintain mitochondrial integrity, reduce the release of mitochondrial death signals, and attenuate mitochondrially regulated cell death responses of various types [11,304,305]. Mitochondria-targeted H_2S donors have been developed by coupling DTT (1,2-Dithiole-3-thione) with a phosphonium salt. One of such donors, AP39 ([(10-oxo-10-(4-(3-thioxo-3H-1,2-dithiol-5yl) phenoxy)decyl) triphenyl phosphonium bromide]) (Fig. 6A and Table 1), has shown promising activities [306]. Compared with standard and nonmitochondria-targeted donors such as GYY4137 and ADT-OH, AP39 significantly inhibited the oxidative stress-induced toxicity. It also attenuated the loss of cellular bioenergetics during oxidative stress and protected against oxidative mitochondrial DNA damage. These results suggest that mitochondria targeting is important.

There are several reports that H_2S can reduce visceral pain [53,307,308]. Based on this, gIcare Pharma has developed an H_2S-releasing salt of trimebutine (GIC-1001, Fig. 6B and

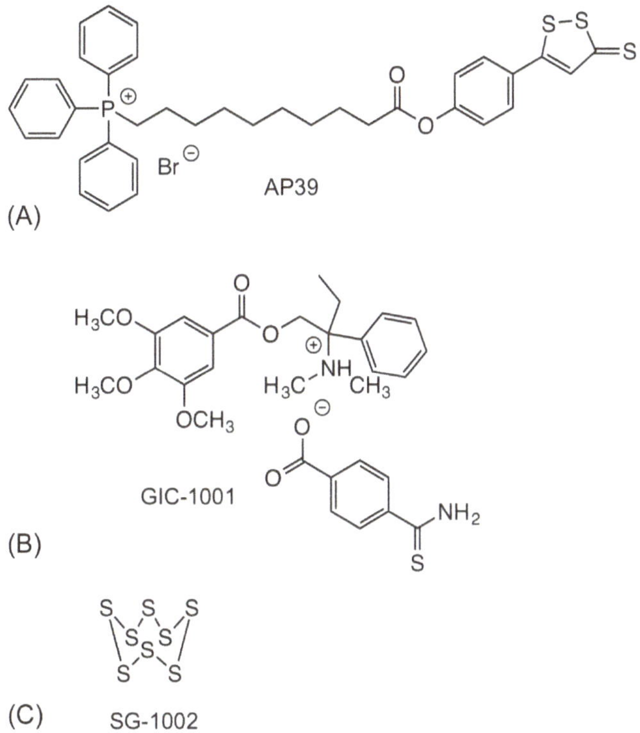

FIG. 6 (A) AP39, a mitochondria-targeting H_2S donor; (B) GIC-1001, Trimebutine salt; (C) SG-1002, polyvalent sulfur.

Table 1). Trimebutine is an opioid antispasmodic that has been used to treat a range of gastrointestinal conditions, including irritable bowel syndrome, for over 40 years. Shown to be safe in a phase-I clinical trial (ClinicalTrials.gov identifier, NCT01738425), this agent is now in phase-II clinical trials as a precolonoscopy analgesic (ClinicalTrials.gov identifier, NCT01926444 and NCT02276768).

Several H_2S-based therapeutics target disorders that are characterized by oxidative stress and associated tissue injury [11]. Based on these observations, SulfaGENIX (in New Orleans, Louisiana, the United States) is developing zerovalent sulfur (SG-1002, Fig. 6C and Table 1) as a medicinal food, with the initial aim of targeting heart failure [55]. Preclinical studies in relevant animal models of heart failure have confirmed that SG-1002 is effective in decreasing infarct size, improving cardiac function, increasing angiogenesis, decreasing inflammation, and downregulating oxidative stress after infarction. A phase-I trial of SG-1002 (ClinicalTrials.gov identifier, NCT02278276) evaluated doses of 200, 400, and 800 mg/day in healthy volunteers and reported dose-dependent increases in plasma H_2S levels, and only minor adverse effects. A second study has been designed (ClinicalTrials.gov identifier, NCT01989208) to evaluate the ability of SG-1002 to elevate plasma H_2S levels (in effect, reversing the defects in circulating H_2S) and to reduce markers of oxidative stress and heart failure.

ATB-429 (Table 1) is an H_2S-releasing derivative of mesalamine and was developed with the notion that the H_2S released from the compound would provide additional antiinflammatory and ulcer-healing effects to those of mesalamine [309], which is used for treatment of Crohn's disease and ulcerative colitis. In one study, mice with hapten-induced colitis were treated orally twice daily with vehicle, mesalamine, or ATB-429 [310]. After 3 days of treatment, there was a significant reduction in the colitis disease activity score and in tissue granulocyte numbers (MPO activity), which was not seen with mesalamine.

CONCLUDING REMARKS AND FUTURE DIRECTIONS

A major challenge for designing novel anticancer drugs is the generation of compounds with improved efficacy, lower side effects, and potential synergism with currently available chemotherapeutic drugs. All of the compounds reviewed in this chapter have the potential of being developed and formulated for chemotherapeutic applications against cancer. NO and H_2S appear to have major roles in modulating cancer therapy resistance, and this in itself holds a great deal of promise in improving patient outcomes; thus, targeted cancer therapy with NO- and H_2S-based compounds is a feasible proposition. A major challenge in terms of drug development is whether to focus on chemoprevention or on chemotherapy. While chemoprevention is ideal, its achievement with drugs is an arduous task, as safety concerns are of the highest paramount importance and testing for this via clinical trials may prove to be an impossibility. However, chemotherapy of an already developed cancer takes a different dimension with much more options. For example, phase-I clinical trials with NO-aspirin (NCX-4016) developed by NicOx (Sophia Antipolis, France) confirmed its improved GI safety compared with aspirin and its efficacy to inhibit platelet aggregation [311–313]. However, its development as a chemopreventive drug in patients who are at high risk of developing colorectal cancer was terminated due to the putative genotoxicity of a potential metabolite [314]. Although the genotoxicity of this metabolite has not been confirmed by others or for that matter for the intact molecule, nevertheless, the development of NCX-4016 as a chemopreventive agent was halted. Had the focus been on chemotreatment as a single agent alone or in combination with established chemotherapeutic agents, the outcome may have been quite different. Another NO-releasing NSAID, NO-naproxen (naproxcinod), which was being evaluated for the treatment of knee and hip osteoarthritis [315–319], has completed phase-III clinical trials. However, concerns by the FDA about its GI and CV safety for long-term use did not lead to the approval of the NDA application in 2010 without further studies [320].

An exciting class of compounds is the NOSH-NSAIDs, which are dual NO- and H2S-donating agents. NO and H_2S share many similar actions but may interact together giving rise to a transient highly reactive species, which may ultimately be responsible for the actions. Organ-specific delivery of H_2S is something that appears to be quite achievable in the near future and, with that, the possibility of disease-specific H_2S donors. The selective inhibitors of the various H_2S-generating enzymes, such as CSE, offer yet another avenue for therapeutic applications and development. No doubt the next decade will be quite exciting as we move forward in translating from the bench to the bedside.

Acknowledgment

Supported in part by NIH Grant R24 DA018055.

Conflict of Interest: No potential conflicts of interest were disclosed.

References

[1] Ignarro LJ, Buga GM, Wood KS, Byrns RE, Chaudhuri G. Endothelium-derived relaxing factor produced and released from artery and vein is nitric oxide. Proc Natl Acad Sci U S A 1987;84:9265–9.

[2] Hattori Y, Kasai K, Gross SS. NO suppresses while peroxynitrite sustains NF-kappaB: a paradigm to rationalize cytoprotective and cytotoxic actions attributed to NO. Cardiovasc Res 2004;63:31–40.

[3] Katsuyama K, Shichiri M, Marumo F, Hirata Y. NO inhibits cytokine-induced iNOS expression and NF-kappaB activation by interfering with phosphorylation and degradation of IkappaB-alpha. Arterioscler Thromb Vasc Biol 1998;18:1796–802.

[4] Hogaboam CM, Befus AD, Wallace JL. Modulation of rat mast cell reactivity by IL-1 beta. Divergent effects on nitric oxide and platelet-activating factor release. J Immunol 1993;151:3767–74.

[5] Huang JQ, Sridhar S, Chen Y, Hunt RH. Meta-analysis of the relationship between helicobacter pylori seropositivity and gastric cancer. Gastroenterology 1998;114:1169–79.

[6] Naseem KM. The role of nitric oxide in cardiovascular diseases. Mol Aspects Med 2005;26:33–65.

[7] Wallace JL, Miller MJ. Nitric oxide in mucosal defense: a little goes a long way. Gastroenterology 2000;119:512–20.

[8] Wallace JL, Viappiani S, Bolla M. Cyclooxygenase-inhibiting nitric oxide donators for osteoarthritis. Trends Pharmacol Sci 2009;30:112–7.

[9] Abe K, Kimura H. The possible role of hydrogen sulfide as an endogenous neuromodulator. J Neurosci 1996;16:1066–71.

[10] Zhao W, Zhang J, Lu Y, Wang R. The vasorelaxant effect of H(2)S as a novel endogenous gaseous K(ATP) channel opener. EMBO J 2001;20:6008–16.

[11] Elrod JW, Calvert JW, Morrison J, Doeller JE, Kraus DW, Tao L, et al. Hydrogen sulfide attenuates myocardial ischemia-reperfusion injury by preservation of mitochondrial function. Proc Natl Acad Sci U S A 2007;104:15560–5.

[12] Yu XH, Cui LB, Wu K, Zheng XL, Cayabyab FS, Chen ZW, et al. Hydrogen sulfide as a potent cardiovascular protective agent. Clin Chimca Acta 2014;437:78–87.

[13] Zanardo RC, Brancaleone V, Distrutti E, Fiorucci S, Cirino G, Wallace JL. Hydrogen sulfide is an endogenous modulator of leukocyte-mediated inflammation. FASEB J 2006;20:2118–20.

[14] Wallace JL, Ferraz JG, Muscara MN. Hydrogen sulfide: an endogenous mediator of resolution of inflammation and injury. Antioxid Redox Signal 2012;17:58–67.

[15] Moncada S, Higgs A. The L-arginine-nitric oxide pathway. N Engl J Med 1993;329:2002–12.

[16] Ignarro LJ. Endothelium-derived nitric oxide: actions and properties. FASEB J 1989;3:31–6.

[17] Stuehr DJ. Structure-function aspects in the nitric oxide synthases. Annu Rev Pharmacol Toxicol 1997;37:339–59.

[18] Dimmeler S, Fleming I, Fisslthaler B, Hermann C, Busse R, Zeiher AM. Activation of nitric oxide synthase in endothelial cells by Akt-dependent phosphorylation. Nature 1999;399:601–5.

[19] Sessa WC. eNOS at a glance. J Cell Sci 2004;117:2427–9.

[20] Geller DA, Billiar TR. Molecular biology of nitric oxide synthases. Cancer Metastasis Rev 1998;17:7–23.

[21] Alderton WK, Cooper CE, Knowles RG. Nitric oxide synthases: structure, function and inhibition. Biochem J 2001;357:593–615.

[22] Nathan C, Xie QW. Nitric oxide synthases: roles, tolls, and controls. Cell 1994;78:915–8.

[23] Kleinert H, Schwarz PM, Forstermann U. Regulation of the expression of inducible nitric oxide synthase. Biol Chem 2003;384:1343–64.

[24] Michel T, Feron O. Nitric oxide synthases: which, where, how, and why? J Clin Invest 1997;100:2146–52.

[25] Kolios G, Valatas V, Ward SG. Nitric oxide in inflammatory bowel disease: a universal messenger in an unsolved puzzle. Immunology 2004;113:427–37.

[26] Goligorsky MS, Brodsky SV, Noiri E. NO bioavailability, endothelial dysfunction, and acute renal failure: new insights into pathophysiology. Semin Nephrol 2004;24:316–23.
[27] Bukovska G, Kery V, Kraus JP. Expression of human cystathionine beta-synthase in Escherichia coli: purification and characterization. Protein Expr Purif 1994;5:442–8.
[28] Erickson PF, Maxwell IH, Su LJ, Baumann M, Glode LM. Sequence of cDNA for rat cystathionine gamma-lyase and comparison of deduced amino acid sequence with related *Escherichia coli* enzymes. Biochem J 1990;269:335–40.
[29] Shibuya N, Tanaka M, Yoshida M, Ogasawara Y, Togawa T, Ishii K, et al. 3-Mercaptopyruvate sulfurtransferase produces hydrogen sulfide and bound sulfane sulfur in the brain. Antioxid Redox Signal 2009;11:703–14.
[30] Szabo C. Gasotransmitters in cancer: from pathophysiology to experimental therapy. Nat Rev Drug Discov 2016;15:185–203.
[31] Paul BD, Snyder SH. H2S: a novel gasotransmitter that signals by sulfhydration. Trends Biochem Sci 2015;40:687–700.
[32] Sen N, Paul BD, Gadalla MM, Mustafa AK, Sen T, Xu R, et al. Hydrogen sulfide-linked sulfhydration of NF-kappaB mediates its antiapoptotic actions. Mol Cell 2012;45:13–24.
[33] Zhu XY, Liu SJ, Liu YJ, Wang S, Ni X. Glucocorticoids suppress cystathionine gamma-lyase expression and H2S production in lipopolysaccharide-treated macrophages. Cell Mol Life Sci 2010;67:1119–32.
[34] Hine C, Harputlugil E, Zhang Y, Ruckenstuhl C, Lee BC, Brace L, et al. Endogenous hydrogen sulfide production is essential for dietary restriction benefits. Cell 2015;160:132–44.
[35] Krishnan N, Fu C, Pappin DJ, Tonks NK. H2S-Induced sulfhydration of the phosphatase PTP1B and its role in the endoplasmic reticulum stress response. Sci Signal 2011;4:ra86.
[36] Westley AM, Westley J. Biological sulfane sulfur. Anal Biochem 1991;195:63–7.
[37] Searcy DG, Lee SH. Sulfur reduction by human erythrocytes. J Exp Zool 1998;282:310–22.
[38] Kolluru GK, Shen X, Kevil CG. A tale of two gases: NO and H2S, foes or friends for life? Redox Biol 2013;1:313–8.
[39] Bucci M, Papapetropoulos A, Vellecco V, Zhou Z, Pyriochou A, Roussos C, et al. Hydrogen sulfide is an endogenous inhibitor of phosphodiesterase activity. Arterioscler Thromb Vasc Biol 2010;30:1998–2004.
[40] Lancaster Jr. JR. Reactivity and diffusivity of nitrogen oxides in mammalian biology. In: Forman HJ, Fukuto J, Torres M, editors. Signal transduction by reactive oxygen and nitrogen species: pathways and chemical principles. Netherlands: Springer; 2003. p. 53–79.
[41] Hill BG, Dranka BP, Bailey SM, Lancaster Jr. JR, Darley-Usmar VM. What part of NO don't you understand? Some answers to the cardinal questions in nitric oxide biology. J Biol Chem 2010;285:19699–704.
[42] Vasudevan D, Thomas DD. Insights into the diverse effects of nitric oxide on tumor biology. Vitam Horm 2014;96:265–98.
[43] Kashfi K, Duvalsaint PL. Nitric oxide donors and therapeutic applications in cancer. In: Seabra AB, editor. Nitric Oxide Donors: Novel Biomedical Applications. Elsevier; 2017. p. 75–120.
[44] Mustafa AK, Gadalla MM, Sen N, Kim S, Mu W, Gazi SK, et al. H2S signals through protein S-sulfhydration. Sci Signal 2009;2:ra72.
[45] Toohey JI. Sulfur signaling: is the agent sulfide or sulfane? Anal Biochem 2011;413:1–7.
[46] Toohey JI. The conversion of H2S to sulfane sulfur. Nat Rev Mol Cell Biol 2012;13:803.
[47] Toohey JI, Cooper AJ. Thiosulfoxide (sulfane) sulfur: new chemistry and new regulatory roles in biology. Molecules 2014;19:12789–813.
[48] Hara MR, Agrawal N, Kim SF, Cascio MB, Fujimuro M, Ozeki Y, et al. S-nitrosylated GAPDH initiates apoptotic cell death by nuclear translocation following Siah1 binding. Nat Cell Biol 2005;7:665–74.
[49] Tang G, Wu L, Liang W, Wang R. Direct stimulation of K(ATP) channels by exogenous and endogenous hydrogen sulfide in vascular smooth muscle cells. Mol Pharmacol 2005;68:1757–64.
[50] Yang W, Yang G, Jia X, Wu L, Wang R. Activation of KATP channels by H2S in rat insulin-secreting cells and the underlying mechanisms. J Physiol 2005;569:519–31.
[51] Fitzgerald R, DeSantiago B, Lee DY, Yang G, Kim JY, Foster DB, et al. H2S relaxes isolated human airway smooth muscle cells via the sarcolemmal K(ATP) channel. Biochem Biophys Res Commun 2014;446:393–8.
[52] Dawe GS, Han SP, Bian JS, Moore PK. Hydrogen sulphide in the hypothalamus causes an ATP-sensitive K+ channel-dependent decrease in blood pressure in freely moving rats. Neuroscience 2008;152:169–77.

[53] Distrutti E, Sediari L, Mencarelli A, Renga B, Orlandi S, Antonelli E, et al. Evidence that hydrogen sulfide exerts antinociceptive effects in the gastrointestinal tract by activating KATP channels. J Pharmacol Exp Ther 2006;316:325–35.
[54] Medeiros JV, Bezerra VH, Lucetti LT, Lima-Junior RC, Barbosa AL, Tavares BM, et al. Role of KATP channels and TRPV1 receptors in hydrogen sulfide-enhanced gastric emptying of liquid in awake mice. Eur J Pharmacol 2012;693:57–63.
[55] Wallace JL, Wang R. Hydrogen sulfide-based therapeutics: exploiting a unique but ubiquitous gasotransmitter. Nat Rev Drug Discov 2015;14:329–45.
[56] Tang G, Zhang L, Yang G, Wu L, Wang R. Hydrogen sulfide-induced inhibition of L-type Ca^{2+} channels and insulin secretion in mouse pancreatic beta cells. Diabetologia 2013;56:533–41.
[57] Zhang R, Sun Y, Tsai H, Tang C, Jin H, Du J. Hydrogen sulfide inhibits L-type calcium currents depending upon the protein sulfhydryl state in rat cardiomyocytes. PLoS One 2012;7:e37073.
[58] Sekiguchi F, Miyamoto Y, Kanaoka D, Ide H, Yoshida S, Ohkubo T, et al. Endogenous and exogenous hydrogen sulfide facilitates T-type calcium channel currents in Cav3.2-expressing HEK293 cells. Biochem Biophys Res Commun 2014;445:225–9.
[59] Maeda Y, Aoki Y, Sekiguchi F, Matsunami M, Takahashi T, Nishikawa H, et al. Hyperalgesia induced by spinal and peripheral hydrogen sulfide: evidence for involvement of Cav3.2T-type calcium channels. Pain 2009;142:127–32.
[60] Coletta C, Papapetropoulos A, Erdelyi K, Olah G, Modis K, Panopoulos P, et al. Hydrogen sulfide and nitric oxide are mutually dependent in the regulation of angiogenesis and endothelium-dependent vasorelaxation. Proc Natl Acad Sci U S A 2012;109:9161–6.
[61] Wang R. Shared signaling pathways among gasotransmitters. Proc Natl Acad Sci U S A 2012;109:8801–2.
[62] Moccia F, Bertoni G, Pla AF, Dragoni S, Pupo E, Merlino A, et al. Hydrogen sulfide regulates intracellular Ca^{2+} concentration in endothelial cells from excised rat aorta. Curr Pharm Biotechnol 2011;12:1416–26.
[63] Schottenfeld D, Beebe-Dimmer J. Chronic inflammation: a common and important factor in the pathogenesis of neoplasia. CA Cancer J Clin 2006;56:69–83.
[64] Mantovani A, Allavena P, Sica A, Balkwill F. Cancer-related inflammation. Nature 2008;454:436–44.
[65] Kashfi K. Anti-inflammatory agents as cancer therapeutics. Adv Pharmacol 2009;57:31–89.
[66] Thun MJ, Namboodiri MM, Heath Jr. C. Aspirin use and reduced risk of fatal colon cancer. N Engl J Med 1991;325:1593–6.
[67] Benamouzig R, Deyra J, Martin A, Girard B, Jullian E, Piednoir B, et al. Daily soluble aspirin and prevention of colorectal adenoma recurrence: one-year results of the APACC trial. Gastroenterology 2003;125:328–36.
[68] Rothwell PM, Price JF, Fowkes FG, Zanchetti A, Roncaglioni MC, Tognoni G, et al. Short-term effects of daily aspirin on cancer incidence, mortality, and non-vascular death: analysis of the time course of risks and benefits in 51 randomised controlled trials. Lancet 2012;379:1602–12.
[69] Takkouche B, Regueira-Mendez C, Etminan M. Breast cancer and use of nonsteroidal anti-inflammatory drugs: a meta-analysis. J Natl Cancer Inst 2008;100:1439–47.
[70] Yiannakopoulou E. Aspirin and NSAIDs for breast cancer chemoprevention. Eur J Cancer Prev 2015;24:416–21.
[71] de Pedro M, Baeza S, Escudero MT, Dierssen-Sotos T, Gomez-Acebo I, Pollan M, et al. Effect of COX-2 inhibitors and other non-steroidal inflammatory drugs on breast cancer risk: a meta-analysis. Breast Cancer Res Treat 2015;149:525–36.
[72] Cui XJ, He Q, Zhang JM, Fan HJ, Wen ZF, Qin YR. High-dose aspirin consumption contributes to decreased risk for pancreatic cancer in a systematic review and meta-analysis. Pancreas 2014;43:135–40.
[73] Shebl FM, Sakoda LC, Black A, Koshiol J, Andriole GL, Grubb R, et al. Aspirin but not ibuprofen use is associated with reduced risk of prostate cancer: a PLCO study. Br J Cancer 2012;107:207–14.
[74] Sun L, Yu S. Meta-analysis: non-steroidal anti-inflammatory drug use and the risk of esophageal squamous cell carcinoma. Dis Esophagus 2011;24:544–9.
[75] Liao LM, Vaughan TL, Corley DA, Cook MB, Casson AG, Kamangar F, et al. Nonsteroidal anti-inflammatory drug use reduces risk of adenocarcinomas of the esophagus and esophagogastric junction in a pooled analysis. Gastroenterology 2012;142:442–52.
[76] Becker C, Wilson JC, Jick SS, Meier CR. Non-steroidal anti-inflammatory drugs and the risk of head and neck cancer: a case-control analysis. Int J Cancer 2015;137:2424–31.

[77] Baandrup L, Kjaer SK, Olsen JH, Dehlendorff C, Friis S. Low-dose aspirin use and the risk of ovarian cancer in Denmark. Ann Oncol 2015;26:787–92.
[78] Trabert B, Ness RB, Lo-Ciganic WH, Murphy MA, Goode EL, Poole EM, et al. Aspirin, nonaspirin nonsteroidal anti-inflammatory drug, and acetaminophen use and risk of invasive epithelial ovarian cancer: a pooled analysis in the Ovarian Cancer Association Consortium. J Natl Cancer Inst 2014;106:431.
[79] Clouser MC, Roe DJ, Foote JA, Harris RB. Effect of non-steroidal anti-inflammatory drugs on non-melanoma skin cancer incidence in the SKICAP-AK trial. Pharmacoepidemiol Drug Saf 2009;18:276–83.
[80] Elmets CA, Viner JL, Pentland AP, Cantrell W, Lin HY, Bailey H, et al. Chemoprevention of nonmelanoma skin cancer with celecoxib: a randomized, double-blind, placebo-controlled trial. J Natl Cancer Inst 2010;102:1835–44.
[81] Daugherty SE, Pfeiffer RM, Sigurdson AJ, Hayes RB, Leitzmann M, Schatzkin A, et al. Nonsteroidal antiinflammatory drugs and bladder cancer: a pooled analysis. Am J Epidemiol 2011;173:721–30.
[82] Nicastro HL, Grubbs CJ, Margaret Juliana M, Bode AM, Kim MS, Lu Y, et al. Preventive effects of NSAIDs, NO-NSAIDs, and NSAIDs plus difluoromethylornithine in a chemically induced urinary bladder cancer model. Cancer Prev Res (Phila) 2014;7:246–54.
[83] Thun MJ, Henley SJ, Patrono C. Nonsteroidal anti-inflammatory drugs as anticancer agents: mechanistic, pharmacologic, and clinical issues. J Natl Cancer Inst 2002;94:252–66.
[84] Cui S, Reichner JS, Mateo RB, Albina JE. Activated murine macrophages induce apoptosis in tumor cells through nitric oxide-dependent or -independent mechanisms. Cancer Res 1994;54:2462–7.
[85] Brune B, von Knethen A, Sandau KB. Nitric oxide (NO): an effector of apoptosis. Cell Death Differ 1999;6:969–75.
[86] Thomsen LL, Miles DW, Happerfield L, Bobrow LG, Knowles RG, Moncada S. Nitric oxide synthase activity in human breast cancer. Br J Cancer 1995;72:41–4.
[87] Glynn SA, Boersma BJ, Dorsey TH, Yi M, Yfantis HG, Ridnour LA, et al. Increased NOS2 predicts poor survival in estrogen receptor-negative breast cancer patients. J Clin Invest 2010;120:3843–54.
[88] Basudhar D, Somasundaram V, de Oliveira GA, Kesarwala A, Heinecke JL, Cheng RY, et al. NOS2-derived NO drives multiple pathways of breast cancer progression. Antioxid Redox Signal 2016. http://dx.doi.org/10.1089/ars.2016.6813 [ahead of print].
[89] Klotz T, Bloch W, Volberg C, Engelmann U, Addicks K. Selective expression of inducible nitric oxide synthase in human prostate carcinoma. Cancer 1998;82:1897–903.
[90] Masri FA, Comhair SA, Koeck T, Xu W, Janocha A, Ghosh S, et al. Abnormalities in nitric oxide and its derivatives in lung cancer. Am J Respir Crit Care Med 2005;172:597–605.
[91] Gallo O, Masini E, Morbidelli L, Franchi A, Fini-Storchi I, Vergari WA, et al. Role of nitric oxide in angiogenesis and tumor progression in head and neck cancer. J Natl Cancer Inst 1998;90:587–96.
[92] Ambs S, Merriam WG, Bennett WP, Felley-Bosco E, Ogunfusika MO, Oser SM, et al. Frequent nitric oxide synthase-2 expression in human colon adenomas: implication for tumor angiogenesis and colon cancer progression. Cancer Res 1998;58:334–41.
[93] Lagares-Garcia JA, Moore RA, Collier B, Heggere M, Diaz F, Qian F. Nitric oxide synthase as a marker in colorectal carcinoma. Am Surg 2001;67:709–13.
[94] Lechner M, Lirk P, Rieder J. Inducible nitric oxide synthase (iNOS) in tumor biology: the two sides of the same coin. Semin Cancer Biol 2005;15:277–89.
[95] Oronsky B, Fanger GR, Oronsky N, Knox S, Scicinski J. The implications of hyponitroxia in cancer. Transl Oncol 2014;7:167–73.
[96] Fulton D, Gratton JP, Sessa WC. Post-translational control of endothelial nitric oxide synthase: why isn't calcium/calmodulin enough? J Pharmacol Exp Ther 2001;299:818–24.
[97] Ambs S, Hussain SP, Harris CC. Interactive effects of nitric oxide and the p53 tumor suppressor gene in carcinogenesis and tumor progression. FASEB J 1997;11:443–8.
[98] Ridnour LA, Thomas DD, Switzer C, Flores-Santana W, Isenberg JS, Ambs S, et al. Molecular mechanisms for discrete nitric oxide levels in cancer. Nitric oxide Biol Chem 2008;19:73–6.
[99] Li LM, Kilbourn RG, Adams J, Fidler IJ. Role of nitric oxide in lysis of tumor cells by cytokine-activated endothelial cells. Cancer Res 1991;51:2531–5.
[100] Jiang H, Stewart CA, Fast DJ, Leu RW. Tumor target-derived soluble factor synergizes with IFN-gamma and IL-2 to activate macrophages for tumor necrosis factor and nitric oxide production to mediate cytotoxicity of the same target. J Immunol 1992;149:2137–46.

[101] Xiao L, Eneroth PH, Qureshi GA. Nitric oxide synthase pathway may mediate human natural killer cell cytotoxicity. Scand J Immunol 1995;42:505–11.
[102] Fukumura D, Yonei Y, Kurose I, Saito H, Ohishi T, Higuchi H, et al. Role in nitric oxide in Kupffer cell-mediated hepatoma cell cytotoxicity in vitro and ex vivo. Hepatology (Baltimore, MD) 1996;24:141–9.
[103] Wink DA, Vodovotz Y, Laval J, Laval F, Dewhirst MW, Mitchell JB. The multifaceted roles of nitric oxide in cancer. Carcinogenesis 1998;19:711–21.
[104] Ridnour LA, Thomas DD, Donzelli S, Espey MG, Roberts DD, Wink DA, et al. The biphasic nature of nitric oxide responses in tumor biology. Antioxid Redox Signal 2006;8:1329–37.
[105] Xu W, Liu LZ, Loizidou M, Ahmed M, Charles IG. The role of nitric oxide in cancer. Cell Res 2002;12:311–20.
[106] Miller MR, Megson IL. Recent developments in nitric oxide donor drugs. Br J Pharmacol 2007;151:305–21.
[107] Choudhari SK, Chaudhary M, Bagde S, Gadbail AR, Joshi V. Nitric oxide and cancer: a review. World J Surg Oncol 2013;11:118.
[108] Vannini F, Kashfi K, Nath N. The dual role of iNOS in cancer. Redox Biol 2015;6:334–43.
[109] Bhatia M, Sidhapuriwala J, Moochhala SM, Moore PK. Hydrogen sulphide is a mediator of carrageenan-induced hindpaw oedema in the rat. Br J Pharmacol 2005;145:141–4.
[110] Li L, Bhatia M, Zhu YZ, Zhu YC, Ramnath RD, Wang ZJ, et al. Hydrogen sulfide is a novel mediator of lipopolysaccharide-induced inflammation in the mouse. FASEB J 2005;19:1196–8.
[111] Anuar F, Whiteman M, Siau JL, Kwong SE, Bhatia M, Moore PK. Nitric oxide-releasing flurbiprofen reduces formation of proinflammatory hydrogen sulfide in lipopolysaccharide-treated rat. Br J Pharmacol 2006;147:966–74.
[112] Anuar F, Whiteman M, Bhatia M, Moore PK. Flurbiprofen and its nitric oxide-releasing derivative protect against septic shock in rats. Inflamm Res 2006;55:498–503.
[113] Bhatia M, Wong FL, Fu D, Lau HY, Moochhala SM, Moore PK. Role of hydrogen sulfide in acute pancreatitis and associated lung injury. FASEB J 2005;19:623–5.
[114] Tamizhselvi R, Moore PK, Bhatia M. Hydrogen sulfide acts as a mediator of inflammation in acute pancreatitis: in vitro studies using isolated mouse pancreatic acinar cells. J Cell Mol Med 2007;11:315–26.
[115] Whiteman M, Armstrong JS, Chu SH, Jia-Ling S, Wong BS, Cheung NS, et al. The novel neuromodulator hydrogen sulfide: an endogenous peroxynitrite "scavenger"? J Neurochem 2004;90:765–8.
[116] Whiteman M, Cheung NS, Zhu YZ, Chu SH, Siau JL, Wong BS, et al. Hydrogen sulphide: a novel inhibitor of hypochlorous acid-mediated oxidative damage in the brain? Biochem Biophys Res Commun 2005;326:794–8.
[117] Mariggio MA, Minunno V, Riccardi S, Santacroce R, De Rinaldis P, Fumarulo R. Sulfide enhancement of PMN apoptosis. Immunopharmacol Immunotoxicol 1998;20:399–408.
[118] Gilroy DW, Lawrence T, Perretti M, Rossi AG. Inflammatory resolution: new opportunities for drug discovery. Nat Rev Drug Discov 2004;3:401–16.
[119] Hu LF, Wong PT, Moore PK, Bian JS. Hydrogen sulfide attenuates lipopolysaccharide-induced inflammation by inhibition of p38 mitogen-activated protein kinase in microglia. J Neurochem 2007;100:1121–8.
[120] Hu LF, Lu M, Wu ZY, Wong PT, Bian JS. Hydrogen sulfide inhibits rotenone-induced apoptosis via preservation of mitochondrial function. Mol Pharmacol 2009;75:27–34.
[121] Li L, Bhatia M, Moore PK. Hydrogen sulphide—a novel mediator of inflammation? Curr Opin Pharmacol 2006;6:125–9.
[122] Wallace JL. Hydrogen sulfide-releasing anti-inflammatory drugs. Trends Pharmacol Sci 2007;28:501–5.
[123] Aranda E, Lopez-Pedrera C, De La Haba-Rodriguez JR, Rodriguez-Ariza A. Nitric oxide and cancer: the emerging role of S-nitrosylation. Curr Mol Med 2012;12:50–67.
[124] Wink DA, Ridnour LA, Hussain SP, Harris CC. The reemergence of nitric oxide and cancer. Nitric oxide Biol Chem 2008;19:65–7.
[125] Ning S, Bednarski M, Oronsky B, Scicinski J, Saul G, Knox SJ. Dinitroazetidines are a novel class of anticancer agents and hypoxia-activated radiation sensitizers developed from highly energetic materials. Cancer Res 2012;72:2600–8.
[126] Oronsky B, Scicinski J, Ning S, Peehl D, Oronsky A, Cabrales P, et al. RRx-001, a novel dinitroazetidine radiosensitizer. Invest New Drugs 2016;34:371–7.
[127] Bian K, Ghassemi F, Sotolongo A, Siu A, Shauger L, Kots A, et al. NOS-2 signaling and cancer therapy. IUBMB Life 2012;64:676–83.

[128] Burke AJ, Sullivan FJ, Giles FJ, Glynn SA. The yin and yang of nitric oxide in cancer progression. Carcinogenesis 2013;34:503–12.
[129] Wallace JL, Ignarro LJ, Fiorucci S. Potential cardioprotective actions of no-releasing aspirin. Nat Rev Drug Discov 2002;1:375–82.
[130] Wallace JL, Soldato PD. The therapeutic potential of NO-NSAIDs. Fundam Clin Pharmacol 2003;17:11–20.
[131] Kashfi K, Ryan Y, Qiao LL, Williams JL, Chen J, Del Soldato P, et al. Nitric oxide-donating nonsteroidal anti-inflammatory drugs inhibit the growth of various cultured human cancer cells: evidence of a tissue type-independent effect. J Pharmacol Exp Ther 2002;303:1273–82.
[132] Kashfi K, Borgo S, Williams JL, Chen J, Gao J, Glekas A, et al. Positional isomerism markedly affects the growth inhibition of colon cancer cells by nitric oxide-donating aspirin in vitro and in vivo. J Pharmacol Exp Ther 2005;312:978–88.
[133] Nath N, Liu X, Jacobs L, Kashfi K. Flurbiprofen benzyl nitrate (NBS-242) inhibits the growth of A-431 human epidermoid carcinoma cells and targets beta-catenin. Drug Des Devel Ther 2013;7:389–96.
[134] Kashfi K, Rigas B. Non-COX-2 targets and cancer: expanding the molecular target repertoire of chemoprevention. Biochem Pharmacol 2005;70:969–86.
[135] Dunlap T, Chandrasena RE, Wang Z, Sinha V, Wang Z, Thatcher GR. Quinone formation as a chemoprevention strategy for hybrid drugs: balancing cytotoxicity and cytoprotection. Chem Res Toxicol 2007;20:1903–12.
[136] Hulsman N, Medema JP, Bos C, Jongejan A, Leurs R, Smit MJ, et al. Chemical insights in the concept of hybrid drugs: the antitumor effect of nitric oxide-donating aspirin involves a quinone methide but not nitric oxide nor aspirin. J Med Chem 2007;50:2424–31.
[137] Kashfi K, Rigas B. The mechanism of action of nitric oxide-donating aspirin. Biochem Biophys Res Commun 2007;358:1096–101.
[138] Bak AW, McKnight W, Li P, Del Soldato P, Calignano A, Cirino G, et al. Cyclooxygenase-independent chemoprevention with an aspirin derivative in a rat model of colonic adenocarcinoma. Life Sci 1998;62:367–73.
[139] Williams JL, Kashfi K, Ouyang N, del Soldato P, Kopelovich L, Rigas B. NO-donating aspirin inhibits intestinal carcinogenesis in Min (APC(Min/+)) mice. Biochem Biophys Res Commun 2004;313:784–8.
[140] Rao CV, Reddy BS, Steele VE, Wang CX, Liu X, Ouyang N, et al. Nitric oxide-releasing aspirin and indomethacin are potent inhibitors against colon cancer in azoxymethane-treated rats: effects on molecular targets. Mol Cancer Ther 2006;5:1530–8.
[141] Leonetti C, Scarsella M, Zupi G, Zoli W, Amadori D, Medri L, et al. Efficacy of a nitric oxide-releasing nonsteroidal anti-inflammatory drug and cytotoxic drugs in human colon cancer cell lines in vitro and xenografts. Mol Cancer Ther 2006;5:919–26.
[142] Ouyang N, Williams JL, Tsioulias GJ, Gao J, Iatropoulos MJ, Kopelovich L, et al. Nitric oxide-donating aspirin prevents pancreatic cancer in a hamster tumor model. Cancer Res 2006;66:4503–11.
[143] Cena C, Lolli ML, Lazzarato L, Guaita E, Morini G, Coruzzi G, et al. Antiinflammatory, gastrosparing, and antiplatelet properties of new NO-donor esters of aspirin. J Med Chem 2003;46:747–54.
[144] Grosser N, Schroder H. A common pathway for nitric oxide release from NO-aspirin and glyceryl trinitrate. Biochem Biophys Res Commun 2000;274:255–8.
[145] Carini M, Aldini G, Orioli M, Maffei Facino R. In vitro metabolism of a nitroderivative of acetylsalicylic acid (NCX4016) by rat liver: LC and LC-MS studies. J Pharm Biomed Anal 2002;29:1061–71.
[146] Gao J, Kashfi K, Rigas B. In vitro metabolism of nitric oxide-donating aspirin: the effect of positional isomerism. J Pharmacol Exp Ther 2005;312:989–97.
[147] Turnbull CM, Cena C, Fruttero R, Gasco A, Rossi AG, Megson IL. Mechanism of action of novel NO-releasing furoxan derivatives of aspirin in human platelets. Br J Pharmacol 2006;148:517–26.
[148] Williams JL, Borgo S, Hasan I, Castillo E, Traganos F, Rigas B. Nitric oxide-releasing nonsteroidal anti-inflammatory drugs (NSAIDs) alter the kinetics of human colon cancer cell lines more effectively than traditional NSAIDs: implications for colon cancer chemoprevention. Cancer Res 2001;61:3285–9.
[149] Nath N, Vassell R, Chattopadhyay M, Kogan M, Kashfi K. Nitro-aspirin inhibits MCF-7 breast cancer cell growth: effects on COX-2 expression and Wnt/beta-catenin/TCF-4 signaling. Biochem Pharmacol 2009;78:1298–304.
[150] Williams JL, Nath N, Chen J, Hundley TR, Gao J, Kopelovich L, et al. Growth inhibition of human colon cancer cells by nitric oxide (NO)-donating aspirin is associated with cyclooxygenase-2 induction and beta-catenin/

T-cell factor signaling, nuclear factor-kappaB, and NO synthase 2 inhibition: implications for chemoprevention. Cancer Res 2003;63:7613–8.
[151] Williams JL, Ji P, Ouyang N, Liu X, Rigas B. NO-donating aspirin inhibits the activation of NF-kappaB in human cancer cell lines and Min mice. Carcinogenesis 2008;29:390–7.
[152] Pierce EN, Piyankarage SC, Dunlap T, Litosh V, Siklos MI, Wang YT, et al. Prodrugs bioactivated to quinones target NF-kappaB and multiple protein networks: identification of the quinonome. Chem Res Toxicol 2016;29:1151–9.
[153] Gao J, Liu X, Rigas B. Nitric oxide-donating aspirin induces apoptosis in human colon cancer cells through induction of oxidative stress. Proc Natl Acad Sci U S A 2005;102:17207–12.
[154] Nath N, Kashfi K, Chen J, Rigas B. Nitric oxide-donating aspirin inhibits beta-catenin/T cell factor (TCF) signaling in SW480 colon cancer cells by disrupting the nuclear beta-catenin-TCF association. Proc Natl Acad Sci U S A 2003;100:12584–9.
[155] Nath N, Labaze G, Rigas B, Kashfi K. NO-donating aspirin inhibits the growth of leukemic Jurkat cells and modulates beta-catenin expression. Biochem Biophys Res Commun 2005;326:93–9.
[156] Moon RT, Kohn AD, De Ferrari GV, Kaykas A. WNT and beta-catenin signalling: diseases and therapies. Nat Rev Genet 2004;5:691–701.
[157] Cobbs CS, Brenman JE, Aldape KD, Bredt DS, Israel MA. Expression of nitric oxide synthase in human central nervous system tumors. Cancer Res 1995;55:727–30.
[158] Kasper HU, Wolf H, Drebber U, Wolf HK, Kern MA. Expression of inducible nitric oxide synthase and cyclooxygenase-2 in pancreatic adenocarcinoma: correlation with microvessel density. World J Gastroenterol 2004;10:1918–22.
[159] Hara A, Okayasu I. Cyclooxygenase-2 and inducible nitric oxide synthase expression in human astrocytic gliomas: correlation with angiogenesis and prognostic significance. Acta Neuropathol 2004;108:43–8.
[160] Brandao MM, Soares E, Salles TS, Saad ST. Expression of inducible nitric oxide synthase is increased in acute myeloid leukaemia. Acta Haematol 2001;106:95–9.
[161] Ouyang N, Williams JL, Rigas B. NO-donating aspirin isomers downregulate peroxisome proliferator-activated receptor (PPAR){delta} expression in $APC^{min/+}$ mice proportionally to their tumor inhibitory effect: Implications for the role of PPAR{delta} in carcinogenesis. Carcinogenesis 2006;27:232–9.
[162] Hundley TR, Rigas B. Nitric oxide-donating aspirin inhibits colon cancer cell growth via mitogen-activated protein kinase activation. J Pharmacol Exp Ther 2006;316:25–34.
[163] Chaudhary SC, Singh T, Kapur P, Weng Z, Arumugam A, Elmets CA, et al. Nitric oxide-releasing sulindac is a novel skin cancer chemopreventive agent for UVB-induced photocarcinogenesis. Toxicol Appl Pharmacol 2013;268:249–55.
[164] Gao J, Kashfi K, Liu X, Rigas B. NO-donating aspirin induces phase II enzymes in vitro and in vivo. Carcinogenesis 2006;27:803–10.
[165] Rigas B, Sun Y. Induction of oxidative stress as a mechanism of action of chemopreventive agents against cancer. Br J Cancer 2008;98:1157–60.
[166] Fiorucci S, Santucci L, Antonelli E, Distrutti E, Del Sero G, Morelli O, et al. NO-aspirin protects from T cell-mediated liver injury by inhibiting caspase-dependent processing of Th1-like cytokines. Gastroenterology 2000;118:404–21.
[167] Fiorucci S, Santucci L, Cirino G, Mencarelli A, Familiari L, Soldato PD, et al. IL-1 beta converting enzyme is a target for nitric oxide-releasing aspirin: new insights in the antiinflammatory mechanism of nitric oxide-releasing nonsteroidal antiinflammatory drugs. J Immunol 2000;165:5245–54.
[168] Marshall HE, Merchant K, Stamler JS. Nitrosation and oxidation in the regulation of gene expression. FASEB J 2000;14:1889–900.
[169] Chattopadhyay M, Goswami S, Rodes DB, Kodela R, Velazquez CA, Boring D, et al. NO-releasing NSAIDs suppress NF-kappaB signaling in vitro and in vivo through S-nitrosylation. Cancer Lett 2010;298:204–11.
[170] Kashfi K. Nitric oxide-releasing hybrid drugs target cellular processes through S-nitrosylation. Forum Immunopathol Dis Therap 2012;3:97–108.
[171] Solomon DH. Recommendations for use of selective and nonselective nonsteroidal antiinflammatory drugs: an American College of Rheumatology white paper. Arthritis Rheum 2008;59:1058–73.
[172] Funk CD, FitzGerald GA. COX-2 inhibitors and cardiovascular risk. J Cardiovasc Pharmacol 2007;50:470–9.

[173] Boschi D, Cena C, Di Stilo A, Rolando B, Manzini P, Fruttero R, et al. Nitrooxymethyl-substituted analogues of rofecoxib: synthesis and pharmacological characterization. Chem Biodivers 2010;7:1173–82.
[174] Dhawan V, Schwalb DJ, Shumway MJ, Warren MC, Wexler RS, Zemtseva IS, et al. Selective nitros(yl)ation induced in vivo by a nitric oxide-donating cyclooxygenase-2 inhibitor: a NObonomic analysis. Free Radic Biol Med 2005;39:1191–207.
[175] Chowdhury MA, Abdellatif KR, Dong Y, Yu G, Huang Z, Rahman M, et al. Celecoxib analogs possessing a N-(4-nitrooxybutyl)piperidin-4-yl or N-(4-nitrooxybutyl)-1,2,3,6-tetrahydropyridin-4-yl nitric oxide donor moiety: synthesis, biological evaluation and nitric oxide release studies. Bioorg Med Chem Lett 2010;20:1324–9.
[176] Martelli A, Testai L, Anzini M, Cappelli A, Di Capua A, Biava M, et al. The novel anti-inflammatory agent VA694, endowed with both NO-releasing and COX2-selective inhibiting properties, exhibits NO-mediated positive effects on blood pressure, coronary flow and endothelium in an experimental model of hypertension and endothelial dysfunction. Pharmacol Res 2013;78:1–9.
[177] Bechmann N, Kniess T, Kockerling M, Pigorsch A, Steinbach J, Pietzsch J. Novel (pyrazolyl)benzenesulfonamides with a nitric oxide-releasing moiety as selective cyclooxygenase-2 inhibitors. Bioorg Med Chem Lett 2015;25:3295–300.
[178] Abdellatif KR, Chowdhury MA, Velazquez CA, Huang Z, Dong Y, Das D, et al. Celecoxib prodrugs possessing a diazen-1-ium-1,2-diolate nitric oxide donor moiety: synthesis, biological evaluation and nitric oxide release studies. Bioorg Med Chem Lett 2010;20:4544–9.
[179] Basudhar D, Cheng RC, Bharadwaj G, Ridnour LA, Wink DA, Miranda KM. Chemotherapeutic potential of diazeniumdiolate-based aspirin prodrugs in breast cancer. Free Radic Biol Med 2015;83:101–14.
[180] Abdellatif KR, Abdelgawad MA, Labib MB, Zidan TH. Synthesis, cyclooxygenase inhibition, anti-inflammatory evaluation and ulcerogenic liability of novel triarylpyrazoline derivatives as selective COX-2 inhibitors. Bioorg Med Chem Lett 2015;25:5787–91.
[181] Abdellatif KR, Elsaady MT, Abdel-Aziz SA, Abusabaa AH. Synthesis, cyclooxygenase inhibition and anti-inflammatory evaluation of new 1,3,5-triaryl-4,5-dihydro-1H-pyrazole derivatives possessing methanesulphonyl pharmacophore. J Enzyme Inhib Med Chem 2016;31(6):1545–55. http://dx.doi.org/10.3109/14756366.2016.1158168.
[182] Abdellatif KR, Fadaly WA, Ali WA, Kamel GM. Synthesis, cyclooxygenase inhibition, anti-inflammatory evaluation and ulcerogenic liability of new 1,5-diarylpyrazole derivatives. J Enzyme Inhib Med Chem 2016;31(Suppl. 3):54–60. http://dx.doi.org/10.1080/14756366.2016.1201815.
[183] Bakr RB, Azouz AA, Abdellatif KR. Synthesis, cyclooxygenase inhibition, anti-inflammatory evaluation and ulcerogenic liability of new 1-phenylpyrazolo[3,4-d]pyrimidine derivatives. J Enzyme Inhib Med Chem 2016;31(Suppl. 2):6–12. http://dx.doi.org/10.1080/14756366.2016.1186018.
[184] Biava M, Battilocchio C, Poce G, Alfonso S, Consalvi S, Di Capua A, et al. Enhancing the pharmacodynamic profile of a class of selective COX-2 inhibiting nitric oxide donors. Bioorg Med Chem 2014;22:772–86.
[185] Gouda AM, Abdelazeem AH, Arafa el SA, Abdellatif KR. Design, synthesis and pharmacological evaluation of novel pyrrolizine derivatives as potential anticancer agents. Bioorg Chem 2014;53:1–7.
[186] Velazquez CA, Praveen Rao PN, Citro ML, Keefer LK, Knaus EE. O2-acetoxymethyl-protected diazeniumdiolate-based NSAIDs (NONO-NSAIDs): synthesis, nitric oxide release, and biological evaluation studies. Bioorg Med Chem 2007;15:4767–74.
[187] Velazquez C, Praveen Rao PN, Knaus EE. Novel nonsteroidal antiinflammatory drugs possessing a nitric oxide donor diazen-1-ium-1,2-diolate moiety: design, synthesis, biological evaluation, and nitric oxide release studies. J Med Chem 2005;48:4061–7.
[188] Velazquez CA, Chen QH, Citro ML, Keefer LK, Knaus EE. Second-generation aspirin and indomethacin prodrugs possessing an O(2)-(acetoxymethyl)-1-(2-carboxypyrrolidin-1-yl)diazenium-1,2-diolate nitric oxide donor moiety: design, synthesis, biological evaluation, and nitric oxide release studies. J Med Chem 2008;51:1954–61.
[189] Cheng H, Mollica MY, Lee SH, Wang L, Velazquez-Martinez CA, Wu S. Effects of nitric oxide-releasing nonsteroidal anti-inflammatory drugs (NONO-NSAIDs) on melanoma cell adhesion. Toxicol Appl Pharmacol 2012;264:161–6.
[190] Keefer LK. Fifty years of diazeniumdiolate research. From laboratory curiosity to broad-spectrum biomedical advances. ACS Chem Biol 2011;6:1147–55.
[191] Nandurdikar RS, Maciag AE, Cao Z, Keefer LK, Saavedra JE. Diazeniumdiolated carbamates: a novel class of nitric oxide donors. Bioorg Med Chem 2012;20:2025–9.

[192] Paolocci N, Katori T, Champion HC, St John ME, Miranda KM, Fukuto JM, et al. Positive inotropic and lusitropic effects of HNO/NO- in failing hearts: independence from beta-adrenergic signaling. Proc Natl Acad Sci U S A 2003;100:5537–42.

[193] Paolocci N, Saavedra WF, Miranda KM, Martignani C, Isoda T, Hare JM, et al. Nitroxyl anion exerts redox-sensitive positive cardiac inotropy in vivo by calcitonin gene-related peptide signaling. Proc Natl Acad Sci U S A 2001;98:10463–8.

[194] Nagasawa HT, Kawle SP, Elberling JA, DeMaster EG, Fukuto JM. Prodrugs of nitroxyl as potential aldehyde dehydrogenase inhibitors vis-a-vis vascular smooth muscle relaxants. J Med Chem 1995;38:1865–71.

[195] Norris AJ, Sartippour MR, Lu M, Park T, Rao JY, Jackson MI, et al. Nitroxyl inhibits breast tumor growth and angiogenesis. Int J Cancer 2008;122:1905–10.

[196] Basudhar D, Bharadwaj G, Cheng RY, Jain S, Shi S, Heinecke JL, et al. Synthesis and chemical and biological comparison of nitroxyl- and nitric oxide-releasing diazeniumdiolate-based aspirin derivatives. J Med Chem 2013;56:7804–20.

[197] Keefer LK. Broad-spectrum anti-cancer activity of O-arylated diazeniumdiolates. Forum Immunopathol Dis Therap 2010;1:205–18.

[198] Laschak M, Spindler KD, Schrader AJ, Hessenauer A, Streicher W, Schrader M, et al. JS-K, a glutathione/glutathione S-transferase-activated nitric oxide releasing prodrug inhibits androgen receptor and WNT-signaling in prostate cancer cells. BMC Cancer 2012;12:130.

[199] Bico P, Chen CY, Jones M, Erhardt J, Dirr H. Class pi glutathione S-transferase: Meisenheimer complex formation. Biochem Mol Biol Int 1994;33:887–92.

[200] Shami PJ, Saavedra JE, Wang LY, Bonifant CL, Diwan BA, Singh SV, et al. JS-K, a glutathione/glutathione S-transferase-activated nitric oxide donor of the diazeniumdiolate class with potent antineoplastic activity. Mol Cancer Ther 2003;2:409–17.

[201] Shami PJ, Maciag AE, Eddington JK, Udupi V, Kosak KM, Saavedra JE, et al. JS-K, an arylating nitric oxide (NO) donor, has synergistic anti-leukemic activity with cytarabine (ARA-C). Leuk Res 2009;33:1525–9.

[202] Findlay VJ, Townsend DM, Saavedra JE, Buzard GS, Citro ML, Keefer LK, et al. Tumor cell responses to a novel glutathione S-transferase-activated nitric oxide-releasing prodrug. Mol Pharmacol 2004;65:1070–9.

[203] Saavedra JE, Srinivasan A, Buzard GS, Davies KM, Waterhouse DJ, Inami K, et al. PABA/NO as an anticancer lead: analogue synthesis, structure revision, solution chemistry, reactivity toward glutathione, and in vitro activity. J Med Chem 2006;49:1157–64.

[204] Kogias E, Osterberg N, Baumer B, Psarras N, Koentges C, Papazoglou A, et al. Growth-inhibitory and chemosensitizing effects of the glutathione-S-transferase-pi-activated nitric oxide donor PABA/NO in malignant gliomas. Int J Cancer 2012;130:1184–94.

[205] Kiziltepe T, Hideshima T, Ishitsuka K, Ocio EM, Raje N, Catley L, et al. JS-K, a GST-activated nitric oxide generator, induces DNA double-strand breaks, activates DNA damage response pathways, and induces apoptosis in vitro and in vivo in human multiple myeloma cells. Blood 2007;110:709–18.

[206] Shami PJ, Saavedra JE, Bonifant CL, Chu J, Udupi V, Malaviya S, et al. Antitumor activity of JS-K [O2-(2,4-dinitrophenyl) 1-[(4-ethoxycarbonyl)piperazin-1-yl]diazen-1-ium-1,2-diolate] and related O2-aryl diazeniumdiolates in vitro and in vivo. J Med Chem 2006;49:4356–66.

[207] Chakrapani H, Goodblatt MM, Udupi V, Malaviya S, Shami PJ, Keefer LK, et al. Synthesis and in vitro anti-leukemic activity of structural analogues of JS-K, an anti-cancer lead compound. Bioorg Med Chem Lett 2008;18:950–3.

[208] McMurtry V, Saavedra JE, Nieves-Alicea R, Simeone AM, Keefer LK, Tari AM. JS-K, a nitric oxide-releasing prodrug, induces breast cancer cell death while sparing normal mammary epithelial cells. Int J Oncol 2011;38:963–71.

[209] Maciag AE, Holland RJ, Robert Cheng YS, Rodriguez LG, Saavedra JE, Anderson LM, Keefer LK. Nitric oxide-releasing prodrug triggers cancer cell death through deregulation of cellular redox balance. Redox Biol 2013;1:115–24. http://dx.doi.org/10.1016/j.redox.2012.12.002.

[210] Weyerbrock A, Osterberg N, Psarras N, Baumer B, Kogias E, Werres A, et al. JS-K, a glutathione S-transferase-activated nitric oxide donor with antineoplastic activity in malignant gliomas. Neurosurgery 2012;70:497–510.

[211] Maciag AE, Chakrapani H, Saavedra JE, Morris NL, Holland RJ, Kosak KM, et al. The nitric oxide prodrug JS-K is effective against non-small-cell lung cancer cells in vitro and in vivo: involvement of reactive oxygen species. J Pharmacol Exp Ther 2011;336:313–20.

[212] Chung EJ, Hwang SG, Nguyen P, Lee S, Kim JS, Kim JW, et al. Regulation of leukemic cell adhesion, proliferation, and survival by beta-catenin. Blood 2002;100:982–90.

[213] Nath N, Chattopadhyay M, Pospishil L, Cieciura LZ, Goswami S, Kodela R, et al. JS-K, a nitric oxide-releasing prodrug, modulates ß-catenin/TCF signaling in leukemic Jurkat cells: evidence of an S-nitrosylated mechanism. Biochem Pharmacol 2010;80:1641–9.

[214] Chattopadhyay M, Kodela R, Olson KR, Kashfi K. NOSH-aspirin (NBS-1120), a novel nitric oxide- and hydrogen sulfide-releasing hybrid is a potent inhibitor of colon cancer cell growth in vitro and in a xenograft mouse model. Biochem Biophys Res Commun 2012;419:523–8.

[215] Fu J, Liu L, Huang Z, Lai Y, Ji H, Peng S, et al. Hybrid molecule from O2-(2,4-dinitrophenyl)diazeniumdiolate and oleanolic acid: a glutathione S-transferase pi-activated nitric oxide prodrug with selective anti-human hepatocellular carcinoma activity and improved stability. J Med Chem 2013;56:4641–55.

[216] Zhao H, Ning S, Scicinski J, Oronsky B, Knox SJ, Peehl DM. Epigenetic effects of RRx-001: a possible unifying mechanism of anticancer activity. Oncotarget 2015;6:43172–81.

[217] Sikder N, Sikder AK, Bulakh NR, Gandhe BR. 1,3,3-Trinitroazetidine (TNAZ), a melt-cast explosive: synthesis, characterization and thermal behaviour. J Hazard Mater 2004;113:35–43.

[218] Knox SJ, Cannizzo L, Warner K, Wardle R, Velarde S, Ning S. Cyclic nitro compounds, pharmaceutical compositions thereof and uses thereof. United States: RadioRx, Inc., Alliant Techsystems Inc.; 2009.

[219] Vitturi DA, Sun CW, Harper VM, Thrash-Williams B, Cantu-Medellin N, Chacko BK, et al. Antioxidant functions for the hemoglobin beta93 cysteine residue in erythrocytes and in the vascular compartment in vivo. Free Radic Biol Med 2013;55:119–29.

[220] Scicinski J, Oronsky B, Cooper V, Taylor M, Alexander M, Hadar R, et al. Development of methods for the bioanalysis of RRx-001 and metabolites. Bioanalysis 2014;6:947–56.

[221] Scicinski J, Oronsky B, Taylor M, Luo G, Musick T, Marini J, et al. Preclinical evaluation of the metabolism and disposition of RRx-001, a novel investigative anticancer agent. Drug Metab Dispos 2012;40:1810–6.

[222] Ning S, Sekar TV, Scicinski J, Oronsky B, Peehl DM, Knox SJ, et al. Nrf2 activity as a potential biomarker for the pan-epigenetic anticancer agent, RRx-001. Oncotarget 2015;6:21547–56.

[223] Scicinski J, Oronsky B, Ning S, Knox S, Peehl D, Kim MM, et al. NO to cancer: the complex and multifaceted role of nitric oxide and the epigenetic nitric oxide donor, RRx-001. Redox Biol 2015;6:1–8.

[224] Reid T, Oronsky B, Scicinski J, Scribner CL, Knox SJ, Ning S, et al. Safety and activity of RRx-001 in patients with advanced cancer: a first-in-human, open-label, dose-escalation phase 1 study. Lancet Oncol 2015;16:1133–42.

[225] Carter CA, Oronsky B, Caroen S, Scicinski J, Degesys A, Cabrales P, et al. Partial response in an RRx-001-primed patient with refractory small-cell lung cancer after a third introduction of platinum doublets. Case Rep Oncol 2016;9:285–9.

[226] Kim MM, Parmar H, Cao Y, Knox SJ, Oronsky B, Scicinski J, et al. Concurrent whole brain radiotherapy and RRx-001 for melanoma brain metastases. Neuro Oncol 2016;18:455–6.

[227] Carter CA, Oronsky BT, Caroen SZ, Scicinski JJ, Cabrales P, Reid T, et al. Partial response to platinum doublets in refractory EGFR-positive non-small cell lung cancer patients after RRx-001: evidence of episensitization. Case Rep Oncol 2016;9:62–7.

[228] Brzezniak C, Schmitz BA, Peterson PG, Degesys A, Oronsky BT, Scicinski JJ, et al. RRx-001-induced tumor necrosis and immune cell infiltration in an EGFR mutation-positive NSCLC with resistance to EGFR tyrosine kinase inhibitors: a case report. Case Rep Oncol 2016;9:45–50.

[229] Chen KY, Morris JC. Oxidation of sulfide by O2: catalysis and inhibition. J Sanit Eng Div 1972;98:215–27.

[230] Insko MA, Deckwerth TL, Hill P, Toombs CF, Szabo C. Detection of exhaled hydrogen sulphide gas in rats exposed to intravenous sodium sulphide. Br J Pharmacol 2009;157:944–51.

[231] Kaschula CH, Hunter R, Hassan HT, Stellenboom N, Cotton J, Zhai XQ, et al. Anti-proliferation activity of synthetic ajoene analogues on cancer cell-lines. Anticancer Agents Med Chem 2011;11:260–6.

[232] Lee M, Tazzari V, Giustarini D, Rossi R, Sparatore A, Del Soldato P, et al. Effects of hydrogen sulfide-releasing L-DOPA derivatives on glial activation: potential for treating Parkinson disease. J Biol Chem 2010;285:17318–28.

[233] Whiteman M, Perry A, Zhou Z, Bucci M, Papapetropoulos A, Cirino G, et al. Phosphinodithioate and phosphoramidodithioate hydrogen sulfide donors. Handb Exp Pharmacol 2015;230:337–63.

[234] Shukla N, Rossoni G, Hotston M, Sparatore A, Del Soldato P, Tazzari V, et al. Effect of hydrogen sulphide-donating sildenafil (ACS6) on erectile function and oxidative stress in rabbit isolated corpus cavernosum and in hypertensive rats. BJU Int 2009;103:1522–9.

[235] Lee M, Sparatore A, Del Soldato P, McGeer E, McGeer PL. Hydrogen sulfide-releasing NSAIDs attenuate neuroinflammation induced by microglial and astrocytic activation. Glia 2010;58:103–13.

[236] Perrino E, Uliva C, Lanzi C, Soldato PD, Masini E, Sparatore A. New prostaglandin derivative for glaucoma treatment. Bioorg Med Chem Lett 2009;19:1639–42.
[237] Sparatore A, Santus G, Giustarini D, Rossi R, Del Soldato P. Therapeutic potential of new hydrogen sulfide-releasing hybrids. Expert Rev Clin Pharmacol 2011;4:109–21.
[238] Whiteman M, Winyard PG. Hydrogen sulfide and inflammation: the good, the bad, the ugly and the promising. Expert Rev Clin Pharmacol 2011;4:13–32.
[239] Banerjee R, Zou CG. Redox regulation and reaction mechanism of human cystathionine-beta-synthase: a PLP-dependent hemesensor protein. Arch Biochem Biophys 2005;433:144–56.
[240] Kabil O, Vitvitsky V, Xie P, Banerjee R. The quantitative significance of the transsulfuration enzymes for H2S production in murine tissues. Antioxid Redox Signal 2011;15:363–72.
[241] Kabil O, Weeks CL, Carballal S, Gherasim C, Alvarez B, Spiro TG, et al. Reversible heme-dependent regulation of human cystathionine beta-synthase by a flavoprotein oxidoreductase. Biochemistry 2011;50:8261–3.
[242] Li L, Whiteman M, Guan YY, Neo KL, Cheng Y, Lee SW, et al. Characterization of a novel, water-soluble hydrogen sulfide-releasing molecule (GYY4137): new insights into the biology of hydrogen sulfide. Circulation 2008;117:2351–60.
[243] Whiteman M, Li L, Rose P, Tan CH, Parkinson DB, Moore PK. The effect of hydrogen sulfide donors on lipopolysaccharide-induced formation of inflammatory mediators in macrophages. Antioxid Redox Signal 2010;12:1147–54.
[244] Lee ZW, Zhou J, Chen CS, Zhao Y, Tan CH, Li L, et al. The slow-releasing hydrogen sulfide donor, GYY4137, exhibits novel anti-cancer effects in vitro and in vivo. PLoS One 2011;6:e21077.
[245] Lu S, Gao Y, Huang X, Wang X. GYY4137, a hydrogen sulfide (H(2)S) donor, shows potent anti-hepatocellular carcinoma activity through blocking the STAT3 pathway. Int J Oncol 2014;44:1259–67.
[246] Lee ZW, Teo XY, Tay EY, Tan CH, Hagen T, Moore PK, et al. Utilizing hydrogen sulfide as a novel anti-cancer agent by targeting cancer glycolysis and pH imbalance. Br J Pharmacol 2014;171:4322–36.
[247] Putz VR, Johnson BL, Setzer JV. A comparative study of the effects of carbon monoxide and methylene chloride on human performance. J Environ Pathol Toxicol 1979;2:97–112.
[248] Fagin J, Bradley J, Williams D. Carbon monoxide poisoning secondary to inhaling methylene chloride. Br Med J 1980;281:1461.
[249] Peers C, Boyle JP, Scragg JL, Dallas ML, Al-Owais MM, Hettiarachichi NT, et al. Diverse mechanisms underlying the regulation of ion channels by carbon monoxide. Br J Pharmacol 2015;172:1546–56.
[250] Peers C, Lefer DJ. Emerging roles for gasotransmitters. Exp Physiol 2011;96:831–2.
[251] Nassour I, Kautza B, Rubin M, Escobar D, Luciano J, Loughran P, et al. Carbon monoxide protects against hemorrhagic shock and resuscitation-induced microcirculatory injury and tissue injury. Shock 2015;43:166–71.
[252] Kashfi K, Olson KR. Biology and therapeutic potential of hydrogen sulfide and hydrogen sulfide-releasing chimeras. Biochem Pharmacol 2012;24:019.
[253] Kashfi K. Anti-cancer activity of new designer hydrogen sulfide-donating hybrids. Antioxid Redox Signal 2014;20:831–46.
[254] Wallace JL, Dicay M, McKnight W, Martin GR. Hydrogen sulfide enhances ulcer healing in rats. FASEB J 2007;21:4070–6.
[255] Li L, Rossoni G, Sparatore A, Lee LC, Del Soldato P, Moore PK. Anti-inflammatory and gastrointestinal effects of a novel diclofenac derivative. Free Radic Biol Med 2007;42:706–19.
[256] Wallace JL, Caliendo G, Santagada V, Cirino G, Fiorucci S. Gastrointestinal safety and anti-inflammatory effects of a hydrogen sulfide-releasing diclofenac derivative in the rat. Gastroenterology 2007;132:261–71.
[257] Wallace JL, Caliendo G, Santagada V, Cirino G. Markedly reduced toxicity of a hydrogen sulphide-releasing derivative of naproxen (ATB-346). Br J Pharmacol 2010;159:1236–46.
[258] Liu L, Cui J, Song CJ, Bian JS, Sparatore A, Soldato PD, et al. H(2)S-releasing aspirin protects against aspirin-induced gastric injury via reducing oxidative stress. PLoS One 2012;7:e46301.
[259] Chattopadhyay M, Kodela R, Nath N, Barsegian A, Boring D, Kashfi K. Hydrogen sulfide-releasing aspirin suppresses NF-kappaB signaling in estrogen receptor negative breast cancer cells in vitro and in vivo. Biochem Pharmacol 2012;83:723–32.
[260] Chattopadhyay M, Kodela R, Nath N, Dastagirzada YM, Velazquez-Martinez CA, Boring D, et al. Hydrogen sulfide-releasing NSAIDs inhibit the growth of human cancer cells: a general property and evidence of a tissue type-independent effect. Biochem Pharmacol 2012;83:715–22.

[261] Chattopadhyay M, Kodela R, Nath N, Street CR, Velazquez-Martinez CA, Boring D, et al. Hydrogen sulfide-releasing aspirin modulates xenobiotic metabolizing enzymes in vitro and in vivo. Biochem Pharmacol 2012;83:733–40.
[262] Elsheikh W, Blackler RW, Flannigan KL, Wallace JL. Enhanced chemopreventive effects of a hydrogen sulfide-releasing anti-inflammatory drug (ATB-346) in experimental colorectal cancer. Nitric Oxide Biol Chem 2014;41:131–7.
[263] De Cicco P, Panza E, Ercolano G, Armogida C, Sessa G, Pirozzi G, et al. ATB-346, a novel hydrogen sulfide-releasing anti-inflammatory drug, induces apoptosis of human melanoma cells and inhibits melanoma development in vivo. Pharmacol Res 2016;114:67–73.
[264] Paul-Clark M, Elsheikh W, Kirkby N, Chan M, Devchand P, Agbor TA, et al. Profound chemopreventative effects of a hydrogen sulfide-releasing NSAID in the $APC^{Min/+}$ mouse model of intestinal tumorigenesis. PLoS One 2016;11:e0147289.
[265] Hanif R, Pittas A, Feng Y, Koutsos MI, Qiao L, Staiano-Coico L, et al. Effects of nonsteroidal anti-inflammatory drugs on proliferation and on induction of apoptosis in colon cancer cells by a prostaglandin-independent pathway. Biochem Pharmacol 1996;52:237–45.
[266] Ekundi-Valentim E, Mesquita FP, Santos KT, de Paula MA, Florenzano J, Zanoni CI, et al. A comparative study on the anti-inflammatory effects of single oral doses of naproxen and its hydrogen sulfide (H2S)-releasing derivative ATB-346 in rats with carrageenan-induced synovitis. Med Gas Res 2013;3:24.
[267] Karin M. NF-kappaB and cancer: mechanisms and targets. Mol Carcinog 2006;45:355–61.
[268] Li L, Salto-Tellez M, Tan CH, Whiteman M, Moore PK. GYY4137, a novel hydrogen sulfide-releasing molecule, protects against endotoxic shock in the rat. Free Radic Biol Med 2009;47:103–13.
[269] Nakshatri H, Bhat-Nakshatri P, Martin DA, Goulet Jr. RJ, Sledge Jr. GW. Constitutive activation of NF-kappaB during progression of breast cancer to hormone-independent growth. Mol Cell Biol 1997;17:3629–39.
[270] Biswas DK, Cruz AP, Gansberger E, Pardee AB. Epidermal growth factor-induced nuclear factor kappa B activation: a major pathway of cell-cycle progression in estrogen-receptor negative breast cancer cells. Proc Natl Acad Sci U S A 2000;97:8542–7.
[271] Romieu-Mourez R, Landesman-Bollag E, Seldin DC, Traish AM, Mercurio F, Sonenshein GE. Roles of IKK kinases and protein kinase CK2 in activation of nuclear factor-kappaB in breast cancer. Cancer Res 2001;61:3810–8.
[272] Patel NM, Nozaki S, Shortle NH, Bhat-Nakshatri P, Newton TR, Rice S, et al. Paclitaxel sensitivity of breast cancer cells with constitutively active NF-kappaB is enhanced by IkappaBalpha super-repressor and parthenolide. Oncogene 2000;19:4159–69.
[273] Weldon CB, Burow ME, Rolfe KW, Clayton JL, Jaffe BM, Beckman BS. NF-kappa B-mediated chemoresistance in breast cancer cells. Surgery 2001;130:143–50.
[274] Nakshatri H, Goulet Jr. RJ. NF-kappaB and breast cancer. Curr Probl Cancer 2002;26:282–309.
[275] Bonizzi G, Karin M. The two NF-kappaB activation pathways and their role in innate and adaptive immunity. Trends Immunol 2004;25:280–8.
[276] Ghosh S, Karin M. Missing pieces in the NF-kappaB puzzle. Cell 2002;109(Suppl.):S81–96.
[277] Switzer CH, Cheng RY, Ridnour LA, Murray MC, Tazzari V, Sparatore A, et al. Dithiolethiones inhibit NF-kappaB activity via covalent modification in human estrogen receptor-negative breast cancer. Cancer Res 2012;72:2394–404.
[278] Kodela R, Nath N, Chattopadhyay M, Nesbitt DE, Velazquez-Martinez CA, Kashfi K. Hydrogen sulfide-releasing naproxen suppresses colon cancer cell growth and inhibits NF-kappaB signaling. Drug Des Devel Ther 2015;9:4873–82.
[279] Dalle-Donne I, Giustarini D, Colombo R, Rossi R, Milzani A. Protein carbonylation in human diseases. Trends Mol Med 2003;9:169–76.
[280] Circu ML, Aw TY. Reactive oxygen species, cellular redox systems, and apoptosis. Free Radic Biol Med 2010;48:749–62.
[281] Mustacich D, Powis G. Thioredoxin reductase. Biochem J 2000;346(Pt 1):1–8.
[282] Arner ES, Holmgren A. The thioredoxin system in cancer. Semin Cancer Biol 2006;16:420–6.
[283] Lincoln DT, Ali Emadi EM, Tonissen KF, Clarke FM. The thioredoxin-thioredoxin reductase system: over-expression in human cancer. Anticancer Res 2003;23:2425–33.
[284] Zhang H, Zhi L, Moore PK, Bhatia M. Role of hydrogen sulfide in cecal ligation and puncture induced sepsis in the mouse. Am J Physiol Lung Cell Mol Physiol 2006;290:L1193–201.

[285] Matthews JR, Wakasugi N, Virelizier JL, Yodoi J, Hay RT. Thioredoxin regulates the DNA binding activity of NF-kappa B by reduction of a disulphide bond involving cysteine 62. Nucleic Acids Res 1992;20: 3821–30.

[286] Chiang JM, Chou YH, Chen TC, Ng KF, Lin JL. Nuclear beta-catenin expression is closely related to ulcerative growth of colorectal carcinoma. Br J Cancer 2002;86:1124–9.

[287] Lu D, Zhao Y, Tawatao R, Cottam HB, Sen M, Leoni LM, et al. Activation of the Wnt signaling pathway in chronic lymphocytic leukemia. Proc Natl Acad Sci U S A 2004;101:3118–23.

[288] Gan Z, Chattopadhyay M, Kodela R, Boring D, Crowell J, Kashfi K. Hydrogen sulfide-releasing aspirin inhibits the growth of leukemic Jurkat cells and.modulates β-catenin expression. In: American Association for Cancer Research, 102nd annual meeting, April 2–6, Orlando, FL; 2011.

[289] Bellosillo B, Pique M, Barragan M, Castano E, Villamor N, Colomer D, et al. Aspirin and salicylate induce apoptosis and activation of caspases in B-cell chronic lymphocytic leukemia cells. Blood 1998;92:1406–14.

[290] Kodela R, Chattopadhyay M, Kashfi K. NOSH-aspirin: a novel nitric oxide-hydrogen sulfide-releasing hybrid: a new class of anti-inflammatory pharmaceuticals. ACS Med Chem Lett 2012;3:257–62.

[291] Kodela R, Chattopadhyay M, Kashfi K. Synthesis and biological activity of NOSH-naproxen (AVT-219) and NOSH-sulindac (AVT-18A) as potent anti-inflammatory agents with chemotherapeutic potential. Med Chem Commun 2013;4.

[292] Kodela R, Chattopadhyay M, Velazquez-Martinez CA, Kashfi K. NOSH-aspirin (NBS-1120), a novel nitric oxide- and hydrogen sulfide-releasing hybrid has enhanced chemo-preventive properties compared to aspirin, is gastrointestinal safe with all the classic therapeutic indications. Biochem Pharmacol 2015;98:564–72.

[293] Chattopadhyay M, Kodela R, Duvalsaint PL, Kashfi K. Gastrointestinal safety, chemotherapeutic potential, and classic pharmacological profile of NOSH-naproxen (AVT-219) a dual NO- and H2S-releasing hybrid. Pharmacol Res Perspect 2016;4:e00224.

[294] Kashfi K, Chattopadhyay M, Kodela R. NOSH-sulindac (AVT-18A) is a novel nitric oxide- and hydrogen sulfide-releasing hybrid that is gastrointestinal safe and has potent anti-inflammatory, analgesic, antipyretic, anti-platelet, and anti-cancer properties. Redox Biol 2015;6:287–96.

[295] Fonseca MD, Cunha FQ, Kashfi K, Cunha TM. NOSH-aspirin (NBS-1120), a dual nitric oxide and hydrogen sulfide-releasing hybrid, reduces inflammatory pain. Pharmacol Res Perspect 2015;3:e00133.

[296] Lee M, McGeer E, Kodela R, Kashfi K, McGeer PL. NOSH-aspirin (NBS-1120), a novel nitric oxide and hydrogen sulfide releasing hybrid, attenuates neuroinflammation induced by microglial and astrocytic activation: a new candidate for treatment of neurodegenerative disorders. Glia 2013;61:1724–34.

[297] Chattopadhyay M, Velazquez CA, Pruski A, Nia KV, Abdellatif KR, Keefer LK, et al. Comparison between 3-nitrooxyphenyl acetylsalicylate (NO-ASA) and O2-(acetylsalicyloxymethyl)-1-(pyrrolidin-1-yl)diazen-1-ium-1,2-diolate (NONO-ASA) as safe anti-inflammatory, analgesic, antipyretic, antioxidant prodrugs. J Pharmacol Exp Ther 2010;335:443–50.

[298] Cortese-Krott MM, Fernandez BO, Kelm M, Butler AR, Feelisch M. On the chemical biology of the nitrite/sulfide interaction. Nitric Oxide Biol Chem 2015;46:14–24.

[299] Filipovic MR, Miljkovic J, Nauser T, Royzen M, Klos K, Shubina T, et al. Chemical characterization of the smallest S-nitrosothiol, HSNO; cellular cross-talk of H2S and S-nitrosothiols. J Am Chem Soc 2012;134:12016–27.

[300] Cortese-Krott MM, Kuhnle GG, Dyson A, Fernandez BO, Grman M, DuMond JF, et al. Key bioactive reaction products of the NO/H2S interaction are S/N-hybrid species, polysulfides, and nitroxyl. Proc Natl Acad Sci U S A 2015;112(34):E4651–60. http://dx.doi.org/10.1073/pnas.1509277112.

[301] Szabo C, Papapetropoulos A. Hydrogen sulphide and angiogenesis: mechanisms and applications. Br J Pharmacol 2011;164:853–65.

[302] Whiteman M, Le Trionnaire S, Chopra M, Fox B, Whatmore J. Emerging role of hydrogen sulfide in health and disease: critical appraisal of biomarkers and pharmacological tools. Clin Sci (Lond) 2011;121:459–88.

[303] Wang R. Physiological implications of hydrogen sulfide: a whiff exploration that blossomed. Physiol Rev 2012;92:791–896.

[304] Bouillaud F, Blachier F. Mitochondria and sulfide: a very old story of poisoning, feeding, and signaling? Antioxid Redox Signal 2011;15:379–91.

[305] Sun WH, Liu F, Chen Y, Zhu YC. Hydrogen sulfide decreases the levels of ROS by inhibiting mitochondrial complex IV and increasing SOD activities in cardiomyocytes under ischemia/reperfusion. Biochem Biophys Res Commun 2012;421:164–9.

[306] Szczesny B, Modis K, Yanagi K, Coletta C, Le Trionnaire S, Perry A, et al. AP39, a novel mitochondria-targeted hydrogen sulfide donor, stimulates cellular bioenergetics, exerts cytoprotective effects and protects against the loss of mitochondrial DNA integrity in oxidatively stressed endothelial cells in vitro. Nitric Oxide Biol Chem 2014;41:120–30.

[307] Distrutti E, Sediari L, Mencarelli A, Renga B, Orlandi S, Russo G, et al. 5-Amino-2-hydroxybenzoic acid 4-(5-thioxo-5H-[1,2]dithiol-3yl)-phenyl ester (ATB-429), a hydrogen sulfide-releasing derivative of mesalamine, exerts antinociceptive effects in a model of postinflammatory hypersensitivity. J Pharmacol Exp Ther 2006;319:447–58.

[308] Ekundi-Valentim E, Santos KT, Camargo EA, Denadai-Souza A, Teixeira SA, Zanoni CI, et al. Differing effects of exogenous and endogenous hydrogen sulphide in carrageenan-induced knee joint synovitis in the rat. Br J Pharmacol 2010;159:1463–74.

[309] Chan MV, Wallace JL. Hydrogen sulfide-based therapeutics and gastrointestinal diseases: translating physiology to treatments. Am J Physiol Gastrointest Liver Physiol 2013;305:G467–73.

[310] Fiorucci S, Orlandi S, Mencarelli A, Caliendo G, Santagada V, Distrutti E, et al. Enhanced activity of a hydrogen sulphide-releasing derivative of mesalamine (ATB-429) in a mouse model of colitis. Br J Pharmacol 2007;150:996–1002.

[311] Fiorucci S, Santucci L, Gresele P, Faccino RM, Del Soldato P, Morelli A. Gastrointestinal safety of NO-aspirin (NCX-4016) in healthy human volunteers: a proof of concept endoscopic study. Gastroenterology 2003;124:600–7.

[312] Fiorucci S, Santucci L, Wallace JL, Sardina M, Romano M, del Soldato P, et al. Interaction of a selective cyclooxygenase-2 inhibitor with aspirin and NO-releasing aspirin in the human gastric mucosa. Proc Natl Acad Sci U S A 2003;100:10937–41.

[313] Fiorucci S, Mencarelli A, Meneguzzi A, Lechi A, Renga B, del Soldato P, et al. Co-administration of nitric oxide-aspirin (NCX-4016) and aspirin prevents platelet and monocyte activation and protects against gastric damage induced by aspirin in humans. J Am Coll Cardiol 2004;44:635–41.

[314] NicOx SA. NicOx provides an update on NCX-4016. Sophia Antipolis, France. http://www.nicox.com/wp-content/uploads/pr2007061800en.pdf?4051a2; 2007.

[315] Schnitzer TJ, Hochberg MC, Marrero CE, Duquesroix B, Frayssinet H, Beekman M. Efficacy and safety of naproxcinod in patients with osteoarthritis of the knee: a 53-week prospective randomized multicenter study. Semin Arthritis Rheum 2011;40:285–97.

[316] Schnitzer TJ, Kivitz A, Frayssinet H, Duquesroix B. Efficacy and safety of naproxcinod in the treatment of patients with osteoarthritis of the knee: a 13-week prospective, randomized, multicenter study. Osteoarthritis Cartilage 2010;18:629–39.

[317] Karlsson J, Pivodic A, Aguirre D, Schnitzer TJ. Efficacy, safety, and tolerability of the cyclooxygenase-inhibiting nitric oxide donator naproxcinod in treating osteoarthritis of the hip or knee. J Rheumatol 2009;36:1290–7.

[318] Schnitzer TJ, Kivitz AJ, Lipetz RS, Sanders N, Hee A. Comparison of the COX-inhibiting nitric oxide donator AZD3582 and rofecoxib in treating the signs and symptoms of Osteoarthritis of the knee. Arthritis Rheum 2005;53:827–37.

[319] Lohmander LS, McKeith D, Svensson O, Malmenas M, Bolin L, Kalla A, et al. A randomised, placebo controlled, comparative trial of the gastrointestinal safety and efficacy of AZD3582 versus naproxen in osteoarthritis. Ann Rheum Dis 2005;64:449–56.

[320] NicOx SA. FDA provides complete response letter to NicOx's new drug application for naproxcinod. http://www.nicox.com/wp-content/uploads/pr2010072200en.pdf?aeb643; 2010.

CHAPTER

8

Nitric Oxide-Based Anticancer Therapeutics: The New Technologies of the Nanoparticles

Federica Vannini[*,†], *Chiara Lenzi*[‡], *Valter Lubrano*[§]

[*]Institute of Life Sciences, Sant'Anna School of Advanced Studies, Pisa, Italy

[†]Institute of Clinical Physiology, CNR, Pisa, Italy

[‡]University of Pisa, Pisa, Italy

[§]Fondazione G. Monasterio, CNR-Regione Toscana, Pisa, Italy

Abstract

In the study of new drug molecules for the treatment of cancer, nitric oxide has received much attention for its tumoricidal role. In addition to its direct cytotoxic effect, NO may enhance the efficiency of classical chemotherapeutic drugs, resulting in a promising strategy against tumorigenesis and tumor progression. However, the clinical therapeutic potential of NO is limited by the short half-life, the concentration, and the biological microenvironment with which it interacts. Particular attention should be paid to the concentrations, as according to some authors, low NO levels could promote the growth of cancer cells. The development of a nanoparticle-based delivery system is aimed to the selective release of NO and drugs at the target tumor site, increasing their stability and biodistribution and preserving their therapeutic dosage. This review summarizes the current knowledge of NO-based anticancer treatments and their association with chemotherapy, focusing on the benefit of using nanotechnology.

Abbreviations

CDDP	*cis*-diamminedichloridoplatinum
DOX	doxorubicin
NO	nitric oxide
NONOates	*N*-diazeniumdiolate
NOS	nitric oxide synthase
NSAIDs	nonsteroidal antiinflammatory drugs
RSNO	*S*-nitrosothiols

Nitric Oxide (Donor/Induced) in Chemosensitization, Volume 1
http://dx.doi.org/10.1016/B978-0-12-811020-1.00008-9

INTRODUCTION

Nitric oxide (NO) is a short-lived endogenously produced gas that acts as a signaling molecule in the body. Ignarro et al. and Palmer et al. simultaneously identified NO in 1987 as the endothelium-derived relaxing factor, the main relaxation factor of vessels in smooth muscle cells [1,2].

NO is a ubiquitous free-radical gas that plays a key role in various physiological and pathological processes. It is produced by several different cell types, and it is an important regulator and mediator of numerous processes including smooth muscle relaxation, neurotransmission, and murine macrophage-mediated cytotoxicity for microbes and tumor cells [3–5]. NO is synthesized from L-arginine by the enzyme nitric oxide synthase (NOS) (Fig. 1).

In the active form, NOS is a dimeric enzyme consisting of an N-terminal oxidative domain and a C-terminal reductive domain, both linked to a calmodulin-binding site (CaM), on which the enzymatic activity depends [6]. Several cofactors are also required for the NOS activity, such as nicotinamide adenine dinucleotide phosphate (NADPH), flavin adenine dinucleotide (FAD), flavin mononucleotide (FMN), and (6*R*-)5,6,7,8-tetrahydro-l-biopterin (BH_4). In fact, from the reductive domain, electrons are transferred via NADPH-FAD-FMN to the heme prosthetic group of the oxidative domain where O_2 reduction and NO formation take place. There are at least three distinct isoforms of NOS present in human cells, which have been isolated and cloned. Two enzymes, the endothelial and the neuronal isoforms encoded by eNOS and nNOS genes, respectively, are constantly present and termed constitutive NOS (cNOS). The third isoform, encoded by the *iNOS* gene, is inducible and expressed in response to immunostimulants, for example, in hepatocytes [7] (Fig. 2).

Murine macrophages express high levels of iNOS mRNA and protein after stimulation with various agents such as lipopolysaccharide (LPS) and cytokines (including interferon-γ (IFN-γ), interleukin-l (IL-l), and tumor necrosis factor (TNF)) [8]. It has been reported that human monocytes/macrophages are capable of producing NO [9].

In relation to these immunologic characteristics, in recent years, NO has raised much interest in the study of tumor therapy. Some authors have shown the production of NO in various cancer [10,11]. The investigations suggested that NO can act sometimes promoting cancer cell proliferation and in other cases acting with a tumoricidal effect [12–14]. These lines of evidence have suggested the importance of the development of new therapies in which the effectiveness of NO could be exploited. An important implication in cancer is the regulatory role of the tumor suppressor gene *p53* that shows a negative feedback loop on iNOS [15,16].

In the presence of various types of cancer, the cells that are most frequently involved in the production of NO for tumoricidal activity are macrophages, natural killer cells, Kupffer cells, and endothelial cells. Some studies have shown that NO has a cytotoxic and/or cytostatic action on cancer cells [17,18]. An example is the stimulus of NO on aconitase and ribonucleotide reductase, which are implicated in the indirect cytostatic/cytotoxic effects by causing the suppression of DNA synthesis [19]. A large and prolonged production of NO actives caspases, induces the mitochondrial cytochrome c release in the cytosol and upregulates of p53, and alters the Bcl2 family, and thus, behaving like a preapoptotic modulator [20].

Some investigators have suggested that high levels of NO activates eliminate metastases [21]. Indeed, Baritaki et al. have shown that high levels of NO derived from the NO donor DETA-NONOate inhibited the epithelial-mesenchymal transition (EMT) and reversed both

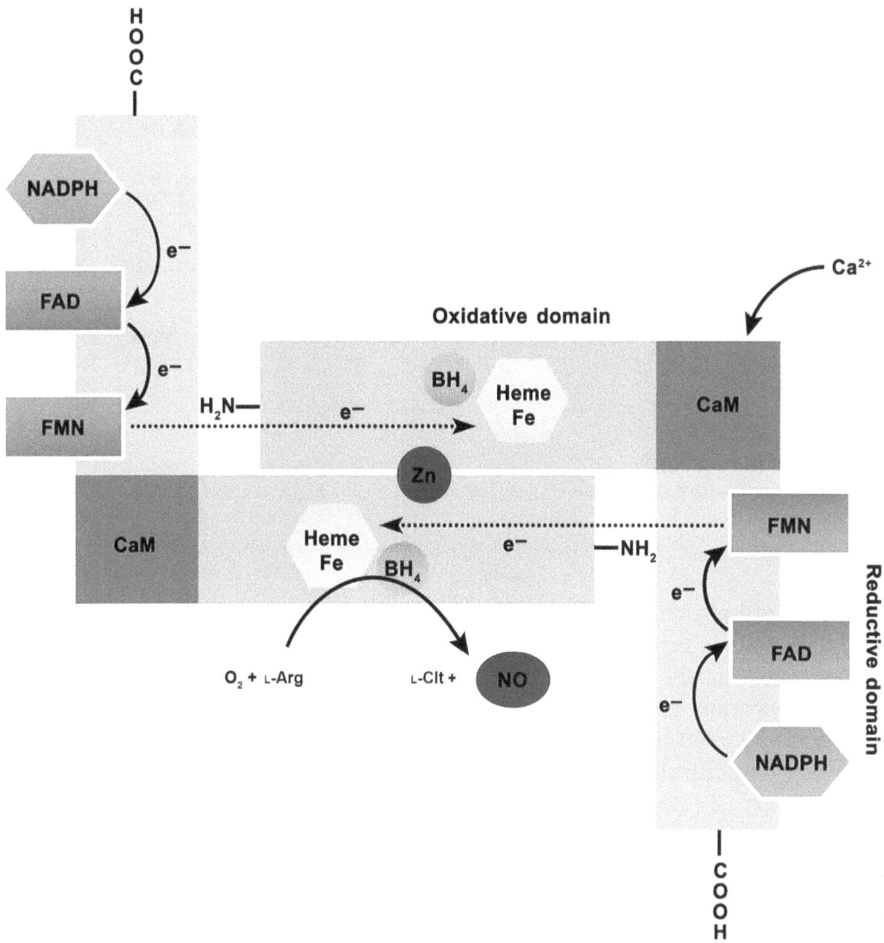

FIG. 1 Structure of NOS enzyme. Nitric oxide synthase (NOS) is an enzyme that oxidizes arginine to citrulline and generates NO. It is formed by an N-terminal oxidative domain, consisting of a heme binding site, (6*R*-)5,6,7,8-tetrahydro-l-biopterin (BH4), and L-arginine and a C-terminal reductive domain, containing binding sites for nicotinamide adenine dinucleotide phosphate (NADPH), flavin adenine dinucleotide (FAD), and flavin mononucleotide (FMN). The two domains are linked to a calmodulin binding site (CaM), required for the catalytic reaction. Electrons flow from the reductive domain to the oxidative domain. A Zn(2+) ion stabilizes the dimeric structure of NOS.

the mesenchymal phenotype and the invasive properties of human prostate metastatic cells [22]. Although most experiments that were performed in vitro proposed the tumoricidal role of NO [7,23,24], such findings have not been reported in cancer patients. It has been suggested that NO concentrations found in solid tumors are insufficient to produce apoptosis and other tumoricidal effects and are likely to facilitate angiogenesis and tumor dissemination [25,26]. In addition, whether NO has an inhibitory or stimulatory effect on the cancer, the process initially depends on the concentration of NO achieved and also on other factors such as the type of the cell exposed, the redox state, the final intracellular concentration, and the duration of exposure in the tumor bed. Once the cancer has begun, NO seems to play a

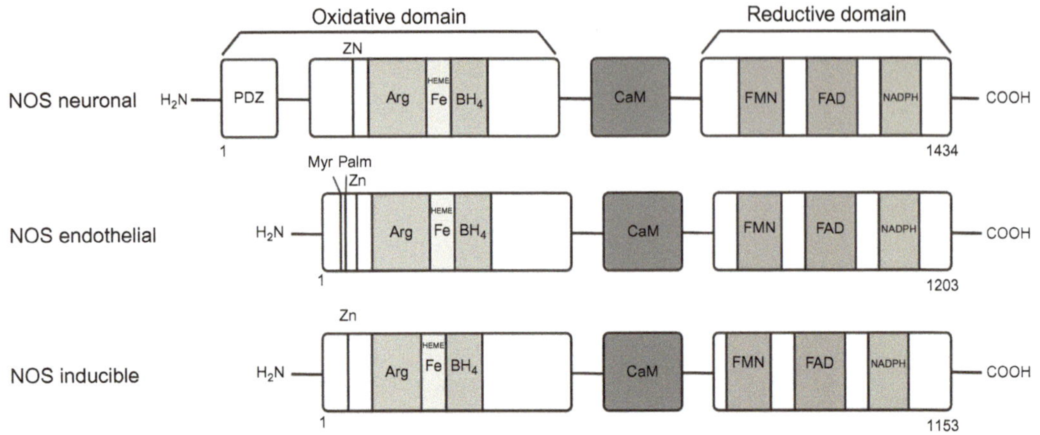

FIG. 2 Isoforms of human nNOS, eNOS, and iNOS. In mammals, there are three main isoforms of nitric oxide synthase that are termed neuronal NOS (nNOS), endothelial NOS (eNOS), and inducible NOS (iNOS). The oxidative and reductive domains, linked to a calmodulin-binding domain, are presented in all three isoforms. Although they probably have an ancestral *NOS* gene in common, they lightly differ in structure and function.

protumoral role rather than an antitumoral one as the concentration required to cause tumor cell cytotoxicity cannot be achieved by the cancer cells [26]. However, considering the cytostatic and/or cytotoxic effects of NO, strategies are being developed to manipulate NO levels in the tumor environment for a therapeutic gain. Viewing this biphasic NO effect on cancer cells, investigators took into account the critical concentrations of this factor by studying various strategies for manipulating the NO concentration in vivo, that is, the production and the exogenous delivery of NO into the tumor cells.

iNOS GENE THERAPY

One of the first mechanisms thought for the delivery of NO was to transfer the iNOS-encoding cDNA sequences into the cancer cells with retroviral and adenoviral vectors. Modulation of cellular production of NO provides a more localized therapeutic effect on the target site [27]. However, this type of gene therapy raises some fears about the danger to the host [28]. Indeed, the vector combined with the gene could invade not only the intended target but also the many other organs, triggering an activation of innate immunity and inducing a systemic inflammatory response, as unfortunately it has happened in some cases. Given the difficulty of a successful gene transfer, the gene therapy is usually performed with high doses of the viral vector, but this involves the risk that we have just described. Further studies into the precise mechanisms and efficacy and toxicity of this kind of therapy are required.

NO RELEASING DRUGS

Donors of NO drugs should result in a sustained release and a wide range of half-life, with predictable estimated doses. These drugs must simultaneously exert the inhibition of angiogenesis and metastasis and the stimulation of apoptosis, according to their concentration and the type and the stage of the cancer [29]. In addition, they are emerging as potential antioncogene agents to overcome the resistance of cancer cells to conventional therapeutic agents. Indeed, Yasuda et al. demonstrated the possible role of nitroglycerine as the alternative for the management of the resistant tumors [30,31]. Moreover, NO donors were observed to sensitize cancer cells to chemotherapy and immunotherapy. However, the potentials of these drugs are limited by the systemic effect exerted by NO, like vasodilation with a sudden decrease of blood pressure and an accumulation of toxic and mutagenic compounds [32,33]. The need to use the antiproliferative efficacy with reduced side effects of NO in vivo has led to the development of NO-drug hybrid.

NO donors like ester nitrates, furoxans, benzofuroxans, *N*-diazeniumdiolate (NONOates), *S*-nitrosothiols (RSNO), and metal complexes that can be generated showed a NO release benefit with maintaining the activity of the native drug. NO combined with nonsteroidal antiinflammatory drugs (NO-NSAIDs), in which a NO-releasing moiety is covalently attached, showed promising results in patients affected by colorectal cancer [34].

Recent studies indicate the possibility of using the NO-drug hybrid in various types of cancer cells. For example, Fraix et al. induced the photogeneration of NO with a molecule containing two chromogen compounds. These two chromogenic units absorb in distinct regions of the visible spectrum, and their individual photochemical properties are conserved in the molecular conjugate. The compound, irradiated with green light, was tested on melanoma cancer cells that resulted in a strong amplification of mortality with respect for the structurally correlated model compound deprived of the NO release capacity [35].

Novel NO-releasing prodrug hybrids, like *O*(2)-(2,4-dinitrophenyl)diazeniumdiolate and oleanolic acid (OA), are activated by glutathione-*S*-transferase π that is overexpressed in many tumor cells. These compounds tend to release a quantity of NO much higher in cancer cells as compared with normal cells than showing a strong in vitro antiproliferative activity, largely also observable in vivo [36].

Novel compounds capable of releasing both NO and hydrogen sulfide (H_2S) combine the cytotoxic and antineoplastic activities of NO-NSAIDs with the H_2S-dependent gastrointestinal defense mechanism. Nitric oxide-hydrogen sulfide (NOSH-sulindac) inhibited the growth of all cancer cell lines studied, thus amplifying significantly the cell mortality compared with simple sulindac, which is a nonsteroidal antiinflammatory drug (NSAID) of the arylalkanoic acid. NOSH-sulindac inhibited cell proliferation, induced apoptosis, and caused G_2/M cell cycle block [37]. In a similar manner, NOSH-aspirin and NOSH-naproxen were found to be more potent in inhibiting human cancer cell growth than their traditional counterparts, aspirin and naproxen, respectively, both in vitro and in xenograft mouse models and showing no signs of gastrointestinal and renal damages [38–40]. The promising data obtained from these studies have prompted researchers to synthesize molecules capable of releasing a more optimal amount of NO at the right time and at the right site.

EMERGING ANTICANCER THERAPIES: THE NANOPARTICLE TECHNOLOGY

Given the limited solubility of NO in aqueous solutions, nanoparticles have been attracting increasing attention in cancer therapy as novel vehicles for the delivery of NO into biological systems, because they show many advantages compared with classical NO donors.

The hydrophobicity, the size, and the chemical structure of the particle affect the pharmacokinetic properties of NO donors and the tunability of NO release. In fact, high levels of NO can be loaded into nanomaterials and released directly into a specific target after stimulation with light, temperature, pH, or a magnetic field. In this way, the biodistribution of NO can be altered in a controlled manner, avoiding the release of low levels of NO that could be tumorigenic. Another reported advantage of using nanomaterials is the possibility to combine the cytotoxicity of NO donors with classical chemotherapy agents, such as cisplatin and doxorubicin (DOX). The systemic side effects associated with a nonselective drug distribution can be overcome by the nanotechnology (Fig. 3).

NO-Releasing Nanomaterials

NO donors can be incorporated into or chemically linked to the nanoparticles, which release the endogenous NO to a specific destination site. Based on the biochemical structure of the nanomaterial, nanoparticles can be classified into polymeric nanocarriers (including polymeric nanoparticles, micelles, dendrimers, and hydrogels), liposomes, silica nanomaterials, metallic nanoparticles, and quantum dots [41].

Polymeric nanoparticles are nontoxic carriers that can be easily biodegraded in the body. They are usually composed of polylactic acid (PLA), polyglycolic acid (PGA), or polylactic-*co*-glycolic acid (PLGA) [42]. Duan and colleagues have developed a multiarm polymeric

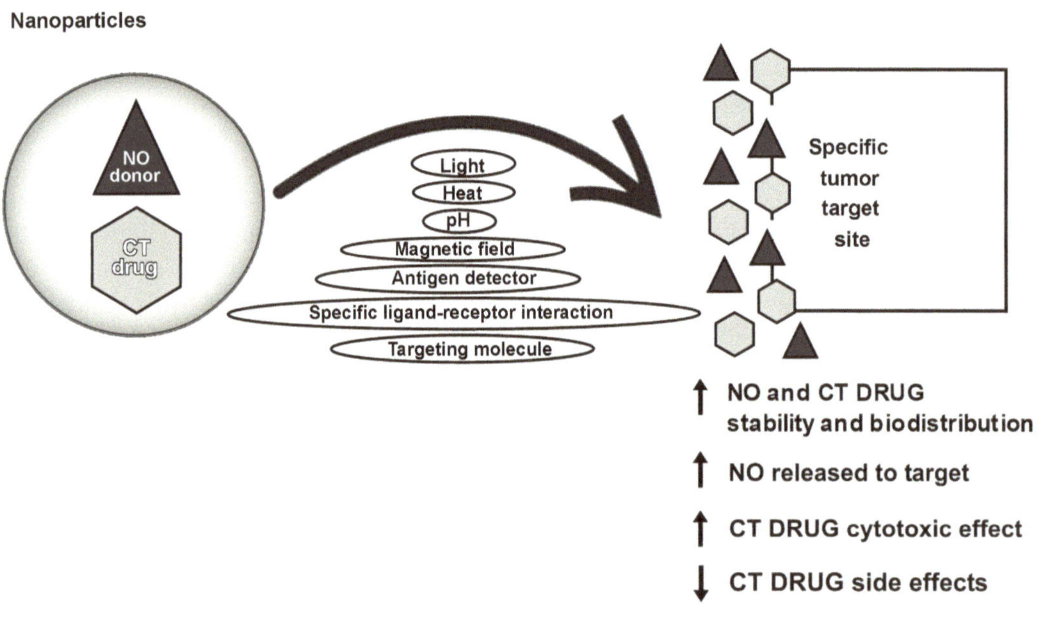

FIG. 3 Nanoparticles as novel delivery systems for NO donors and chemotherapeutic drugs in cancer treatment. The stability and biodistribution of NO and chemotherapeutic (CT) drugs may be increased by using a delivery system based on nanotechnology. In this way, high levels of NO are released from a NO donor directly to a specific tumor target site. The consequent NO cytotoxic effect is able to enhance the efficacy of chemotherapeutic drugs, such as cisplatin or doxorubicin, and reduce the drug side effects.

nanocarrier (poly-(6-*O*-methacryloyl-D-galactose)) to deliver NO to xenografts of human head and neck squamous carcinoma (HNSCC) and compared its efficacy to classical JS-K anticancer therapy [43]. They found a 50% tumor growth inhibition and a fivefold tumor size reduction in mice treated with NONOate-multiarm polymer nanocarriers compared with mice treated with the NO-donating prodrug JS-K (O^2-(2,4-dinitrophenyl) 1-[(4-ethoxycarbonyl) piperazin-1-yl] diazen-1-ium-1,2-diolate). In addition, using this polymeric nanocarrier, the average survival time of the animals was extended to 7 weeks.

To develop a more controlled NO delivery system, Kanayama et al. encapsulated a NO photodonor into the hydrophobic core of PEGylated polymer micelles [44]. The photoinduced NO-mediated antitumor effect was registered in HeLa cells. Recently, a similar photocontrolled NO delivery method was applied in combination with a fluorescent dye to human melanoma cancer cells and human skin carcinoma cells by Sortino and colleagues [45,46]. The NO photodonor included in the nanocarrier, formed by a cyclodextrin-based polymer, was released after light excitation, leading to cancer cell death. In addition, the imaging of the tumor cells was detected by the presence of the fluorophore.

Friedman et al. proposed a new nanodelivery platform based on the combination of glassy matrices and hydrogels that is able to release therapeutic sustained NO levels [47]. Currently, no studies have been reported on the use of dendrimers as polymeric carriers for NO-based anticancer treatments.

Another class of nanocarriers is represented by liposomes. In particular, thermosensitive liposomes were investigated for their ability to deliver the NO donor into the tumor site, where heat is generated [48,49]. In fact, Tai et al. demonstrated that the pH inside the liposome decreased by inducing heat, leading to a rapid NO release from the zwitterionic NONOate encapsulated in the nanocarrier. Silica nanoparticles are nontoxic porous structures derived from sol-gels, usually prepared with alkoxysilanes.

Stevensen et al. found an increased antitumor activity of the silica nanomaterial-derived NO against ovarian tumor compared with NO released from the pyrrolidine-NO donor (PYRRO/NO). Indeed, by using silica nanoparticles, the NO release was prolonged, and thus, the NO donor was preserved until reaching the target site. Preferential cytotoxicity for tumor versus nontumor ovarian cells was enhanced by increasing the nanoparticle size (from $d = 90 \pm 10$ to 350 ± 50 nm), which influenced the NO delivery volume [50]. In this way, the destabilization of the mitochondrial function and apoptosis was stimulated. In addition, NO-releasing nanoparticles containing 3-methylaminopropyltrimethoxysilane (MAP3) resulted in a less cytotoxic activation to normal cells than *N*-(6-aminohexyl)aminopropyltrimethoxysilane (AHAP3) silica nanoparticles, due to the absence of primary amine. The investigator demonstrated that the size and the surface properties of the particle influence the efficiency of the NO release [50].

Metallic nanoparticles are biocompatible delivery vehicles with potential applications in cancer treatment. Among them, gold nanoparticles are the best studied. They are derived from the reaction with tetrachloroaurate ($AuCl^{4-}$) and sodium citrate or sodium borohydride [51]. The properties of the gold nanoparticle surface are dependent on the presence of a stabilizing agent, such as alkanethiol or 2-mercapto-5-nitro benzimidazole (MNBI). In fact, Sudhesh and colleagues demonstrated that MNBI-coated gold nanoparticles with their small sizes, water solubility, and stability were able to release high levels of NO, after visible-light irradiation, in order to induce cervical cancer cell death with a cytotoxic effect comparable to cisplatin [52]. Application of a magnetic field can also be used to guide NO-releasing iron oxide nanoparticles to a specific tumor site [41].

Another class of nanoparticles that is receiving increasing attention is the quantum dots. These fluorescent nanocrystals can release rapidly high amounts of NO into tumor cells by photodynamic therapy and improving in vitro and in vivo imaging of the cancer cells [41]. Thus, nanoparticle technologies exhibit a high efficiency as delivery system and may be successfully used to combine anticancer therapeutics and also used for diagnostic applications.

The Combination of NO Nanodelivery System and Chemotherapy

NO donors were shown to enhance the efficacy of chemotherapeutic treatments both in vitro and in vivo [41,53], and they are able to reverse chemotherapeutic drug resistance in tumor cells [54,55,32].

Huerta et al. examined the chemosensitizing effect of DETANONOate on metastatic SW620 and nonmetastatic SW480 colon carcinoma cell lines that were resistant to cisplatin (CDDP). DETANONOate pretreatment (1000 μM) for 8 h followed by CDDP administration (5 μg/ml) induced a 6.6-fold and 5.6-fold increase for apoptosis in SW480 and SW620 compared with control, respectively [56]. Additional studies revealed a reduction of 36% in tumor size in nude mice bearing SW620 tumor xenografts treated with the DETANONOate and CDDP combination, while it was not observed with these compounds, where they were used alone [56].

The capability of NO to augment the cytotoxicity of DOX was analyzed by Evig and coworkers on the MCF-7 breast cancer cell line [27]. Treatment with diethylamine-NO donor (DEANO) prior to DOX showed a two- to threefold increased of cytotoxicity compared with control.

The same experiment was also conducted using cells transduced with iNOS, revealing an eightfold increased of cytotoxicity compared with control. Although highly efficient, the effect of the NO donor on cytotoxicity was less enhanced compared with iNOS gene therapy because the exogenous NO may not have reached critical intercellular sites at an effective concentration to sensitize the cells to chemotherapeutic agents.

Thus, recent studies have aimed at developing novel nanodelivery systems to overcome this limitation. Duong et al. demonstrated that the nanoparticle-based NO delivery enhanced the response to cisplatin in the treatment of neuroblastoma cells [57]. In fact, a neuroblastoma BE(2)-C cell line was incubated for 8 h with a NO donor, *S*-nitrosoglutathione (GSNO), encapsulated in polymeric nanoparticles and subsequently treated with cisplatin. The results showed that NO-pretreated cells were more sensitive to chemotherapy than the non-NO-treated ones. The IC50 of cisplatin was increased fivefold by pretreatment with a NO-releasing nanoparticle. In addition, the synergistic cytotoxic effect of NO nanodelivery and cisplatin was not observed in noncancerous fibroblast cells (MRC-5) [57].

Munaweera and colleagues confirmed the capability of NO to sensitize non-small-cell lung cancer (H596 and A549 NSCLC) to cisplatin [58]. In fact, the cytotoxicity induced by NO- and cisplatin-loaded amine mesoporous silica nanoparticles (NO-Si-DETA-cisplatin-AMS) to tumor cells was significantly higher than that observed for treatments with either of the two molecules alone. In addition, NO-Si-DETA-cisplatin-AMS did not increase the toxicity toward normal lung cell lines (WI-38 and BEAS-2B) compared with cisplatin-releasing silica nanoparticle [58].

A method for further improving the efficiency of the nanodelivery system for cancer therapy was proposed by Wang et al. Their approach is based on the development of a codelivery system that let the encapsulation of both NO and anticancer agent in the same nanoparticles. An organic phase containing *S*-nitroso (SNO) polysilsesquioxane was mixed with DOX containing solvent-water. That interaction caused nanoparticle formation by precipitation and simultaneous drug encapsulation. After lyophilization in water, SNODOX nanoparticles were generated. The concentrations of DOX and NO in the spherical nanoparticles (50–100 nm) were 1.76 and 3.71 μg/mg, respectively. The highest drug encapsulation was found using DMSO as solvent. Interestingly, the DOX release was pH-dependent, unlike NO. This property makes SNODOX a suitable nanoplatform for a localized delivery of NO and anticancer drugs in the acidic tumor environment [59].

Another efficient NO donor/chemotherapeutic agent codelivery nanoplatform was developed by Fan and coworkers [60]. A monomethoxy-polyethylene glycol-poly-lactic-*co*-glycolic acid (mPEG-PLGA) nanoparticle was loaded with *N,N*′-di-sec-butyl-*N,N*′-dinitroso-1,4-phenylenediamine (BNN6), a photoresponsive NO donor, together with DOX. After 2 min of UV light exposure, the NO release from BNN6 caused the nanoparticles disruption with a subsequent DOX release. In vitro analysis revealed that mPEG-PLGA-BNN6-DOX increased the cytotoxicity toward the multidrug resistance ovarian cancer cell line (OVCAR-8/ADR) compared with the treatment with mPEG-PLGA-DOX [60].

These results confirm that NO is able to reverse the multidrug resistance, increasing the potency of the chemotherapy. In addition, the data highlight the great efficacy of the nanoparticle technology as a delivery system that leads to a controlled interaction between NO donors and chemotherapeutic agents.

CONCLUSION

NO is a small gaseous molecule with physiological relevance and has a short half-life in the body of about 5 min, due to its reactions with hemoglobin, molecular oxygen, and superoxide [43,57].

The NO concentration highly exerts its biological effects on tumor growth, resulting in protumorigenic or antitumorigenic activity. High levels of NO can also make cancer cells more susceptible to conventional chemotherapeutic drugs, such as cisplatin and DOX, enhancing the therapeutic outcome of the treatments and overcoming multidrug resistance.

To improve the NO stability and biodistribution, several NO-releasing drugs have been proposed, such as NO donors, RSNO and NONOates. However, the systemic administration of the drug may limit the delivery of NO to a specific tumorigenic tissue leading to a nontherapeutic low dosage of NO. To overcome this challenge, NO-releasing drugs can be conjugated to nanoparticles, as a novel NO delivery system for cancer treatment. Due to their favorable physicochemical properties, nanoparticles increase the stability of NO donors, prolong blood circulation time, and preserve NO efficacy, leading to a controlled release into a specific tumor site. In addition, lower doses of chemotherapeutic agents may be used, since an efficient NO nanodelivery can sensitize tumor cells to the drug effect. So the combination of NO donors and chemotherapeutic agents with nanomaterials represents a promising anticancer strategy that is worth further exploration.

Acknowledgment

The authors are grateful to Lucrecia Mota Garcia for her editing support.
Conflict of Interest: No potential conflicts of interest were disclosed.

References

[1] Ignarro LJ, Buga GM, Wood KS, Byrns RE, Chaudhuri G. Endothelium-derived relaxing factor produced and released from artery and vein is nitric oxide. Proc Natl Acad Sci U S A 1987;11:9265–9.
[2] Palmer RMJ, Ferrige AG, Moncada S. Nitric oxide release accounts for the biological activity of endothelial-derived relaxing factor. Nature 1987;11:524–6.
[3] Hibbs J. Synthesis of nitric oxide from L-arginine: a recently discovered pathway induced by cytokines with antitumour and antimicrobial activity. Res Immunol 1991;142:565–9.
[4] Moncada S, Higgs A. The L-arginine-nitric oxide pathway. N Engl J Med 1993;329(27):2002–12.
[5] Nathan C. Nitric oxide as a secretory product of mammalian cells. FASEB J 1992;6(12):3051–64.
[6] Förstermann U, Sessa WC. Nitric oxide synthases: regulation and function. Eur Heart J 2012;33(7):829–37.
[7] Moncada S, Palmer RMJ, Higgs EA. Nitric oxide: physiology, pathology, and pharmacology. Pharmacol Rev 1991;43:109–42.
[8] Jorens PG, Matthys KE, Bult H. Modulation of nitric oxide synthase activity in macrophages. Mediators Inflamm 1995;4(2):75–89.
[9] Reiling N, Ulmer AJ, Duchrow M, Ernst M, Flad HD, Hauschildt S. Nitric oxide synthase: mRNA expression of different isoforms in human monocytes/macrophages. Eur J Immunol 1994;24(8):1941–4.
[10] Thomson LL, Lawton FG, Knowles RG, Basley JE, Riversomoreno V, Moncada S. NO synthase activity in human gynecological cancer. Cancer Res 1994;11:1352–4.
[11] Taysi S, Uslu C, Akcay F, Sutbeyaz MY. MDA and nitric oxide in the plasma of patients with advanced laryngeal cancer. Surg Today 2003;11(9):651–4.
[12] Vannini F, Kashfi K, Nath N. The dual role of iNOS in cancer. Redox Biol 2015;6:334–43.
[13] Reveneau S, Arnould L, Jolimoy G, Hilpert S, Lejeune P, Saint-Giorgio V, et al. Nitric oxide synthase in human breast cancer is associated with tumor grade, proliferation rate, and expression of progesterone receptor. Lab Invest 1999;11:1215–25.
[14] Prazma J, Pertrusz P, Mims W, Ball SS, Weissler MC. Immunohistochemical characterization of NOS activity in squamous cell carcinoma of head and neck. Otolaryngol Head Neck Surg 1995;11:541–9.
[15] Chen YK, Hsue SS, Lin LM. Correlation between inducible nitric oxide synthase and p53 expression for DMBA-induced hamster buccal-pouch carcinomas. Oral Dis 2003;9(5):227–34.
[16] Ambs S, Merriam WG, Ogunfusika MO, Bennett WP, Ishibe N, Hussain SP, et al. p53 and vascular endothelial growth factor regulate tumor growth of NOS2-expressing human carcinoma cells. Nat Med 1998;12:1371–6.
[17] Shang ZJ, Li JR, Li ZB. Effects of exogenous nitric oxide on oral squamous cell carcinoma: an in vitro study. J Oral Maxillofac Surg 2002;11(8):905–10.
[18] Harada K, Supriatno, Kawaguchi S, Tomitaro O, Yoshida H, Sato M. Overexpression of iNOS gene suppresses the tumorigenicity and metastasis of oral cancer cells. In Vivo 2004;11(4):449–55.
[19] Lepoivre M, Flaman JM, Henry Y. Early loss of the tyrosyl radical in ribonucleotide reductase of adenocarcinoma cells producing nitric oxide. J Biol Chem 1992;267:22994–3000.
[20] Chung HT, Pae HO, Choi BM, Billiar TR, Kim YM. Nitric oxide as a bioregulator of apoptosis. Biochem Biophys Res Commun 2001;282(5):1075–9.
[21] Aranda E, López-Pedrera C, De La Haba-Rodríguez JR, Rodríguez-Ariza A. Nitric oxide and cancer: the emerging role of S nitrosylation. Curr Mol Med 2012;12:50–67.

[22] Baritaki S, Huerta-Yepez S, Sahakyan A, Karagiannides I, Bakirtzi K, Jazirehi A, et al. Mechanisms of nitric oxide-mediated inhibition of EMT in cancer: inhibition of the metastasis-inducer Snail and induction of the metastasis-suppressor RKIP. Cell Cycle 2010;9:4931–40.

[23] Zhao SF, Tong XY, Zhu FD. Nitric oxide induces oral squamous cell carcinoma cells apoptosis with p53 accumulation. Oral Oncol 2005;41(8):785–90.

[24] Brennan PA, Palacios-Callender M, Umar T, Tant S, Langdon JD. Expression of type 2 nitric oxide synthase and p21 in oral squamous cell carcinoma. Int J Oral Maxillofac Surg 2002;31(2):200–5.

[25] Gallo O, Emanuela M, Lucia M, Franchi A, Fini-storchi I, Vergari WA, et al. Role of NO in angiogenesis and tumor progression in head and neck cancer. J Natl Cancer Inst 1998;90:587–96.

[26] Korde Choudhari S, Sridharan G, Gadbail A, Poornima V. Nitric oxide and oral cancer: a review. Oral Oncol 2012;48:475–83.

[27] Evig CB, Kelley EE, Weydert CJ, Chu Y, Buettner GR, Burns CP. Endogenous production and exogenous exposure to nitric oxide augment doxorubicin cytotoxicity for breast cancer cells but not cardiac myoblasts. Nitric Oxide 2004;10(3):119–29.

[28] Lehrman S. Virus treatment questioned after gene therapy death. Nature 1999;401(6753):517–8.

[29] Huerta S, Chilka S, Bonavida B. Nitric oxide donors: novel cancer therapeutics. Int J Oncol 2008;33:909–27 [Review].

[30] Yasuda H, Yamaya M, Nakayama K, Sasaki T, Ebihara S, Kanda A, et al. Randomized phase II trial comparing nitroglycerin plus vinorelbine and cisplatin with vinorelbine and cisplatin alone in previously untreated stage IIIB/IV non-small-cell lung cancer. J Clin Oncol 2006;24:688–94.

[31] Yasuda H, Nakayama K, Watanabe M, Suzuki S, Fuji H, Okinaga S, et al. Nitroglycerin treatment may enhance chemosensitivity to docetaxel and carboplatin in patients with lung adenocarcinoma. Clin Cancer Res 2006;12:6748–57.

[32] Bonavida B, Baritaki S, Huerta-Yepez S, Vega MI, Chatterjee D, Yeung K. Novel therapeutic applications of nitric oxide donors in cancer: roles in chemo- and immunosensitization to apoptosis and inhibition of metastases. Nitric Oxide 2008;19(2):152–7.

[33] Lee DH, Pfeifer GP. Mutagenesis induced by the nitric oxide donor sodium nitroprusside in mouse cells. Mutagenesis 2007;22:63–7.

[34] Rigas B. Novel agents for cancer prevention based on nitric oxide. Biochem Soc Trans 2007;35:136–8.

[35] Fraix A, Blangetti M, Guglielmo S, Lazzarato L, Marino N, Cardile V, et al. Light-tunable generation of singlet oxygen and nitric oxide with a bichromophoric molecular hybrid: a bimodal approach to killing cancer cells. ChemMedChem 2016;11(12):1371–9.

[36] Fu J, Liu L, Huang Z, Lai Y, Ji H, Peng S, et al. Hybrid molecule from O^2-(2,4-dinitrophenyl)diazeniumdiolate and oleanolic acid: a glutathione *S*-transferase π-activated nitric oxide prodrug with selective anti-human hepatocellular carcinoma activity and improved stability. J Med Chem 2013;56(11):4641–55.

[37] Kashfi K, Chattopadhyay M, Kodela R. NOSH-sulindac (AVT-18A) is a novel nitric oxide- and hydrogen sulfide-releasing hybrid that is gastrointestinal safe and has potent anti-inflammatory, analgesic, antipyretic, anti-platelet, and anti-cancer properties. Redox Biol 2015;6:287–96.

[38] Vannini F, MacKessack-Leitch AC, Eschbach EK, Chattopadhyay M, Kodela R, Kashfi K. Synthesis and anti-cancer potential of the positional isomers of NOSH-aspirin (NBS-1120) a dual nitric oxide and hydrogen sulfide releasing hybrid. Bioorg Med Chem Lett 2015;25(20):4677–82.

[39] Kodela R, Chattopadhyay M, Velázquez-Martínez CA, Kashfi K. NOSH-aspirin (NBS-1120), a novel nitric oxide- and hydrogen sulfide-releasing hybrid has enhanced chemo-preventive properties compared to aspirin, is gastrointestinal safe with all the classic therapeutic indications. Biochem Pharmacol 2015;98(4):564–72.

[40] Chattopadhyay M, Kodela R, Duvalsaint PL, Kashfi K. Gastrointestinal safety, chemotherapeutic potential, and classic pharmacological profile of NOSH-naproxen (AVT-219) a dual NO- and H2S-releasing hybrid. Pharmacol Res Perspect 2016;4(2):e00224.

[41] Seabra AB, de Lima R, Calderón M. Nitric oxide releasing nanomaterials for cancer treatment: current status and perspectives. Curr Top Med Chem 2015;15(4):298–308.

[42] Saraiva J, Marotta-Oliveira SS, Cicillini SA, Eloy Jde O, Marchetti JM. Nanocarriers for nitric oxide delivery. J Drug Deliv 2011;2011:936438.

[43] Duan S, Cai S, Yang Q, Forrest ML. Multi-arm polymeric nanocarrier as a nitric oxide delivery platform for chemotherapy of head and neck squamous cell carcinoma. Biomaterials 2012;33(11):3243–53.

[44] Kanayama N, Yamaguchi K, Nagasaki Y. PEGylated polymer micelle-based nitric oxide (NO) photodonor with NO-mediated antitumor activity. Chem Lett 2010;39:10081009.

[45] Deniz E, Kandoth N, Fraix A, Cardile V, Graziano AC, Lo Furno D, et al. Photoinduced fluorescence activation and nitric oxide release with biocompatible polymer nanoparticles. Chemistry 2012;18(49):15782–7.

[46] Kirejev V, Kandoth N, Gref R, Ericson MB, Sortino S. A polymer-based nanodevice for the photoregulated release of NO with two-photon fluorescence reporting in skin carcinoma cells. J Mater Chem B 2014;2: 1190–5.

[47] Friedman AJ, Han G, Navati MS, Chacko M, Gunther L, Alfieri A, et al. Sustained release nitric oxide releasing nanoparticles: characterization of a novel delivery platform based on nitrite containing hydrogel/glass composites. Nitric Oxide 2008;19(1):12–20.

[48] Lin QY, Yang HQ, Xie SS, Wang YH, Ye Z, Chen SQ. Detecting early breast tumour by finite element thermal analysis. J Med Eng Technol 2009;33(4):274–80.

[49] Tai LA, Wang YC, Yang CS. Heat-activated sustaining nitric oxide release from zwitterionic diazeniumdiolate loaded in thermo-sensitive liposomes. Nitric Oxide 2010;23(1):60–4.

[50] Stevens EV, Wells A, Shin JH, Liu J, Der CJ, Schoenfisch MH. Nitric oxide-releasing silica nanoparticle inhibition of ovarian cancer cell growth. Mol Pharm 2010;7(3):775–85.

[51] Liang H, Nacharaju P, Friedman A, Friedman JM. Nitric oxide generating/releasing materials. Future Sci OA 2015;1(1):149.

[52] Sudhesh P, Tamilarasan K, Arumugam P, Berchmans S. Nitric oxide releasing photoresponsive nanohybrids as excellent therapeutic agent for cervical cancer cell lines. ACS Appl Mater Interfaces 2013;5(17):8263–6.

[53] Mocellin S, Bronte V, Nitti D. Nitric oxide, a double edged sword in cancer biology: searching for therapeutic opportunities. Med Res Rev 2007;27(3):317–52.

[54] Riganti C, Miraglia E, Viarisio D, Costamagna C, Pescarmona G, Ghigo D, et al. Nitric oxide reverts the resistance to doxorubicin in human colon cancer cells by inhibiting the drug efflux. Cancer Res 2005;65(2):516–25.

[55] Bratasz A, Weir NM, Parinandi NL, Zweier JL, Sridhar R, Ignarro LJ, et al. Reversal to cisplatin sensitivity in recurrent human ovarian cancer cells by NCX-4016, a nitro derivative of aspirin. Proc Natl Acad Sci U S A 2006;103(10):3914–9.

[56] Huerta S, Baay-Guzman G, Gonzalez-Bonilla CR, Livingston EH, Huerta-Yepez S, Bonavida B. In vitro and in vivo sensitization of SW620 metastatic colon cancer cells to CDDP-induced apoptosis by the nitric oxide donor DETANONOate: involvement of AIF. Nitric Oxide 2009;20:182–94.

[57] Duong HT, Kamarudin ZM, Erlich RB, Li Y, Jones MW, Kavallaris M, et al. Intracellular nitric oxide delivery from stable NO-polymeric nanoparticle carriers. Chem Commun (Camb) 2013;49(39):4190–2.

[58] Munaweera I, Shi Y, Koneru B, Patel A, Dang MH, Di Pasqua AJ, et al. Nitric oxide- and cisplatin-releasing silica nanoparticles for use against non-small cell lung cancer. J Inorg Biochem 2015;153:23–31.

[59] Wang MR, Chiu SJ, Chou HC, Hu TM. An efficient S-NO-polysilsesquioxane nano-platform for the co-delivery of nitric oxide and an anticancer drug. Chem Commun (Camb) 2015;51(86):15649–52.

[60] Fan J, He Q, Liu Y, Zhang F, Yang X, Wang Z, et al. Light-responsive biodegradable nanomedicine overcomes multidrug resistance via NO-enhanced chemosensitization. ACS Appl Mater Interfaces 2016;8(22):13804–11.

CHAPTER

9

Nitric Oxide in Rectal Cancer: From Mice to Patients

*Jonathan Imran**, *Sergio Huerta**,†

*University of Texas Southwestern Medical Center, Dallas, TX, United States

†VA North Texas Health Care Center, Dallas, TX United States

Abstract

The wide response to treatment in rectal cancer to neoadjuvant chemoradiation (CRT) provides a unique opportunity to study the role of agents to overcome resistance. Up to one-fourth of patients treated with CRT experience no observable response to the same treatment where similarly treated patients might achieve a complete regression of the tumor. We have developed in vitro and in vivo models for the study of radioresistance. We have employed several agents to overcome tumor resistance and found that nitric oxide was one of the most potent radiosensitizers. We extended our experience in vivo and undertook a phase I dose-escalation trial with the goal of determining the safety of utilizing nitric oxide from nitroglycerine transdermal patches in addition to conventional treatment in patients with rectal cancer. The present discussion addresses our models of rectal cancer and the results of our radiosensitizing strategies with nitric oxide from mice to patients.

Abbreviations

CRT	chemoradiation
FDA	Food and Drug Administration
5-FU	5-fluorouracil
GTN	glyceryl trinitrate
IR	irradiation
NO	nitric oxide
NSCLC	nonsmall-cell lung cancer
NONOates	*N*-diazeniumdiolates
pCR	pathological complete response

Nitric Oxide (Donor/Induced) in Chemosensitization, Volume 1
http:/dx.doi.org/10.1016/B978-0-12-811020-1.00009-0

INTRODUCTION

In the United States, 39,220 Americans were affected by rectal cancer in 2016. Rectal and colon cancer claimed 49,190 lives that same year [1]. The current standard of care for patients with stage II (cT3-cT4, N0) and III (any T and N+) rectal cancer is a trimodality approach: preoperative chemoradiation (CRT) (ionizing radiation + chemotherapy), followed by surgical resection of the tumor (total mesorectal excision) and adjuvant therapy [2]. Thus, in contrast to patients affected by colon cancer, patients with rectal cancer receive preoperative CRT [2]. The original intended goal of this strategy was to downstage tumors such that the operative intervention would spare the rectal sphincter and lead to a decrease in locoregional recurrence.

Unfortunately, this treatment leads to a wide range in tumor response from complete regression after surgery (pathological complete response, pCR) in 21% of the patients, to no response in 25% of similarly treated subjects, to patients with a partial response in 54% of the cases (Fig. 1) [3,4]. Our findings are consistent with other reports [5]. It is currently known that patients who achieve a pCR have superior oncological outcomes compared with patients who only have a partial or no response to CRT [6]. It is, therefore, clinically logical to seek strategies to maximize the number of patients who achieve a pCR.

It is currently unclear of the particular characteristics of the tumor or the host that dictate this variability in the response to CRT. A rational approach to overcome this problem is combination treatment. However, while multiple combinations to conventional drugs for the management of colorectal cancer such as 5-fluorouracil (5-FU), oxaliplatin, irinotecan, cetuximab, and/or bevacizumab in various combinations with ionizing radiation have been tested, none has had a substantial change in the observed rate of pCR [7–9]. New strategies in drug development are clearly needed in this endeavor.

Over the past three decades, there has been considerable debate over the role of nitric oxide (NO) in oncology. A dual role of tumor formation has been assigned to this diatomic molecule. It has been reported that at low concentrations, NO might be tumorigenic, while at high concentrations, an antioncogenic property is elicited [10–12].

In the present review, we summarize the radiosensitizing properties of NO in vitro and in vivo from the NO donor: DETANONOate. We also report the extensions of our preclinical

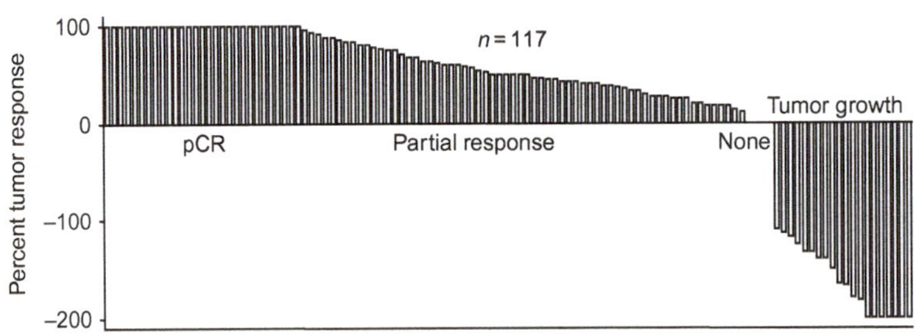

FIG. 1 Response to chemoradiation. The response to chemoradiation is wide. Each bar on the X-axis represents an individual patient. The Y-axis represents the clinical response to preoperative CRT. *Modified with permission from Huerta S, Hrom J, Gao X, Saha D, Anthony T, Reinhart H, et al. Tissue microarray constructs to predict a response to chemoradiation in rectal cancer. Dig Liver Dis 2010;42:679–84.*

observations to patients affected with rectal cancer receiving conventional CRT. In addition to conventional treatment, patients with rectal cancer received NO from nitroglycerine patches in a dose-escalation phase I clinical trial.

IN VIVO AND IN VITRO MODELS FOR THE STUDY OF RECTAL CANCER

The observed wide response to the same treatment in rectal cancer provides a unique opportunity to examine pathways that might lead to overcoming this resistance. In order to examine pathways that might lead to resistance and ways to overcome it, there has to be a model to study the observed resistance to CRT in a reproducible manner. A logical and practical approach is to investigate the response to treatment in vitro. We have examined six colorectal cancer cell lines to determine the response to radiation treatment, HT-29, DLD-1, SW480, SW620, HCT-116, and SW837 [13]. These cell lines were obtained from the American Tissue Culture Collection, and the following characteristics of the cells are available on their website and the literature: HT-29 cells were derived from a 44-year-old Caucasian woman with colorectal adenocarcinoma. All of these cells have mutations of the *p53* gene, except for the HCT-116 cells (p53-Wt). HT-29 cells have mutations of both alleles of the *p53* gene (p53-null) [14]. Both SW480 and SW620 cell lines were derived from the same patient. The SW480 cell line was established from a primary Dukes' stage II colon adenocarcinoma from a 50-year-old Caucasian male, while the SW620 cell line was cultured from a lymph node metastasis from the same patient at a later time. The DLD-1 cell line was established from an adult male with adenocarcinoma of the colon. The SW837 cell line was derived from a 53-year-old Caucasian male with rectal cancer. HCT-116 cells were cultured from an adult male with colon cancer. HCT-116 cells have microsatellite instability [13].

When these cell lines are exposed to an equivalent dose of ionizing radiation as that given to patients, a variable response to ionizing radiation is observed (Fig. 2). Thus, this response is similar to that observed in patients. When these cells are placed in nude mice as xenografts and then the mice are treated with similar doses of ionizing radiation as that given to humans, a similar response as that in vitro is observed (Fig. 3). A model with xenografts in an orthotropic fashion was also established for the study of rectal cancer [15]. A practical approach in the understanding of resistance to treatment is to take the most radioresistant cell line (HT-29) and the most radiosensitive cell line (HCT-116) and determine their genotypic differences (reviewed in Ref. [13]; Fig. 4).

GENOTYPIC DIFFERENCES BETWEEN HT-29 AND HCT-116 CELLS

Our results have been previously reported [13] (Fig. 5) briefly:

1. HT-29 cells expressed lower levels of DNA-PKcs and Ku proteins compared with HCT-116 [16].
2. Cell cycle kinetics demonstrated that the radiosensitive HCT-116 cells substantially accumulate in the G-2 phase of the cell cycle. HT-29 cells proceed through the cell cycle [17,18].

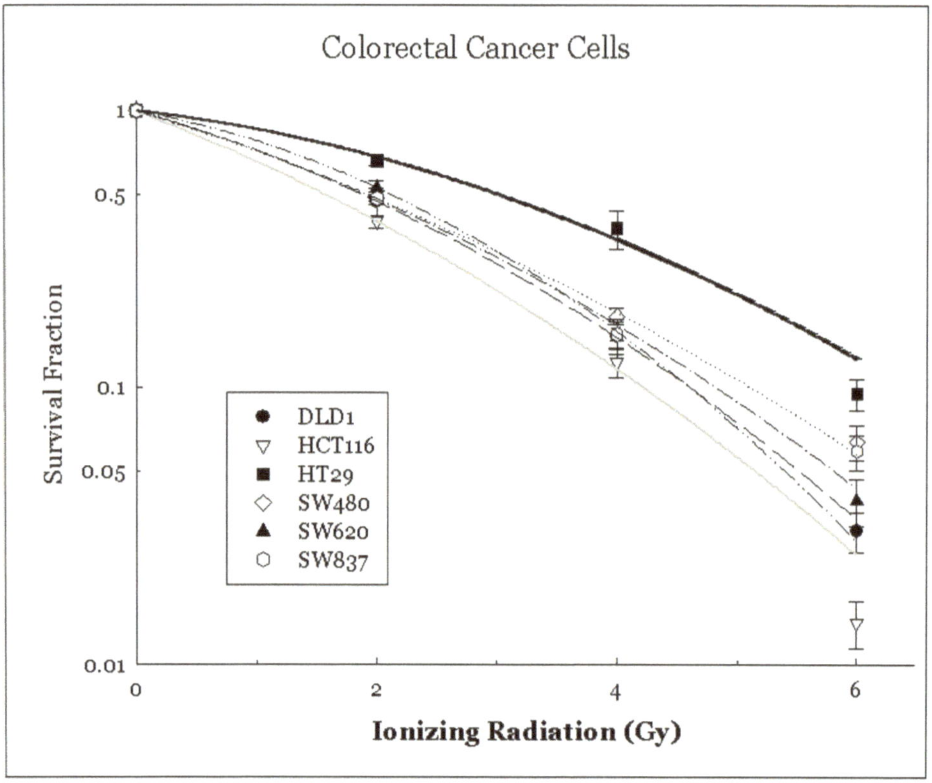

FIG. 2 The clonogenic survival assay. Clonogenic survival assay of six colorectal cancer cell lines revealed a wide range in response to ionizing radiation. *Reproduced with permission from Gao X, Saha D, Kapur P, Anthony T, Livingston EH, Huerta S. Radiosensitization of HT-29 cells and xenografts by the nitric oxide donor DETANONOate. J Surg Oncol 2009;100:149–58.*

3. Apoptosis analysis in HCT-116 versus HT-29 cells showed overexpression of p21 and decreased expression of p53, Bax, Bcl-2, and survivin [14].

Our observations show that there is not a unifying pathway to explain the radioresistance that could be applicable to other models or that is consistent with other reports in the literature [13]. While no genotypic characteristics in our model have been successful, this model can also be used to attempt to identify novel radiosensitizers that might be possibly used in clinical trials. We have been more productive with this endeavor compared with attempts at elucidating unifying pathways leading to radioresistance. Taking the most radioresistant cells and xenografts, the obvious question to ask is whether we can make it respond to treatment and respond to ionizing radiation with a phenotype similar to the HCT-116 cells. We have employed a number of radiosensitizing agents in this model such as (1) the smac mimetic JP-1201 [19], (2) the NO donor JS-K [20], (3) and the NO donor DETANONOate [14]. Of these, the most potent agent has been DETANONOate.

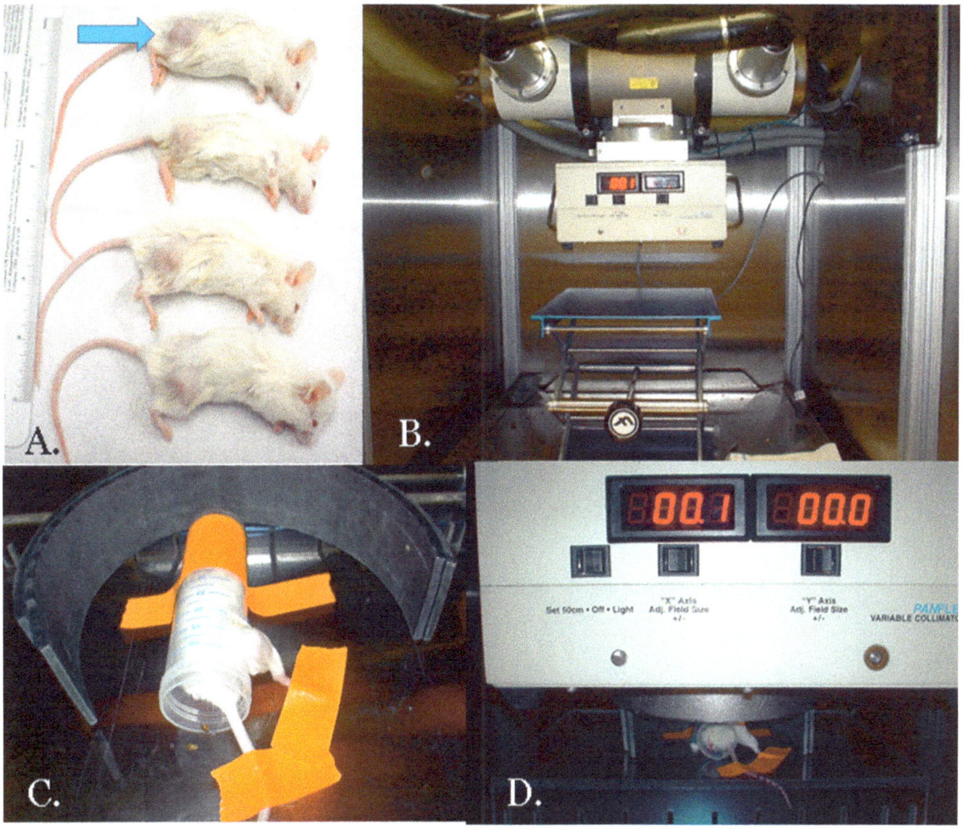

FIG. 3 Response in tumor xenografts. Nude mice were injected with colorectal cancer cells and developed xenografts (A, arrow). An ionizing radiation machine for small animals delivers radiation in similar doses as those given to patients affected with rectal cancer (panel B). Only the xenografts are treated while shielding the rest of the mice with lead (panels C and D). After treatment, xenografts display the same response at that observed in cells.

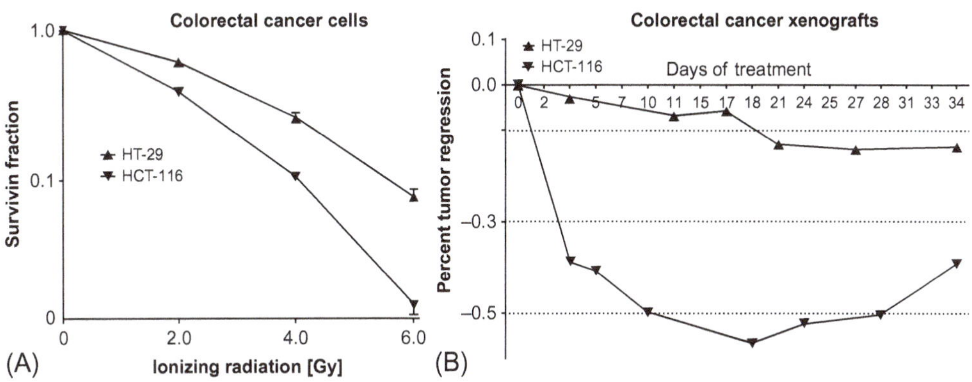

FIG. 4 In vitro and in vivo comparison of cell survival. (A) Survival of cells following treatment in vitro with various doses of ionizing radiation. The most radiosensitive cell line (HCT-116) is compared with the radioresistant HT-29 cell line. (B) In vivo tumor regression of tumor cell xenografts following treatment with ionizing radiation. The same response is observed in xenografts. *Reprinted with permission from Ramzan Z, Nassri AB, Huerta S. Genotypic characteristics of resistant tumors to pre-operative ionizing radiation in rectal cancer. World J Gastrointest Oncol 2014;6:194–210.*

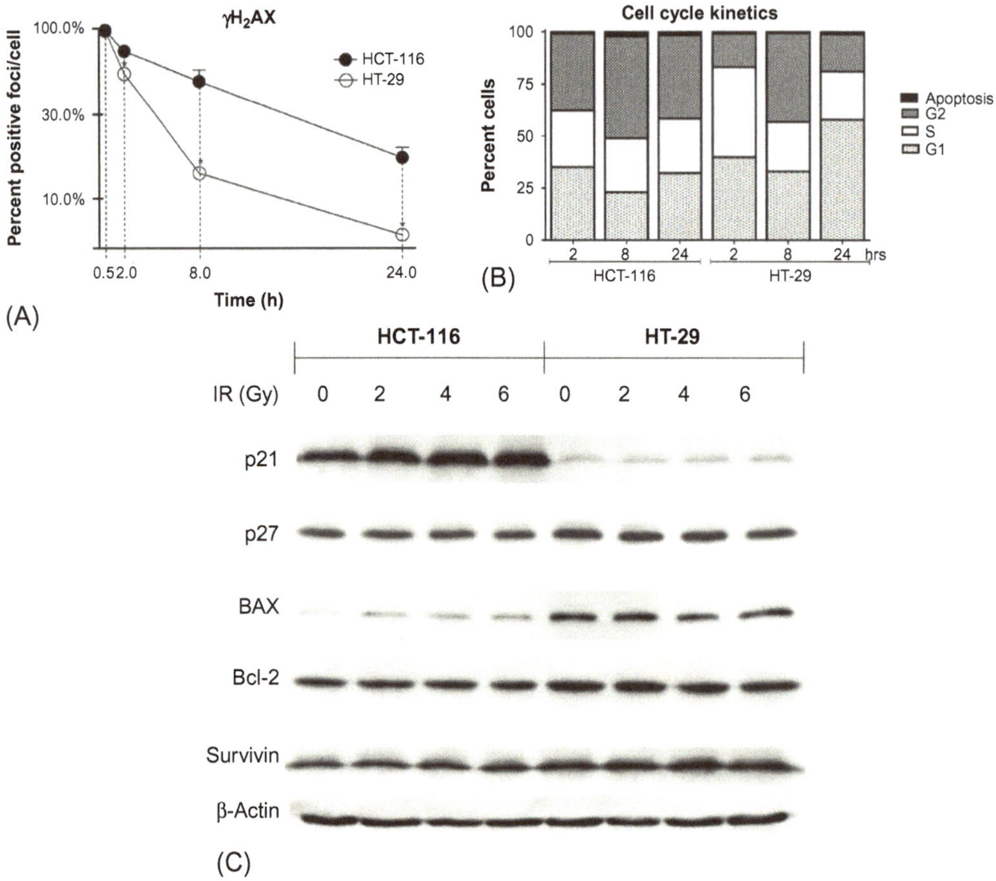

FIG. 5 Tumor cell death. (A) Analysis of different phenotypic properties of HCT-116 and HT-29 cell lines. HCT-116 cells undergo more DNA damage when treated with the same dose of IR compared with HT-29 cells [13]. (B) Cell cycle kinetics of HCT-116 and HT-29 cells treated with 2.0 Gy IR. There is an accumulation of cells in G2 in HCT-116 cells. HT-29 cells continue through the cell cycle in spite of receiving the same dose of IR [13]. (C) Genotypic differences between HCT-116 and HT-29 cells with regard to CDKIs, proapoptotic BAX, antiapoptotic Bcl-2, and survivin. There was a substantial augmentation of p21 with increasing doses of IR compared with the almost undetectable levels of p21 in HT-29 cells. Paradoxically, the levels of BAX were higher in the HT-29 (p53-null) cells compared with the HCT-116 (p53-Wt) cells. Both Bcl-2 and survivin displayed more expression in the HT-29 cells [14].

NO IN CANCER

The role of NO in carcinogenesis was placed in the international forefront after the first International Conference on Nitric Oxide and Cancer (NO Cancer) in 2007 in Paris. This small diatomic molecule has been shown to have a pivotal role in carcinogenesis. However, this role has not been without controversy as reports have split its activity as tumorigenic and tumoricidal (reviewed in Refs. [10,21]). There is currently a general understanding that at low concentrations, NO is tumorigenic and, at high concentrations, it has antioncogenic properties [11,12,21] (Fig. 6).

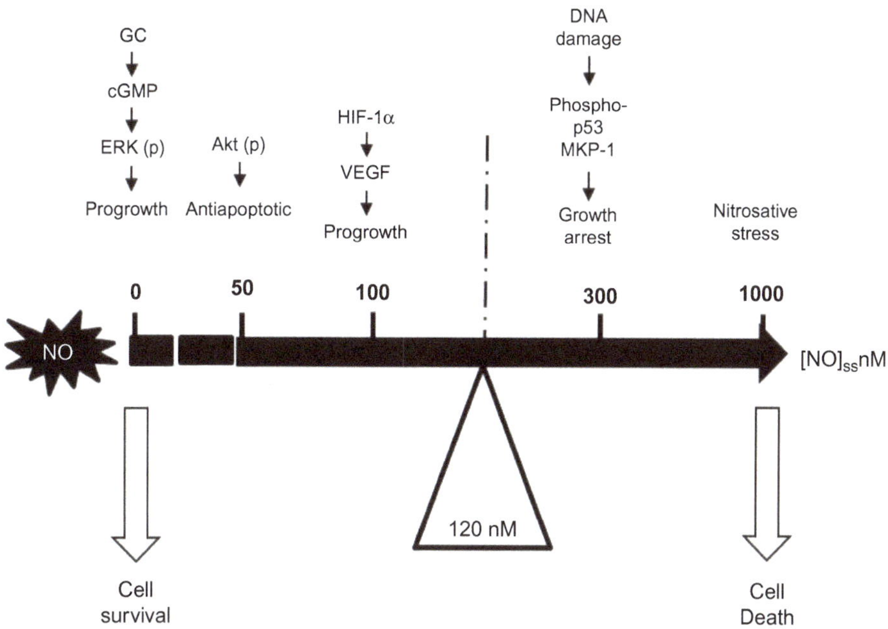

FIG. 6 Threshold concentration levels for the antitumor versus protumor growth of NO with potential mechanisms of action. *Modified with permission from Huerta S. Nitric oxide for cancer therapy. Future Sci OA 2016;1:1–9.*

NO is a highly reactive radical, and it is challenging to predict its biological effects in a given system. NO donors are, therefore, extremely useful in the study of the role of NO in carcinogenesis. NO donors are capable of producing a sustained released with a wide range of half-lives. With a predictable estimated dose, NO donors have become useful tools to study the biological properties of NO in vitro and in vivo models of carcinogenesis.

Multiple NO donors have been used to study the role of NO in carcinogenesis. Of these donors, the group belonging to the class of the *N*-nitroso compounds is of particular interest in the area of oncology. The *N*-nitroso compounds can be divided broadly into nitrosamines, hydroxy-*N*-nitrosamines, nitrosoimines, and diazeniumdiolates. The *N*-diazeniumdiolates (NONOates), in particular DETANONOate, has been studied in our model of rectal cancer (Fig. 7).

DETANONOate IN CANCER THERAPY

Of the many compounds belonging to the diazeniumdiolates, DETANONOate is a good compound to use in oncology therapy (Fig. 7G). It is a stable compound; it has a long half-life and closely mimics sustained endogenous NO release for over 20 h. The role of DETANONOate as an anticancer compound has previously been reported [10]. DETANONOate induces stasis and cell cycle arrest in the human breast cancer cells MDA-MB-231 [22] and apoptosis in the breast cancer cell line MDA-MB-468 [23]. Resistance to doxorubicin has also been overcome by DETANONOate [24]. DETANONOate has been able to act as a chemoimmune-sensitizer in many forms of cancer. DETANONOate reversed the

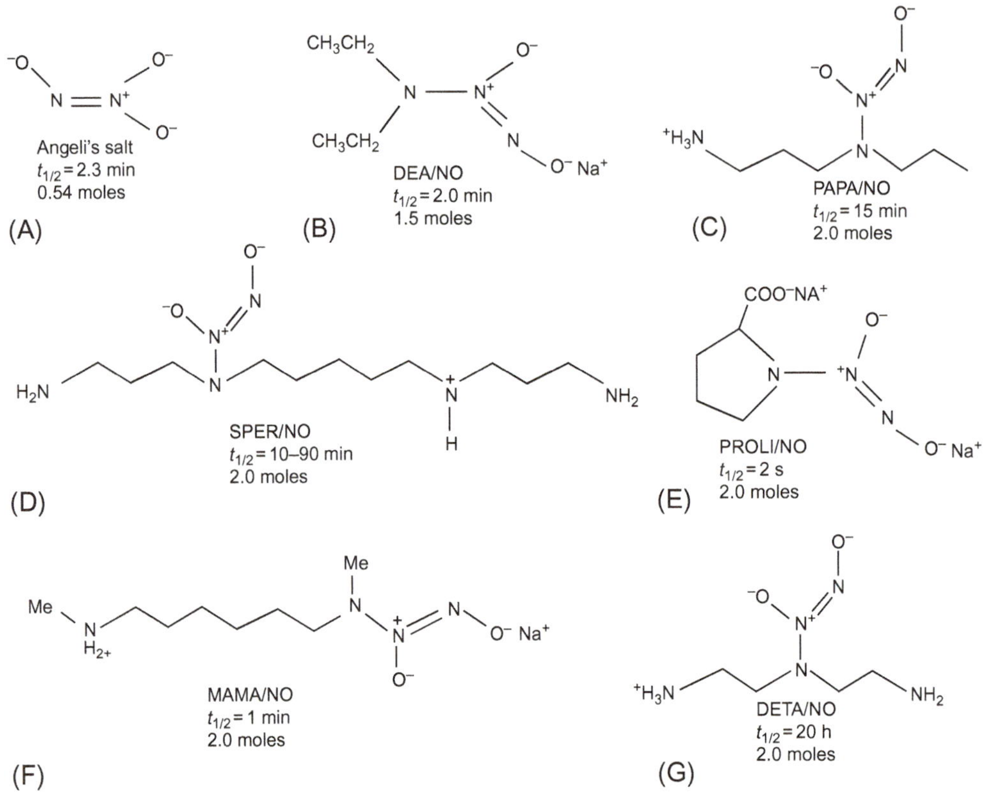

FIG. 7 The *N*-diazeniumdiolates (NONOates). The half-lives ($t_{1/2}$) are also shown at a pH of 7.4 and at a temperature of 37°C. *Reprinted with permission from Huerta S, Chilka S, Bonavida B. Nitric oxide donors: novel cancer therapeutics. Int J Oncol 2008;33:909–27 [Review].*

hypoxia-mediated resistance to chemotherapeutic agents in human breast carcinoma MDA-MB-231 cells and in B16F10 mouse melanoma cells [25].

DETANONOate enhanced cisplatin-mediated toxicity in Chinese hamster V79 lung fibroblasts and in head and in neck squamous cell carcinoma cells [26,27]. Mice receiving DETANONOate in addition to cisplatin had a significant reduction of xenografts and an increased survival compared with mice treated with cisplatin or vehicle alone [28]. In prostate cancer, DETANONOate immunosensitized PC-3 cells to TRAIL- and FasL-mediated apoptosis [29]. DETANONOate effectively chemosensitized SW620 colorectal cancer cells and xenografts to cisplatin-induced apoptosis [30]. These powerful properties of DETANONOate in cancer treatment led us to investigate its role in rectal cancer.

DETANONOate IN RECTAL CANCER

For this analysis, we utilized the most radioresistant of our cell lines (HT-29) to determine if radioresistance could be overcome by adding NO delivered from DETANONOate. Our hypothesis was that DETANONOate could act as a potent radiosensitizer in cells pretreated with this NO donor and could result in tumor reduction in SCID mice with established HT-29 xenografts. We found that DETANONOate effectively radiosensitized HT-29 cells and caused cell death, and we were able to observe a radiosensitive phenotype similar to the one displayed by HCT-116 cells (the cells with the most radiosensitive phenotype) [14].

Pretreatment of HT-29 cells with DETANONOate (1000 μM) prior to irradiation (IR) demonstrated a significant inhibition compared with untreated cells following clonogenic studies at 2.0 and 4.0 Gy (47.0% and 15.0%, respectively). This response was similar in vivo. Mice were treated with 2.0 Gy IR over a period of five days (10.0 Gy) provided directly to the tumor. We found that the IR group had a significant reduction in tumor growth (37.6%) compared with control. However, the group that had been pretreated with DETANONOate prior to IR had a more substantial reduction of tumor growth (70.1%) compared with IR alone (Fig. 8).

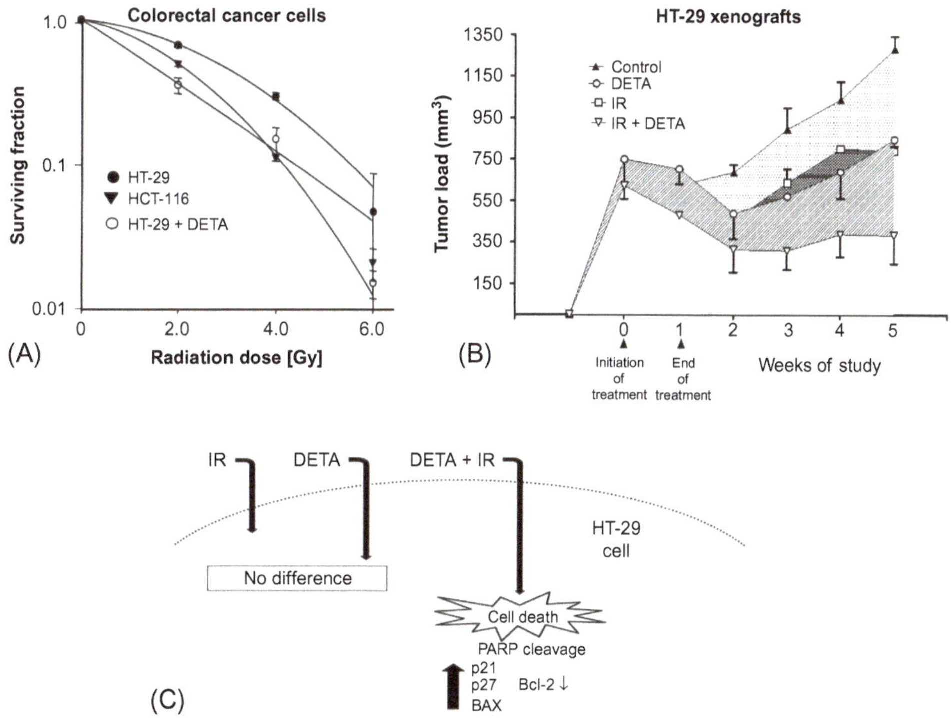

FIG. 8 Characterization of the colorectal cancer cell line HT-29 left untreated or pretreated with DETANONOate. The phenotype of cells pretreated with DETANONOate was similar to the most radiosensitive HCT-116 cells (panel A). DETANONOate was also a potent radiosensitizer in vivo (panel B). Cell death was associated with apoptosis as shown by the increase in PARP cleavage in an IR dose-response fashion with a fixed dose of DETANONOate (1000 μM). This response was associated with an increase in the levels of the cyclin-dependent kinase inhibitors (p21 and p27) and an increase in BAX expression with an ease in Bcl-2 (panel C). *With permission from Gao X, Saha D, Kapur P, Anthony T, Livingston EH, Huerta S. Radiosensitization of HT-29 cells and xenografts by the nitric oxide donor DETANONOate. J Surg Oncol 2009;100:149–58.*

MECHANISM OF ACTION OF DETANONOate IN RECTAL CANCER IN VITRO

In vitro, we examined the mechanism of action leading to the radiosensitizing activity of DETANONOate. Our results showed that radiosensitization was associated with a significant response in apoptosis as shown by a pronounced increased in the cleaved form of PARP-1 compared with untreated cells. This effect in apoptosis was accompanied with an increase of cyclin-dependent kinase inhibitors p21 and p27 and an increase in the proapoptotic BAX. A decrease in the levels of antiapoptotic Bcl-2 was also observed (Fig. 8; panel C) [14].

In a separate study, we showed that DETANONOate led to radiosensitization of HT-29 cells via attenuation of HIF-1alpha [31]. Thus, while the mechanisms of action of the radiosensitizing properties of NO remain unclear, improving the hypoxic environment in rectal cancer might facilitate the therapeutic effect of ionizing radiation leading to cell cycle arrest and apoptotic cell death [32,33].

NO FROM NITROGLYCERIN IN PATIENTS WITH RECTAL CANCER

Extending the experience of the treatment of colon cancer in the adjuvant or metastatic setting, chemotherapeutic agents such as 5-FU [34], oxaliplatin [35], irinotecan [36], bevacizumab, and cetuximab [37] have been used in rectal cancer in addition to ionizing radiation in the neoadjuvant setting. However, none of these have been shown to have a significant difference in response (assessed by the rate of pCR) compared with 5-FU [9]. In fact, in some cases, toxicity has been shown to be more pronounced without the added benefits of additional chemotherapeutic agents such as oxaliplatin [35].

Because of the lack of a substantial difference in increasing the rate of pCR, our group has focused on the development of novel radiosensitizing agents [9,14,19,20]. NO from DETANONOate was the most powerful radiosensitizers in our in vitro and in vivo models of rectal cancer. However, DETANONOate is not an approved Food and Drug Administration (FDA) compound to be used in humans, and systemic administration could lead to severe hypotension from the sustained release of NO. For low rectal cancers, it might be possible to proceed with local injection of a compound. While feasible in xenografts [30], in human tumors, this strategy remains to be tested. Further, FDA approval would need to be obtained if compounds such as DETANONOate were to be used.

Thus, the issue is whether there is an NO donor that could be utilized in humans in combination with conventional treatments. One of the oldest NO donors in clinical use is the class of NO donors belonging to the organic nitrates [10,21]. In fact, the organic nitrate glyceryl trinitrate (GTN) is notorious for its ability to relieve the pain from vasoconstriction of the cardiac vessels and has been used for the management of angina for nearly half a century [10,21]. This clinical ability of GTN has demonstrated safety in clinical use in humans. The biological half-life of GTN is short (1–4 s), but its metabolites might be active for up to 40 min [38]. In spite of the fact that the biological activity of this compound is not limited to the coronary vessels, the side effect profile of GTN is relatively mild and includes headaches and hypotension. In order to have a system with a sustained release of NO and its metabolites from GTN,

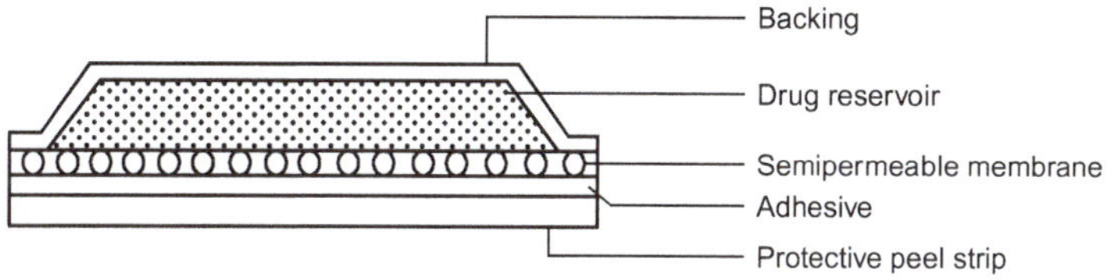

FIG. 9 Cross-sectional view of NTG patches.

GTN patches were developed and have been used by physicians for the management of angina since the 1980s [38] (Fig. 9). Thus, NO patches are NO donors that are currently in clinical use for chest pain. Can NTG patches be useful in oncology?

NO from NTG was shown to radiosensitize breast cancer cells and prostate cancer cells and established prostate cancer xenografts [39]. Further, mice with prostate cancer xenografts that received NTG patches in combination with chemotherapy had reduced tumor load compared with mice that received chemotherapy alone [40].

In humans, Yasuda et al. have shown that NTG patches had therapeutic properties in patients with nonsmall-cell lung cancer (NSCLC) [32,33]. Yasuda et al. conducted a phase II clinical trial inclusive of 120 patients with NSCLC. Patients were randomly assigned to receive chemotherapy and transdermal NTG patches versus chemotherapy and placebo. The response to chemotherapy was significantly higher in patients that received NTG patches compared to placebo. Further, the side effects experienced with transdermal patches were minimal [32,33]. Based on these findings, our group elected to investigate the radiosensitizing properties of transdermal patches in patients with rectal cancer receiving conventional treatments. We, therefore, performed an open-label, nonrandomized, multicohort, dose-escalation, phase I trial with the primary endpoint to evaluate the safety, tolerability, feasibility, dose-limiting toxicity, and maximum tolerated dose of topical nitroglycerin in addition to 5-FU and radiation therapy for neoadjuvant treatment of locoregionally advanced operable rectal cancer. A secondary endpoint was the rate of pCR [36].

In this study, patients were assigned to three sequential cohorts of escalating dose levels of commercially available NTG patches (0.2, 0.4, and 0.6 mg/h), each cohort consisted of three patients in a model previously described [41]. All patients received radiation therapy, 45–50 Gy in 25–28 fractions to the pelvis along with continuous infusion 5-FU 225 mg/m^2/day for the duration of the radiation therapy as per the institutional protocol.

The first cohort of patients received 0.2 mg/h NTG transdermal patch every day for the duration of preoperative CRT. The transdermal patches were placed in the buttock or upper thigh for 2 h prior to each radiation treatment (from Monday to Friday) for 12 h. The following cohorts were escalated only if no dose-limiting toxicity was observed.

A total of 13 patients were enrolled in this trial (age of 59.4 $\pm$ 2.5 (SEM) years old (range = 42–70 years old)). Toxicities were mild to moderate and manageable. Four patients developed asymptomatic grade III lymphopenia during the CRT that resolved promptly upon completion. A complete response was observed in two patients (2/12). Thus, pCR was 17%.

In this open-label, nonrandomized, multicohort, dose-escalation, phase I trial, we demonstrated that NTG patches in combination with standard CRT are safe, tolerable, and feasible at the maximum dose given (0.6 mg/h). Toxicity was similar to patients receiving standard neoadjuvant therapy.

CONCLUSION

We have established in vitro and in vivo models for the study of a response to CRT in rectal cancer. We studied a number of strategies to overcome the observed radioresistance and found that NO from DETANONOate was the most powerful radiosensitizer. Mice with established rectal cancer xenografts had a significant reduction in tumor load when treated with the combination of ionizing radiation and DETANONOate. We extended these observations in mice into patients. NO from NTG transdermal patches was safe at the highest doses utilized (0.6 mg/h) when it was combined with conventional treatments. These observations should serve as the foundation to proceed with phase II trials to determine the efficacy of NTG patches with increasing the number of patients who are able to achieve a pCR.

Acknowledgment

We would like to acknowledge the VA North Texas Health Care System for allowing us to conduct this research study.

Conflict of Interest: No potential conflicts of interest were disclosed.

References

[1] Siegel RL, Miller KD, Jemal A. Cancer statistics, 2016. CA Cancer J Clin 2016;66(1):7–30.
[2] Huerta S, Murray B, Olson C, Patel P, Anthony T. Current evidence-based opinions in the management of adenocarcinoma of the rectum. Indian J Surg 2009;71(6):356–62.
[3] Huerta S. New directions in the preoperative management of adenocarcinoma of the rectum. Crit Rev Oncog 2012;7:393–6.
[4] Huerta S, Hrom J, Gao X, Saha D, Anthony T, Reinhart H, et al. Tissue microarray constructs to predict a response to chemoradiation in rectal cancer. Dig Liver Dis 2010;42:679–84.
[5] Minsky BD, Roedel C, Valentini V. Combined modality therapy for rectal cancer. Cancer J 2010;6:253–61.
[6] Maas M, Nelemans PJ, Valentini V, et al. Long-term outcome in patients with a pathological complete response after chemoradiation for rectal cancer: a pooled analysis of individual patient data. Lancet Oncol 2010;11:835–44.
[7] Huerta S. Critical concepts in the oncogenesis of adenocarcinoma of the rectum. Crit Rev Oncog 2012;17:323–9.
[8] Huerta S. Rectal cancer and importance of chemoradiation in the treatment. Adv Exp Med Biol 2010;685:124–33.
[9] Huerta S. Radiosensitizing agents for the management of rectal cancer. Anticancer Drugs 2011;22:305–7.
[10] Huerta S, Chilka S, Bonavida B. Nitric oxide donors: novel cancer therapeutics. Int J Oncol 2008;33:909–27 [Review].
[11] Ridnour LA, Thomas DD, Switzer C, Flores-Santana W, Isenberg JS, Ambs S, et al. Molecular mechanisms for discrete nitric oxide levels in cancer. Nitric Oxide 2008;19:73–6.
[12] Wink DA, Ridnour LA, Hussain SP, Harris CC. The reemergence of nitric oxide and cancer. Nitric Oxide 2008;19:65–7.
[13] Ramzan Z, Nassri AB, Huerta S. Genotypic characteristics of resistant tumors to pre-operative ionizing radiation in rectal cancer. World J Gastrointest Oncol 2014;6:194–210.
[14] Gao X, Saha D, Kapur P, Anthony T, Livingston EH, Huerta S. Radiosensitization of HT-29 cells and xenografts by the nitric oxide donor DETANONOate. J Surg Oncol 2009;100:149–58.
[15] Huerta S, Gao X, Saha D. Murine orthotopic model for the assessment of chemoradiotherapeutic interventions in rectal cancer. Anticancer Drugs 2011;22:371–6.
[16] Gao X, Meyer J, Huerta S. Role of DNA-PKcs, Ku80 and Bax in radioresistance of HT-29 cells and xenografts. J Surg Res 2014;186:683.
[17] Huerta S, Gao X, Saha D. Mechanisms of resistance to ionizing radiation in rectal cancer. Expert Rev Mol Diagn 2009;9:469–80.

[18] Huerta S, Gao X, Dineen S, Kapur P, Saha D, Meyer J. Role of p53, Bax, p21, and DNA-PKcs in radiation sensitivity of HCT-116 cells and xenografts. Surgery 2013;154:143–51.
[19] Huerta S, Gao X, Livingston EH, Kapur P, Sun H, Anthony T. In vitro and in vivo radiosensitization of colorectal cancer HT-29 cells by the smac mimetic JP-1201. Surgery 2010;148:346–53.
[20] Huerta S, Gao X. Radiosensitization of colorectal HT-29 cells and xenografts by the nitric oxide donor JS-K. J Surg Res 2012;172:304.
[21] Huerta S. Nitric oxide for cancer therapy. Future Sci OA 2016;1:1–9.
[22] Pervin S, Singh R, Chaudhuri G. Nitric oxide-induced cytostasis and cell cycle arrest of a human breast cancer cell line (MDA-MB-231): potential role of cyclin D1. Proc Natl Acad Sci U S A 2001;98:3583–8.
[23] Pervin S, Singh R, Freije WA, Chaudhuri G. MKP-1-induced dephosphorylation of extracellular signal-regulated kinase is essential for triggering nitric oxide-induced apoptosis in human breast cancer cell lines: implications in breast cancer. Cancer Res 2003;63:8853–60.
[24] Muir CP, Adams MA, Graham CH. Nitric oxide attenuates resistance to doxorubicin in three-dimensional aggregates of human breast carcinoma cells. Breast Cancer Res Treat 2006;96:169–76.
[25] Matthews NE, Adams MA, Maxwell LR, Gofton TE, Graham CH. Nitric oxide-mediated regulation of chemosensitivity in cancer cells. J Natl Cancer Inst 2001;93:1879–85.
[26] Azizzadeh B, Yip HT, Blackwell KE, Horvath S, Calcaterra TC, Buga GM, et al. Nitric oxide improves cisplatin cytotoxicity in head and neck squamous cell carcinoma. Laryngoscope 2001;111:1896–900.
[27] Wink DA, Cook JA, Christodoulou D, Krishna MC, Pacelli R, Kim S, et al. Nitric oxide and some nitric oxide donor compounds enhance the cytotoxicity of cisplatin. Nitric Oxide 1997;1:88–94.
[28] Perrotta C, Bizzozero L, Falcone S, Rovere-Querini P, Prinetti A, Schuchman EH, et al. Nitric oxide boosts chemoimmunotherapy via inhibition of acid sphingomyelinase in a mouse model of melanoma. Cancer Res 2007;67:7559–64.
[29] Huerta-Yepez S, Vega M, Jazirehi A, Garban H, Hongo F, Cheng G, et al. Nitric oxide sensitizes prostate carcinoma cell lines to TRAIL-mediated apoptosis via inactivation of NF-kappa B and inhibition of Bcl-xl expression. Oncogene 2004;23:4993–5003.
[30] Huerta S, Baay-Guzman G, Gonzalez-Bonilla CR, Livingston EH, Huerta-Yepez S, Bonavida B. In vitro and in vivo sensitization of SW620 metastatic colon cancer cells to CDDP-induced apoptosis by the nitric oxide donor DETANONOate: involvement of AIF. Nitric Oxide 2009;20:182–94.
[31] Harris K, Gao X, Huerta-Yepez S, Kapul P, Huerta S. HIF-1-alpha and resistance to ionizing radiation in rectal cancer. Surg Curr Res 2014;4:200.
[32] Yasuda H, Nakayama K, Watanabe M, Suzuki S, Fuji H, Okinaga S, et al. Nitroglycerin treatment may enhance chemosensitivity to docetaxel and carboplatin in patients with lung adenocarcinoma. Clin Cancer Res 2006;12:6748–57.
[33] Yasuda H, Yamaya M, Nakayama K, Sasaki T, Ebihara S, Kanda A, et al. Randomized phase II trial comparing nitroglycerin plus vinorelbine and cisplatin with vinorelbine and cisplatin alone in previously untreated stage IIIB/IV non-small-cell lung cancer. J Clin Oncol 2006;24:688–94.
[34] Patel PA. Evolution of 5-fluorouracil-based chemoradiation in the management of rectal cancer. Anticancer Drugs 2011;22:311–6.
[35] Huerta S, Hrom J. Oxaliplatin as a radiosensitizing agent in rectal cancer. Anticancer Drugs 2011;22:317–23.
[36] Illum H, Wang DH, Dowell JE, Hittson WJ, Torrisi JR, Meyer J, et al. Phase I dose-escalation trial of nitroglycerin in addition to 5-fluorouracil and radiation therapy for neoadjuvant treatment of operable rectal cancer. Surgery 2015;158:460–5.
[37] Glynne-Jones R, Mawdsley S, Harrison M. Antiepidermal growth factor receptor radiosensitizers in rectal cancer. Anticancer Drugs 2011;22:330–40.
[38] Parker JD, Parker JO. Nitrate therapy for stable angina pectoris. N Engl J Med 1998;338:520–31.
[39] Frederiksen LJ, Sullivan R, Maxwell LR, Macdonald-Goodfellow SK, Adams MA, Bennett BM, et al. Chemosensitization of cancer in vitro and in vivo by nitric oxide signaling. Clin Cancer Res 2007;13:2199–206.
[40] Frederiksen LJ, Siemens DR, Heaton JP, Maxwell LR, Adams MA, Graham CH. Hypoxia induced resistance to doxorubicin in prostate cancer cells is inhibited by low concentrations of glyceryl trinitrate. J Urol 2003;170:1003–7.
[41] Ji Y, Wang SJ. Modified toxicity probability interval design: a safer and more reliable method than the 3+3 design for practical phase I trials. J Clin Oncol 2013;31:1785–91.

CHAPTER

10

Chemoprotective and Chemosensitizing Effects of Nitric Oxide and Other Biologically Active Gases in Breast Cancer Chemotherapy: Potential Implications

Suvajit Sen, Kirsten Jensen, Kathleen Brennan, Sivakumar Ramadoss, Gautam Chaudhuri

David Geffen School of Medicine at UCLA, Los Angeles, CA, United States

Abstract

Since the discovery in 1987 that vascular endothelial cells are able to synthesize nitric oxide (NO) from L-arginine, the existence of this biochemical pathway in many other cell types has been thoroughly documented, and its relevance in biology has become apparent. NO, a free-radical gas, synthesized by a family of isoenzymes called NO synthases, was shown to play a key role as a cell-signaling molecule in the vascular nervous and immune systems. It also exhibited a cytostatic/cytotoxic property at higher concentrations as generated by activated macrophages and endothelial cells. The role of NO in cancer biology and more so in breast cancer, however, started to be elucidated only in 1995 by Thomsen et al. (1995). In this chapter, we will review some of the key aspects of endogenous NO as produced in breast cancer cells and breast tumors. We will allude to the early studies that described NO production in breast tumor tissues and breast cancer cells and point to some of its proposed functions in breast cancer homeostasis. We will discuss how these findings could be of therapeutic benefit and finally end with some information on how a mechanistic cross talk between endogenously produced NO and another gaseous signaling molecule, hydrogen sulfide (H_2S), in breast cancer cells may open up new opportunities for intervention. We will also discuss one potential mechanism of infiltration of tumor-associated macrophages in breast cancer and its clinical significance.

Nitric Oxide (Donor/Induced) in Chemosensitization, Volume 1
http://dx.doi.org/10.1016/B978-0-12-811020-1.00010-7

Abbreviations

DETA	diethylenetriamine
H_2S	hydrogen sulfide
L-NAME	L-N^G-nitroarginine methyl ester
NO	nitric oxide
NOHA	*N*-hydroxy-L-arginine
ODQ	1H-[1,2,4]oxadiazolo[4,3-a]quinoxalin-1-one

ACTIVITY OF NO SYNTHASE AND ITS CELLULAR LOCALIZATION IN HUMAN BREAST TUMORS

In 1995, Thomsen et al. [1] first demonstrated that NO biosynthesis and NO synthase activity were high for tumor tissues obtained from primary cancers compared with benign lesions and normal breast tissues in which the activity was low or undetectable. Tissues were collected from 34 breast tumors. Normal breast tissues were collected from six reduction mammoplasty specimens, which served as a control group.

NO biosynthesis was found to be significantly greater for tissue explants from grade III compared with grade II invasive ductal carcinomas. The biochemical observations of calcium-independent enzyme activity and the generation of nitrite and nitrate during culture of viable tumor tissue explants suggested the presence of inducible NO synthase (iNOS).

Immunohistochemical studies revealed the presence of iNOS predominantly in macrophages present in the breast tumor tissues. Interestingly, *breast tumor cells* were not labeled with antibodies against iNOS. The authors thus concluded that the generation of NO in breast tumors was most likely by infiltrated macrophages. They hypothesized that NO derived from breast tumor might cause cytostasis/cytotoxicity or it might also increase blood flow and promote angiogenesis supporting the positive correlation between NOS activity and higher-grade tumors.

In 1997, Dueñas-Gonzalez et al. [2] analyzed the expression of iNOS in 22 primary breast tumors and investigated its association to other gene products related to the metastatic ability of tumor cells. They observed that the expression of iNOS in breast cancer tissues positively correlated with the presence of axillary lymph node metastases. These data suggested that NO synthesis and the resulting increase in blood flow to the tumor might play a role in the facilitation of tumor metastasis. Thus, so far, it was known that breast tissues produced NO, most likely by iNOS present in tumor-infiltrated macrophages or other stromal cells, and that it stimulated tumor growth and metastasis. However, whether breast cancer epithelial cells derived from tumors produced NO was still not completely established.

To resolve this, Walter et al. [3] attempted to investigate the cellular distribution of NOS isoforms in a series of benign and malignant epithelial cells derived from breast tumors and normal breast tissues. Immunohistochemistry revealed that in samples of benign disease, the number of iNOS-positive epithelial cells was about 69% ($n=50$). In samples of grade II invasive ductal breast carcinomas, the number of iNOS-positive epithelial cells was about 62% compared with about 12% in samples of grade III carcinomas ($P<0.0001$). iNOS protein was also identifiable in most of the epithelial cells of normal breast tissue ($n=4$). In contrast, eNOS was found to be restricted to vascular endothelial cells in all the specimens studied. Since the presence of tumor-cell-derived iNOS protein was inversely related to the tumor's metastatic potential, the authors concluded that endogenous tumor-cell-derived iNOS expression might have an inhibitory effect on the metastatic processes in breast cancer.

Furthermore, Vakkala et al. [4] studied the expression of iNOS and its association to apoptosis and angiogenesis in 43 in situ and 68 invasive breast carcinomas. It was observed that the proportion of cases with iNOS-positive tumor cells increased in in situ carcinomas from grade I to grade III (20%, 46.2%, and 73.3%, respectively). In invasive ductal carcinomas, there were more cases with iNOS-positive tumor cells than with in situ carcinomas ($P=0.04$). Carcinomas with both iNOS-positive tumor and stromal cells had a higher apoptotic index ($P=0.02$). A higher number of iNOS-positive stromal cells are associated with metastatic disease ($P=0.05$). The authors concluded that breast carcinoma cells in addition to stromal cells expressed iNOS and were capable of producing NO. Carcinomas with iNOS-positive tumor and stromal cells exhibited a higher apoptotic index and increased vascularization that suggested that iNOS contributed to the promotion of apoptosis and angiogenesis in breast carcinoma. The association of the number of iNOS-positive stromal cells with metastatic disease might be attributable to stimulation of angiogenesis, resulting in a higher density and consequently a higher probability of tumor cells to invade. In contrast, the iNOS associated to the tumor cells might promote apoptosis as described in the earlier report.

Nonetheless, this study concluded that iNOS expression was apparent in both stromal cells and tumor-derived epithelial cells and that increased iNOS expression was associated with both tumor growth and a higher apoptotic index.

Taken together, it was thus far apparent that iNOS could be expressed by both stromal cells and tumor-derived epithelial cells in breast cancer. In addition, NO produced by stromal cells could promote tumor growth by enhancing tumor vascularization, whereas iNOS in tumor-derived epithelial cells may induce apoptosis of tumor cells. This paradox warranted mechanistic studies utilizing homogeneous population of breast cancer cells to specifically delineate the role of breast cancer cell-derived NO in cancer homeostasis.

ENDOGENOUS NO MEDIATED MECHANISM(S) IN HUMAN BREAST CANCER CELLS

Studies with human breast cancer cell lines demonstrated the existence of both eNOS and iNOS expressions in these cells. In 1996, Zeillinger et al. [5], for the first time, reported that 16 human breast cancer cell lines expressed eNOS and that its expression positively correlated with that of the estrogen receptor (ER) in 10 out the 16 cell lines investigated. The authors concluded that breast cancer cells could express eNOS and hypothesized that estradiol, the ligand for ER, could stimulate constitutive NO release in the breast tissue.

Alalami et al. [6] investigated the existence of the L-arginine-NO pathway in ZR-75-1 human breast cancer cells. The presence of iNOS in these cells was confirmed by staining with an anti-iNOS antibody. ZR-75-1 cells spontaneously produced nitric oxide (NO), and this production could be significantly ($P<0.001$) enhanced by L-arginine (0.01–10 mM) and was inhibited by L-N^G-nitroarginine methyl ester (L-NAME), a pharmacological inhibitor of NOS. The above studies indicated that human breast cancer cells produced NO from either eNOS or iNOS depending on the specificity of the breast cancer cell. However, the role of NO in these cells was still not known.

Mortensen et al. [7] showed that NO could suppress apoptosis. By reverse transcriptase-polymerase chain reaction and sequencing, they showed that the ER +ve MCF-7, human breast cancer cell, expressed eNOS, but not other NO synthase isoforms. Inhibition of eNOS activity in MCF-7 cells increased tumor cell apoptosis, and this effect was also seen following treatment with an NO scavenger. The authors claimed that the apoptotis-inducing protein p53 could inhibit the expression of eNOS by acting on an eNOS promoter site. These data, for the first time, pointed to a previously unrecognized p53-dependent regulation of eNOS expression.

In 2000, our laboratory [8] screened various human breast cancer cell lines for the presence of arginase (the enzyme that stimulates cell proliferation by inducing polyamine synthesis by converting arginine to ornithine) and NOS. Both arginase and NOS utilize the common substrate arginine. It was observed that two breast cancer cell lines, BT-474 and MDA-MB-468, exhibited relatively high arginase activity and very low eNOS activity. Another cell line, ZR-75-30, had the highest eNOS activity and comparatively low arginase activity. The basal proliferation rates of MDA-MB-468 and BT-474 were found to be higher than the ZR-75-30 cell line as would be expected due to the elevated arginase activity in these cells as polyamines directly stimulate growth.

Furthermore, *N*-hydroxy-L-arginine (NOHA), a stable intermediate product formed during the conversion of L-arginine to NO, inhibited the proliferation of the high arginase-expressing MDA-MB-468 cells and induced apoptosis after 48 h. NOHA arrested these cells in the S phase, increased the expression of p21 and reduced the levels of spermine, which is a polyamine. Interestingly, these effects of NOHA were not observed in the ZR-75-30 cell line, which expressed high eNOS and relatively low arginase. The effects of NOHA were antagonized in the presence of L-ornithine (500 μM), which further suggested that the arginase pathway was important for cell proliferation in these cells. Inhibition of the arginase pathway led to the depletion of intracellular spermine and induction of apoptosis as observed by terminal deoxynucleotidyl transferase (Tdt)-mediated nick end labeling assay and induction of caspase-3. In contrast, the ZR-75-30 cell line maintained its viability and its L-ornithine and spermine levels in the presence of NOHA. We concluded that NOHA had antiproliferative and apoptotic actions on arginase-expressing human breast cancer cells that were independent of NO.

Sen et al. [9] demonstrated that patients with triple-negative breast tumors that expressed NOS had a significantly worse prognosis compared with those that did not express any NOS. We observed that NOS activity was associated with the mitochondria in at least two breast cancer cell lines, ZR-75-30 and BT-474, compared with another NO-producing benign breast epithelial cell line, MCF-12F, in which no significant mitochondrial-associated NOS activity was detected. The rate of proliferation of the malignant cells expressing mitochondrial-associated NOS was decreased in the presence of an inhibitor of NO synthesis, but it had no effect on the normal breast epithelial cells, MCF-12F, which also expressed NOS, but was not associated with the mitochondria. The basal rate of proliferation was not affected by ODQ, an inhibitor of soluble guanylate cyclase, indicating that the effects of the endogenous NO produced by the malignant cell lines on proliferation were cGMP-independent.

Our results [9] also indicated that the mitochondrial-associated NOS activity exhibited by the cancer cell lines ZR-75-30 and BT-474 inhibited cytochrome *c* oxidase, resulting in increased production of hydrogen peroxide (H_2O_2), which inhibited protein phosphatase 2A activity. This resulted in the maintenance of Akt and ERK1/2 in a phosphorylated state, leading to cell proliferation. This study, for the first time, stipulated that endogenous NO could mechanistically regulate growth-promoting effects of another endogenous small molecule H_2O_2 in human breast cancer cells.

Wink et al. [10] demonstrated that pharmacological iNOS inhibition abated tumor growth and metastatic burden of the aggressive ER −ve, MDA-MB-231 human breast cancer cell-derived xenografts in mice. Moreover, low baseline iNOS expression was observed in MDA-MB-231 cells grown in culture, which was diminished by iNOS inhibition, then reaccumulated in the presence of exogenous NO. Similarly, iNOS inhibition blocked MDA-MB-231 cell migration, which was restored to basal levels by the addition of 100 μM diethylenetriamine (DETA)/NO, an NO donor. Importantly, iNOS inhibited cell migration that was significantly enhanced above basal levels by the addition of 300 μM DETA/NO, an NO donor, which alluded to an NO flux-dependent regulation of tumor cell migration. Furthermore, NO donor pretreatment suppressed Taxol-mediated cell killing of MDA-MB-231 cells, indicating the promotion of tumor survival and chemoresistance by iNOS-derived NO. These results suggested a requirement of feed-forward iNOS regulation for the maintenance of a basal NO flux that perpetuates protumorigenic signaling leading to cancer progression and metastases.

Various stimulants present within the tumor microenvironment were found to induce iNOS expression in vitro. Toward this, iNOS was induced in MCF-7, MDA-MB-468, and MDA-MB-231 human breast cancer cells. However, each cell line required different stimulants with varied temporal profiles. Furthermore, iNOS-induced regulation of tumor biomarkers supported iNOS as a biomarker of disease progression in breast cancer patients. In this context, iNOS inhibition in MDA-MB-231 cells decreased the basal levels of proinflammatory cytokines, *IL-8 and TLR4*. Other proinflammatory cytokines such as TIMP1, *S100A8*, and *IL-6* were enhanced by IFN-γ in an NO-dependent manner in these cells.

The levels of NO that activated these pathways were consistent with ~200–400 nM steady-state NO flux. These findings suggested that iNOS could be endogenously activated and that NO donors are a viable resource for understanding the pathological NO effects. The heterogeneity of the stimulation paradigm is reminiscent of that in tumors and supports a role of flux-dependent iNOS-derived NO in disease progression. The elucidation of factors within the tumor microenvironment that regulate iNOS and NO flux-driven tumor progression could lead to more personalized therapeutic options for women whose breast tumors express high iNOS.

THERAPEUTIC IMPLICATIONS

The use of iNOS inhibitors combined with inhibition of upstream and downstream NO targets (i.e., IFN-γ/JAK/STAT, PI3K/AKT, NF-κB, HIF-1α) or the inhibition of microRNAs that regulate human iNOS, such as miRNA-939, could improve clinical outcome of breast cancer patients [10] Moreover, iNOS inhibition suppressed growth and metastatic potential of KRAS mutant MDA-MB-231 cells, which suggested that iNOS inhibition could provide a viable option for patients with currently untreatable RAS mutant tumors [10].

In 2001, Matthews et al. [11] demonstrated that incubation of MDA-MB-231 tumor cells in 1% O_2 maximally increased their resistance to the common chemotherapeutic drugs, doxorubicin, and 5-fluorouracil by 8.5- ($P=0.002$) and 2.3-fold ($P=0.002$), respectively, compared with incubation in 20% O_2. The rapid acquisition of drug resistance after exposure to 1% O_2 could be mimicked by incubating the MDA-MB-231 breast cancer cells for 12 h with the NOS inhibitor *N*(G)-monomethyl-L-arginine (fivefold increase, $P<0.001$). Conversely, replacement of NO activity by the use of the NO-mimetic glyceryl trinitrate (GTN) and DETA NO adduct produced statistically significant attenuations in the development of resistance of 59% ($P<0.001$) and 40% ($P<0.001$), respectively, in MDA-MB-231 cells. The authors concluded that NO-mediated chemosensitivity in tumor cells and hypoxia-induced drug resistance appeared to result, in part, from downstream suppression of endogenous NO production. These results raised the possibility that administration of small doses of NO mimetics could be used as an adjuvant in chemotherapy.

Downregulation of endogenous NO production in human breast cancer cells has also been shown to increase the efficacy of photodynamic treatment [12]. Photodynamic therapy (PDT) employs a photosensitizing agent, molecular oxygen, and visible light to generate reactive species that kill tumor and tumor vasculature cells. NO produced by these cells could be procarcinogenic by inhibiting apoptosis or promoting angiogenesis and tumor growth. Bhowmick et al. [12] assessed whether tumor cells upregulated NO as a cytoprotective measure during PDT. Breast tumor COH-BR1 cells sensitized in their mitochondria with 5-aminolevulinic acid (ALA)-derived protoporphyrin IX died apoptotically after irradiation, while ALA- and light-only controls showed no effects. Immunoblot analysis revealed that iNOS was upregulated > threefold within 4 h after ALA/light treatment, whereas other NOS isoforms were unaffected.

Exposing cells to a NOS inhibitor (L-NAME or 1400 W) during photochallenge enhanced caspase-3/7 activation and apoptotic killing up to two- to threefold while substantially reducing chemiluminescence-assessed NO production, suggesting that NO was cytoprotective. Consistently, the NO scavenger cPTIO enhanced ALA/light-induced caspase-3/caspase-7 activation and apoptotic kill by > 2.5-fold. Moreover, cells could be rescued from 1400 W-exacerbated apoptosis by an exogenous NO donor, spermine NONOate. This was the first reported evidence for increased tumor cell resistance due to iNOS upregulation in a PDT model. These findings indicated that stress-elicited NO in PDT-treated tumors could compromise therapeutic efficacy and suggested NOS-based pharmacological interventions toward preventing this.

NO AND H_2S CROSS TALK AND FUTURE DIRECTIONS

NO and hydrogen sulfide (H_2S) are two gasotransmitters with important physiological functions. NO exerts many different roles: it is a vasorelaxant, an inflammatory mediator, and a neurotransmitter. Similarly, it has been shown that H_2S relaxes blood vessels is also involved in inflammation and is a neurotransmitter [13]. The two gases are also involved in redox signaling reactions, being able to interact with reactive oxygen species (ROS). Interestingly, several pathological conditions such as cardiovascular diseases and neurological disorders are characterized by an imbalance in both the levels of NO and H_2S [13].

All these observations contributed to develop the hypothesis of a possible interaction between the two molecules, especially in the cardiovascular system and during inflammation. While NO biology was very well studied and despite the increasing interest in H_2S biology, the mechanisms of action of H_2S have not been fully elucidated yet [14]. Therefore, there were attempts to characterize some aspects of the cross talk between NO and H_2S signaling pathways, with particular attention given to signaling pathways in the cardiovascular system (e.g., NO synthesis and cGMP production) and in inflammation (e.g., inflammasome complex activation) [15].

By using NO-derived metabolites of pharmacological interest for cardiovascular diseases nitrite (NO_2^-) and *S*-nitrosoalbumin (SNOA), it was possible to study nonenzymatic H_2S-dependent NO synthesis. NO_2^- and SNOA were reduced to NO by H_2S donors (such as the sulfide salt NaSH and the slow releasing GYY 4137), as assessed by electron paramagnetic resonance spectroscopy (EPR) and gas-phase ozone-based chemiluminescence [15].

The reactions were also chemically characterized, and the effects of H_2S-mediated NO synthesis on smooth muscle and endothelial cells were studied. Endogenous H_2S production was shown to significantly increase cGMP synthesis in NO_2^--treated human aortic smooth muscle cells (compared with only NO_2^- treatment). H_2S-mediated NO production from SNOA was also shown to increase nitrosothiol transport through cell membrane and subsequently to increase SNOA antioxidant properties in human microvascular endothelial cells [15,16].

The study of the cross talk between NO and H_2S was also carried forward by characterizing the posttranslational modification (S-nitrosation and sulfhydration) that they cause on key Cys residues of common target proteins. S-nitrosation of albumin was performed using incubation with acidified NO_2^- solutions, and the presence of the modification on the protein was characterized by mass spectrometry. It was possible to show that SNOA had increased antioxidant capacity compared with unmodified albumin controls. Similarly, it was done for H_2S-induced modification [16]. It was possible to demonstrate that H_2S was causing the decysteinylation of albumin samples, inducing an increase in albumin antioxidant properties (DMPD assay) [16].

Finally, the interaction between the two gases was also studied for key inflammatory processes such as the inflammasome complex activation and induction of IL-13 production. Unfortunately, with the assay used (IL-13 ELISA), it was not possible to find a clear interaction between NO and H_2S, but it was possible to clarify some of the mechanisms behind H_2S biological properties. It was indeed demonstrated the potential of H_2S as antiinflammatory mediator by inhibition of inflammasome activation and IL-13 production. In conclusion, with this work, it was possible to demonstrate that H_2S and NO have "overlapping" (at least in part) signaling pathways and that consequently H_2S is able to regulate nonenzymatic NO production in the cardiovascular system. It was also possible to characterize novel posttranslational modifications induced by H_2S, potentially shedding new light on its biological mechanisms of action [16]. The field of gasotransmitter biology is very exciting, and more work is certainly needed to advance our understanding of the key cellular signaling processes that gases like NO and H_2S could modulate.

We reported that human breast cancer cells and tissues expressed cystathionine β-synthase (CBS) an enzyme in the transsulfuration pathway that could catalyze the condensation of homocysteine (Hcy) and cysteine (Cys) to H_2S and cystathionine (CTH) [17]. CBS-derived H_2S was shown to be important in angiogenesis and drug resistance in colon and ovarian cancers, respectively [18]. However, the mechanisms by which cancer-cell-derived H_2S is utilized by cancer cells as a protective agent against host-derived activated macrophages was not yet investigated. Our study investigated the mechanistic role of CBS-derived H_2S in the protection of human breast cancer (HBC) cells against activated macrophages.

HBC-patient-derived tissue arrays and immunoblot analysis of HBC cells exhibited significantly increased levels of CBS when compared with their normal counterparts. This was associated with increased levels of H_2S and CTH. Silencing of CBS in HBC cells caused a significant decrease in the levels of H_2S and CTH but did not affect the growth of these cells per se, in in vitro cultures. However, CBS-silenced cells exhibited significantly reduced growth in the presence of activated macrophages and in xenograft models. This was associated with an increase in the steady-state levels of reactive aldehyde-derived protein adducts. Exogenous addition of H_2S countered the effects of CBS silencing in the presence of macrophages. Conversely, overexpression of CBS in human breast epithelial (HBE) cells (which do not naturally express CBS) protected them from activated macrophages, which were otherwise susceptible to the latter [18].

Despite the fact that HBC cells produce both NO and H_2S, a mechanistic cross talk between these two gasotransmitters remains to be investigated. It was earlier shown by Banerjee et al. [19] that NO could bind to the heme moiety of purified CBS (the enzyme that produces H_2S in HBC cells). The dissociation constant for NO for the full-length enzyme was $281 \pm 50\ \mu M$. Binding of NO was accompanied by inhibition of enzyme activity of CBS. The dissociation constant of CBS for NO was significantly higher than for the NO sensor, guanylate cyclase, which is of the order of 0.25 μM. Thus, it was apparent that inhibition of CBs by NO could only be relevant in cellular compartments or environments where the levels of guanylate cyclase are significantly lower than that of CBS. Moreover, as HBC cells preferentially express the CBS/H_2S pathway, the inhibitory action of NO on CBS would be highly undesired.

However, NO can quench superoxide in HBC cells forming peroxynitrite. Superoxide is known to inhibit heme proteins such as catalase and CBS, and therefore, NO may participate in the removal of superoxide to maintain a high bioactivity of the CBS/H_2S pathway. NO is known to be modifying thiols of proteins and enzymes by the formation of –SNO. This post-translational modification is not yet investigated on CBS and may reveal a mechanistic link between NO and H_2S biosynthesis in HBC.

PATHOPHYSIOLOGIC AND CLINICAL IMPLICATIONS

The immune system is known to play an important role in the tumor environment cross talk. It was initially felt that the accumulation of macrophages at tumor sites especially breast cancer was the body's reaction to suppress progression of cancer by releasing NO from the macrophages as iNOS expressing macrophages have been identified in breast cancer tissue [1]. High tumor infiltration with macrophages has been associated with favorable prognosis in gastric cancer [20], some colorectal carcinomas [20,21] and melanoma [22]. By contrast, it has been reported that large accumulation of macrophages correlates with tumor aggressiveness and predicts poor prognosis in human breast cancer [23,24]. More recently, two types of macrophages have been identified, the M1 and M2 macrophages, which may help in explaining the paradox where accumulation of macrophages to the tumor also known as tumor-associated macrophages or TAMs may actually be beneficial and have a favorable prognosis or be associated with tumor aggressiveness and have a bad prognosis. The M1 phenotype macrophages express numerous proinflammatory mediators including tumor necrosis factor (TNF)-α, IL-1, IL-6, NO, and other reactive nitrogen and oxygen intermediates, which have a strong microbicidal and tumoricidal activity, while the M2 phenotype expresses molecules including resistin-like-α (also known as Fizz1), arginase 1 (Arg1), chitinase 3-like 3 (also known as Ym1), IL-10, and Mrc1 (also known as CD206), which are supposed to be involved in parasite infestation, tissue remodeling, and tumor progression (immunoregulatory functions) [25].

Interestingly, the lactic acid produced within the tumor microenvironment signals macrophages for M2-like polarization [26]. Tumor-derived lactate is uptaken by tumor-associated macrophages (TAMs) through their MCT active transporters on the cell membrane, leading to the transcription of the vascular endothelial growth factor (VEGF) and the L-arginine-metabolizing enzyme arginase 1 (ARG1) genes [26,27]. ARG1 hydrolyzes L-arginine to L-ornithine and urea. ARG1 is expressed in myeloid cells, including TAMs, and can support tumor growth and suppresses antitumor immune responses. Lactic acid has been shown to increase ARG1 expression in macrophages, inhibiting T-cell activation and proliferation [28]. It therefore appears that the extracellular acidic milieu around the cancer cells is responsible for ATP generation in cancer cells by energy produced the proton gradient in addition to signaling macrophages for M2-like polarization, thereby creating an environment for tumor proliferation. Therefore, due to the relative decrease in M1 macrophages at the tumor site, one way to deliver high concentrations of NO to tumor cells could be to use a combination of tamoxifen with a NO donor to target breast cancer tissues, which are estrogen receptor positive [29].

References

[1] Thomsen LL, Miles DW, Happerfield L, Bobrow LG, Knowles RG, Moncada S. Nitric oxide synthase activity in human breast cancer. Br J Cancer 1995;72(1):41.

[2] Dueñas-Gonzalez A, Isales CM, del Mar Abad-Hernandez M, Gonzalez-Sarmiento R, Sangueza O, Rodriguez-Commes J. Expression of inducible nitric oxide synthase in breast cancer correlates with metastatic disease. Mod Pathol 1997;10:645–9.

[3] Tschugguel W, Schneeberger C, Unfried G, Czerwenka K, Weninger W, Mildner M, et al. Expression of inducible nitric oxide synthase in human breast cancer depends on tumor grade. Breast Cancer Res Treat 1999;56(2):143–9.

[4] Vakkala M, Kahlos K, Lakari E, Pääkkö P, Kinnula V, Soini Y. Inducible nitric oxide synthase expression, apoptosis, and angiogenesis in in situ and invasive breast carcinomas. Clin Cancer Res 2000;6(6):2408–16.

[5] Zeillinger R, Tantscher E, Schneeberger C, Tschugguel W, Eder S, Sliutz G, et al. Simultaneous expression of nitric oxide synthase and estrogen receptor in human breast cancer cell lines. Breast Cancer Res Treat 1996;40(2):205–7.

[6] Alalami O, Martin JH. ZR-75-1 human breast cancer cells: expression of inducible nitric oxide synthase and effect of tamoxifen and phorbol ester on nitric oxide production. Cancer Lett 1998;123(1):99–105.

[7] Mortensen K, Skouv J, Hougaard DM, Larsson LI. Endogenous endothelial cell nitric-oxide synthase modulates apoptosis in cultured breast cancer cells and is transcriptionally regulated by p53. J Biol Chem 1999;274(53):37679–84.

[8] Singh R, Pervin S, Karimi A, Cederbaum S, Chaudhuri G. Arginase activity in human breast cancer cell lines: Nω-hydroxy-L-arginine selectively inhibits cell proliferation and induces apoptosis in MDA-MB-468 cells. Cancer Res 2000;60(12):3305–12.

[9] Sen S, Kawahara B, Chaudhuri G. Mitochondrial-associated nitric oxide synthase activity inhibits cytochrome c oxidase: implications for breast cancer. Free Radic Biol Med 2013;57:210–20.

[10] Heinecke JL, Ridnour LA, Cheng RY, Switzer CH, Lizardo MM, Khanna C, et al. Tumor microenvironment-based feed-forward regulation of NOS2 in breast cancer progression. Proc Natl Acad Sci U S A 2014;111(17):6323–8.

[11] Matthews NE, Adams MA, Maxwell LR, Gofton TE, Graham CH. Nitric oxide-mediated regulation of chemosensitivity in cancer cells. J Natl Cancer Inst 2001;93(24):1879–85.
[12] Bhowmick R, Girotti AW. Cytoprotective induction of nitric oxide synthase in a cellular model of 5-aminolevulinic acid-based photodynamic therapy. Free Radic Biol Med 2010;48(10):1296–301.
[13] Li L, Rose P, Moore PK. Hydrogen sulfide and cell signaling. Annu Rev Pharmacol Toxicol 2011;51:169–87.
[14] Wang R. The gasotransmitter role of hydrogen sulfide. Antioxid Redox Signal 2003;5(4):493–501.
[15] Zhao W, Wang R. H2S-induced vasorelaxation and underlying cellular and molecular mechanisms. Am J Physiol Heart Circ Physiol 2002;283(2):H474–80.
[16] Faro L, Letizia M. Nitric oxide and hydrogen sulfide cross-talk: regulation of the vasculature and inflammation [Doctoral dissertation]. Exeter and Plymouth Peninsula Medical School.
[17] Sen S, Kawahara B, Gupta D, Tsai R, Khachatryan M, Roy-Chowdhuri S, et al. Role of cystathionine β-synthase in human breast cancer. Free Radic Biol Med 2015;86:228–38.
[18] Szabo C, Coletta C, Chao C, Módis K, Szczesny B, Papapetropoulos A, et al. Tumor-derived hydrogen sulfide, produced by cystathionine-β-synthase, stimulates bioenergetics, cell proliferation, and angiogenesis in colon cancer. Proc Natl Acad Sci U S A 2013;110(30):12474–9.
[19] Banerjee R, Evande R, Kabil Ö, Ojha S, Taoka S. Reaction mechanism and regulation of cystathionine β-synthase. Biochim Biophys Acta Proteins Proteomics 2003;1647(1):30–5.
[20] Erreni M, Mantovani A, Allavena P. Tumor-associated macrophages (TAM) and inflammation in colorectal cancer. Cancer Microenviron 2011;4:141–54.
[21] Funada Y, Noguchi T, Kikuchi R, Takeno S, Uchida Y. Gabbert HEPrognostic significance of CD8+ T cell and macrophage peritumoral infiltration in colorectal cancer. Oncol Rep 2003;10:309–13.
[22] Piras F, Colombari R, Minerba L, Murtas D, Floris C, Maxia C, et al. The predictive value of CD8, CD4, CD68, and human leukocyte antigen-D related cells in prognosis of cutaneous malignant melanoma with vertical phase. Cancer 2005;104:1246–54.
[23] Tiainen S, Tumelius R, Rilla K, Hämäläinen K, Tammi M, Tammi R, et al. High numbers of macrophages, especially M2-like (CD163-positive), correlate with hyaluronan accumulation and poor outcome in breast cancer. Histopathology 2015;66(6):873–83.
[24] Sousa S, Brion R, Lintunen M, Kronqvist P, Sandholm J, Mönkkönen J, et al. Human breast cancer cells educate macrophages toward the M2 activation status. Breast Cancer Res 2015;17:101.
[25] Gordon S, Martinez FO. Alternative activation of macrophages: mechanism and functions. Immunity 2010;32:593–604.
[26] Colegio OR, Chu NQ, Szabo AL, Chu T, Rhebergen AM, Jairam V, et al. Functional polarization of tumour-associated macrophages by tumour-derived lactic acid. Nature 2014;513(7519):559–63.
[27] Bronte V. Tumor cells hijack macrophages via lactic acid. Immunol Cell Biol 2014;92:647–9.
[28] Ohashi T, Akazawa T, Aoki M, Kuze B, Mizuta K, Ito Y, et al. Dichloroacetate improves immune dysfunction caused by tumor-secreted lactic acid and increases antitumor immunoreactivity. Int J Cancer 2013;133:1107–18.
[29] Pervin S, Singh R, Gau CL, Edamatsu H, Tamanoi F, Chaudhuri G. Potentiation of nitric oxide-induced apoptosis of MDA-MB-468 cells by farnesyltransferase inhibitor: implications in breast cancer. Cancer Res 2001;61 (12):4701–6.

Further Reading

[30] Palmer RM, Ferrige AG, Moncada S. Nitric oxide release accounts for the biological activity of endothelium-derived relaxing factor. Nature 1987;327(6122):524–6.
[31] Moncada SR, Palmer RM, Higgs E. Nitric oxide: physiology, pathophysiology, and pharmacology. Pharmacol Rev 1991;43(2):109–42.
[32] Knowles RG, Moncada S. Nitric oxide synthases in mammals. Biochem J 1994;298(Pt 2):249.
[33] Tayeh MA, Marletta MA. Macrophage oxidation of L-arginine to nitric oxide, nitrite, and nitrate. Tetrahydrobiopterin is required as a cofactor. J Biol Chem 1989;264(33):19654–8.
[34] Dhar G, Sen S, Chaudhuri G. Acid gradient across plasma membrane can drive phosphate bond synthesis in cancer cells: acidic tumor milieu as a potential energy source. PLoS One 2015;10(4):e0124070.

CHAPTER

11

Role of Nitric Oxide in Cancer Stem Cell Regulation and Metastasis

Amruta Manke, Sudjit Luanpitpong†, Liying Wang‡, Yon Rojanasakul§*

*RTI International, Durham, NC, United States

†Mahidol University, Bangkok Thailand

‡National Institute for Occupational Safety and Health, Morgantown, WV, United States

§West Virginia University, Morgantown, WV, United States

Abstract

There is increasing evidence to support the role of cancer stem cells (CSCs) in cancer development, chemoresistance, and relapse; resulting in poor prognosis and high mortality rate of cancer. The tumor microenvironment has been implicated to have an influence on cancer cells in many ways, where cell signaling molecules and mediators including proinflammatory cytokines and reactive species are frequently upregulated. Among them, the concentrations of nitric oxide (NO), a reactive nitrogen species synthesized by many cell types, are found to be upregulated in many cancer microenvironments, and elevated NO levels have been linked to advanced stage and poor survival of cancer patients. A plethora of studies demonstrate the oncogenic activity of NO in tumor progression; however, the underlying mechanisms remain poorly understood. This chapter summarizes and discusses the effects of NO on tumorigenesis and metastasis with a focus on CSC regulation and its effect on tumor progression. NO in the tumor environment appears to have a significant impact on the metastatic behavior of CSCs by affecting hallmarks of cancer including unlimited cell proliferation, evasion of apoptosis, increased cell migration and invasion, and capability of initiating tumor formation. Recent studies have indicated that NO-mediated CSC phenotype occurs through multiple pathways such as induction of target proteins like caveolin-1, expression of CSC markers, and via epithelial-mesenchymal transition. We discuss the emerging evidence documenting the significance of NO in the regulation of CSC phenotype that appears to be of the plausible mechanisms for NO-mediated aggressive behavior of tumor cells. These findings uniquely address the role of NO in CSC-mediated metastasis and can facilitate new drug discovery strategies for cancer treatment.

Nitric Oxide (Donor/Induced) in Chemosensitization, Volume 1
http://dx.doi.org/10.1016/B978-0-12-811020-1.00011-9

Abbreviations

ALDH aldehyde dehydrogenase
CAV-1 caveolin-1
CSC cancer stem cell
ECM extracellular matrix
EMT epithelial-mesenchymal transition
FAK focal adhesion kinase
HIF-1 hypoxia inducible factor-1
MAPK mitogen-activated protein kinase
MMP matrix metalloproteinase
NO nitric oxide
NOS nitric oxide synthase
NF-κB nuclear factor kappa-light-chain-enhancer of activated B cells
AKT protein kinase B
PI3K phosphoinositide 3-kinase
ERK extracellular signal-regulated kinase
VEGF vascular endothelial growth factor
cGMP cyclic guanosine monophosphate
HPV human papillomavirus
PKG protein kinase G

INTRODUCTION

The concept of cancer stem cells (CSCs) has emerged after an identification and characterization of CSC-enriched populations in several different forms of cancer over the last decade. There is accumulating evidence to indicate the role of CSCs in tumor aggressiveness, recurrence, and chemoresistance, which in turn are being reported as causal factors for increased mortality rate in cancer patients [1–3]. The CSC hypothesis assumes hierarchical cellular structure of a tumor, similar to normal regenerative tissue. Tumor cells are heterogeneous in terms of morphology, metabolism, proliferation rate, ability to metastasize, and other features. According to the CSC hypothesis, a subpopulation of highly malignant cancerous cells is responsible for tumor growth and development. Such a subpopulation came to be known as tumor-initiating cells, tumor-propagating cells, or CSCs, and has been identified in many types of cancer [4,5]. CSCs are defined by certain functional traits analogous to physiological stem cells such as unlimited replicative potential, evasion of apoptosis, increased invasion and migration, and ability to initiate tumor formation [3,6,7]. Although detailed knowledge of this cell population remains obscure, the significant impact of this cell population in various forms of cancer has increasingly been reported. CSCs have been identified by their putative markers such as aldehyde dehydrogenase (ALDH) and CD133, and by their cellular traits such as spindle-shaped morphology, colony formation, and other aggressive behaviors [8]. Recent studies suggest that epithelial-mesenchymal transition (EMT), a process characterized by a loss of epithelial markers and a gain of the mesenchymal phenotype, produces cells with CSC features in breast, pancreatic, and colorectal cancers [9]. Notably, several CSC characteristics are highly relevant to metastasis such as motility, invasiveness, and resistance to DNA damage-induced apoptosis [10].

The tumor microenvironment plays a critical role during multistaged tumor progression, particularly local resistance, immune escaping, and distant metastasis. The modulation of CSCs via the tumor niche remains poorly understood. Governed by the surrounding stimuli, the active microenvironment synthesizes a variety of chemokines, cytokines, and growth factors that are frequently upregulated. Such circulating cell signaling molecules and local mediators are capable of inducing rapid reversible changes of CSCs or CSC-like cells within the active tumor microenvironment [11,12]. Among them, several studies have demonstrated the key role of nitric oxide (NO) in the regulation of tumor formation and metastasis. Over the past decade, NO has been proposed as a signaling molecule that regulates cell survival and proliferation in diverse cell types [13,14]. Synthesized by the local stroma such as endothelial, immune, and tumor cells, NO is drastically upregulated in several cancer microenvironments. NO has been shown to modulate major cancer hallmarks including migration, invasion, apoptosis, and anoikis resistance, and angiogenesis, thus, governing the molecular mechanisms and cross talks during various stages of cancer. It has also been shown to affect diverse gene expressions at transcriptional and translational levels. NO is produced by various isoforms of NO synthases (NOSs) that have been found to be expressed in embryonic tissues and recently in stem cells. Therefore, NO is likely to contribute to the regulation of the CSC biology [15].

ROLE OF NITRIC OXIDE IN TUMORIGENESIS

Nitric oxide, a free radical and water-soluble gas produced endogenously, has been shown to regulate numerous physiological and pathological processes. Initially, NO was discovered as a transcellular messenger molecule synthesized by a family of NO synthases, including (1) inducible NOS (iNOS/NOS2), a calcium-independent enzyme that releases large amounts of NO in response to macrophage activation with endotoxin and cytokines, and is involved in cytotoxicity; (2) nNOS/NOS1, a calcium-dependent enzyme discovered in neurons that is involved in neural transmission; and (3) eNOS/NOS3, also a calcium-dependent enzyme that is constitutively expressed, isolated from endothelial cells, and is found in the normal vascular endothelium. The three NOS isoforms catalyze the synthesis of NO in a well-characterized two-step synthetic metabolic pathway whereby L-arginine is first converted to N-gamma-hydroxy-L-arginine and then to L-citrulline and NO. Post synthesis, NO can react with various molecules, resulting in more stable compounds such as S-nitrosothiols, metal adducts, peroxynitrites, and tetrahydrobiopterin [16,17].

Over the past few years, several studies have demonstrated the role of NO in carcinogenesis and tumor progression. Increasing evidence has indicated the involvement of diffusible secondary messengers like NO in the pathophysiology of various types of cancer including the brain, breast, prostate, pancreas, and lung [18]. NO can be derived from tumor cells themselves or from neighboring cells. Due to its lipophilic nature, NO from neighboring cells, for example, endothelial cells in the microvasculature or immune and stromal cells in the tumor, can freely diffuse across cellular membranes and ultimately affect tumor phenotypes and behaviors [19]. NO can mediate its oncogenic activity by damaging DNA, activating oncogenes, regulating apoptosis and metastasis, and inhibiting enzymes for DNA repair and tumor-suppressor genes [18]. Elevated NO levels and upregulation of NOS enzymes are typically found in tumor environments where NO could render cells resistant to death induced by various stimuli [20,21]. Interestingly, NO exerts dichotomous role in tumorigenesis by either inducing cancer progression or halting cancer growth and acting as an anticancer agent. For instance, at lower concentrations, NO promotes angiogenesis thus stimulating tumor progression and subsequently increasing cell proliferation. However, high NO levels induce apoptosis of tumor cells [22]. The variable effects of NO on tumor cells have been shown to be dependent on the localization of NOS and their activity, the duration and concentration of NO, and cellular sensitivity to NO [23,24].

There is accumulating clinical evidence to indicate the role of NO in the progression of several types of cancer. Increased NOS activity is manifested during the progression of endometrial carcinogenesis [25]. A positive correlation has been documented between NO and human papillomavirus (HPV)-induced cervical cancer [26]. Likewise, tumor specimens from head and neck cancer have been reported to have an upregulated expression of NOS in association with enhanced angiogenesis [27]. Clinical data similarly support the role of NO in lung cancer metastasis. Patients with lung cancer exhibited high levels of NO in their lungs, and a positive correlation has been reported between NO, its stable end products nitrite and nitrate, and its generator NOS with advanced cancer staging and poor patient survival. Genetic ablation of NOS suppressed lung tumor formation in mice [28–30]. Moreover, NO has been shown to induce chemoresistance and metastasis in breast cancer cells resulting in poor survival of breast cancer patients [31]. Together, these studies strongly support the role of NO as a signaling molecule in tumor progression although the underlying mechanisms are unclear. As mentioned previously, CSCs are a key determinant of chemoresistance and recurrence; however, our understanding of their regulation is limited. Rapid changes in CSCs may occur due to the presence of circulating signaling mediators such as NO in the tumor microenvironment. Thus, understanding the mechanisms of NO regulation of CSCs is critical in order to address cancer metastasis, chemoresistance, and relapse, which are major therapeutic challenges.

CANCER METASTASIS

The majority of cancer-related deaths has been attributed to cancer metastasis subsequently being a prime target for cancer therapy. The ability of tumor cells to migrate from the primary tumor site to form secondary tumors at distant sites is critical for successful metastasis during tumor progression [32]. Metastasis is a multistep cascade wherein a subpopulation of tumor cells undergo the following in a sequential pattern: (i) cell detachment from the primary tumor, (ii) local invasion of host stroma into the lymphatic and blood circulation (intravasation), (iii) survival of cells in the circulation (anoikis resistance), (iv) adhesion of cells to the capillary walls, (v) cell invasion and penetration out of the circulation (extravasation), and (vi) colonization and formation of secondary tumors [33–35]. The motility required to invade tissues is rendered upon by changes in cell-cell adhesion and cytoskeletal dynamic, which are regulated by signaling molecules such as caveolin-1 (CAV-1) and focal adhesion kinase (FAK) [33].

ROLE OF CSCs IN METASTASIS

The contribution of CSCs to the initiation of metastasis has gained considerable attention due to its potential therapeutic intervention. Several factors support the association between CSCs and enhanced metastatic phenotype. For instance, genetic signatures, cellular phenotype, enhanced migratory capacity conferred upon by EMT, tumor microenvironment, and circulating cell signaling molecules have been reported to contribute toward critical stem cell functions on these cells and hence increased metastasis [36]. Among them, the level of NO has been found to be significantly elevated during cancer development and progression in various tissues including the brain, breast, pancreas, prostate, and lung. Recent discoveries have implicated the role of NO in CSC regulation and associated aggressive cancer behaviors [22].

NO SIGNALING IN CSC REGULATION

The regulation of redox homeostasis entails signaling cascades such as hypoxia-inducible factor-1 (HIF-1), nuclear factor kappa-light-chain-enhancer of activated B cells (NF-κB), phosphoinositide 3-kinase (PI3K), and mitogen-activated protein kinase (MAPK), which control proliferation, metastasis, apoptosis, survival, and inflammation [37–39]. Studies have shown that elevated NO production directly promotes tumorigenesis and metastasis via cancer cell proliferation, migration and invasion, and indirectly by angiogenic factor expression in tumor cells. Moreover, NO has been shown to target NF-κB and promotes metastasis via increased EMT. Exposure to NO activates prosurvival signals such as extracellular signal-regulated kinase (ERK), JNK/SAPK, and p38 kinases among the MAPK family [40,41]. For instance, at NO concentrations of <100 nM, cGMP-dependent protein kinase (PKG) and ERK activation can occur. As NO concentrations increase, protein kinase B (PKB or AKT) is phosphorylated, and in ranges between 300 and 800 nM, HIF-1α and p53 are stabilized [37].

ROLE OF NO IN CELL MIGRATION AND INVASION

One of the critical events during metastasis is the migration of tumor cells away from the primary tumor, and CSCs are likely to be involved in this migration. NO has been shown to modulate the ability of tumor cells to migrate from one location to another, which is relevant to metastasis [42]. Continual exposure to NO has been shown to upregulate the expression of CAV-1, which correlates with the invasive ability and activation of FAK, thus contributing to altered morphology, protrusion, and migration of tumor cells during metastasis. It is likely that NO regulates cell motility and invasion via CAV-1 activation and downstream signaling pathways such as FAK and AKT [43]. Matrix metalloproteinases (MMPs) are responsible for the degradation of extracellular matrix (ECM) during invasion [44]. NO has been found to modulate MMPs expression, thereby regulating tumor cell invasion [45,46]. Additionally, NO reduces cell-cell and cell-ECM adhesion, thus facilitating migration and invasion via inhibition of integrin expression [47]. The process of EMT that is characterized by the loss of epithelial markers and gain of mesenchymal markers has been shown to activate stem phenotype and produce CSCs in breast, pancreatic, and colorectal cancers [48]. Thus, the cross talk between CSCs and EMT may suggest a likely mechanism for NO-mediated cell migration and invasion.

ROLE OF NO IN ANOIKIS RESISTANCE

Cells normally undergo apoptosis following loss of interaction with ECM or neighboring cells, also referred to as anoikis. However, circulating tumor cells acquire "anoikis resistant" phenotype and thereby survive after detachment from the primary tumor site and subsequently travel to distant sites to cause metastasis. Anoikis resistance has been shown to positively correlate with advanced cancer and poor survival of patients [49,50]. Several studies have shown that NO regulates anoikis resistance in tumor cells by modulating prosurvival and apoptotic signaling pathways. For instance, NO-mediated PI3K/AKT signaling suppresses anoikis through nitrosylation of phosphatases and tensin homolog deleted on chromosome 10 (PTEN) [51]. NO has also been reported to upregulate key antiapoptotic proteins such as FLIP and BCL2, thus inducing anoikis resistance in tumor cells [52,53]. Caveolin-1 has been implicated in the metastasis and poor prognosis of lung cancer [54]. NO-mediated nitrosylation of CAV-1 has been shown to induce anoikis resistance in human lung carcinoma cells [20]. In order to metastasize, CSCs from solid tumors have to break free from their primary epithelial sites and resist cell death activation after detachment [33]. Considering the plasticity of CSCs, NO production in CSCs may suppress anoikis through alterations of prosurvival and antiapoptotic pathways, thereby contributing to apoptosis and anoikis resistance.

ROLE OF NO IN ANGIOGENESIS

The angiogenic process involves the formation of new blood vessels from the original vascular bed facilitating intravasation and extravasation of tumor cells and enabling metastasis [55]. NO plays an important role in the process of angiogenesis/lymphangiogenesis where NOS expression is critical for the formation of blood/lymph vessels and their regulation [17]. In patients with hepatocellular carcinoma, iNOS was found to upregulate MMP-9 expression, thereby contributing toward tumor cell angiogenesis, invasion, and metastasis [56]. In head and neck cancer, increased NOS activity is associated with a higher level of angiogenesis [27]. Suppression of eNOS inhibits tumor cells from reaching the draining lymph nodes thus reducing metastasis. There is a positive correlation between the expression of NOS and lymphatic metastasis. eNOS mediates vascular endothelial growth factor (VEGF)-C-induced lymphangiogenesis, suggesting that NO participates in the early steps of lymphatic metastasis [57,58]. CSCs have been reported to express vascular-related factors such as VEGF and CD34 in order to induce neovascular growth and establish an aberrant vascular niche that contributes toward maintenance and proliferation of CSCs [59]. In glioma, CSCs induce angiogenesis by recruiting a substantially higher amount of endothelial progenitor cells (EPCs) and promoting EPC mobilization via stimulation of proangiogenic factors such as VEGF and SDF-1 [60]. In liver carcinoma, high-throughput microarray analyses have identified specific angiogenic factors critical to CSCs such as VEGF and PD-ECGF, and validated their role in promoting tumor angiogenesis [59]. How the vascular niche regulates CSCs to promote tumor angiogenesis remains to be elucidated. One of the possibilities based on the "cancer stem cell niche" hypothesis is the cross talk between circulating NO in the vascular niche and CSCs to promote the angiogenic process. NO stimulates VEGF-C production that contributes toward lymphatic metastasis. CSCs overexpress VEGF to induce angiogenesis, potentially through sprouting of preexisting vasculature or through recruitment of EPCs.

ROLE OF EMT

Another important factor for achieving metastasis is EMT wherein epithelial cells normally interact with the basement membrane to undergo morphological and functional changes leading to a mesenchymal phenotype [61]. It is a multistep process where cells lose their epithelial markers such as E-cadherin and tight junction protein ZO1, and acquire mesenchymal phenotype characterized by the expression of zinc finger protein SNAI1 (SNAIL) and SNAI2 (SLUG), N-cadherin, vimentin, and beta-catenin. Having been detached from the basement membrane, the transformed cells undergo enhanced migration and invasion, and once they enter the circulation and reach new sites they undergo mesenchymal-epithelial transition (MET) and establish new tumors [62]. Long-term NO exposure has been reported to activate EMT in various tumor cell types [63]. NO induces the expression of vimentin and SNAIL via CAV-1-independent mechanism, revealing an alternative route for NO-mediated cell migration. In addition, NO can impair the regulation of MMPs, which is crucial to controlling matrix remodeling and metastasis and thereby contribute toward acquisition of the mesenchymal phenotype [64].

NITRIC OXIDE AND CSCs

Recent studies indicate the role of NO signaling in CSC-induced metastasis and chemoresistance of tumor cells [35]. Decreased survival of estrogen receptor negative (ER−) breast cancer patients has been linked to elevated NOS2 expression in breast tumors. NO has been shown to induce CD44 expression and signal transducer and activator of transcription 3 (STAT3) activation, resulting in the development of stem cell phenotype [65]. Likewise, several studies have implicated the critical role of NO in lung cancer and metastasis; however, the underlying mechanisms remain obscure. Based on the previous findings that NO increased lung cancer aggressiveness, a recent study demonstrated that NO mediated its procarcinogenic effects through CSC-dependent mechanisms [66]. Prolonged exposure to NO was found to upregulate the expression of universal stem cell markers CD133 and ALDH 1A1 in multiple lung cancer cell lines through CAV-1 activation. CAV-1 expression has been linked to increased cell motility and anoikis resistance, thus contributing to the aggressiveness and chemoresistance of tumor cells [66]. These findings suggest that NO-mediated CSC induction could be responsible for increased migration and invasion, anchorage-independent growth, spheroid formation, and anoikis resistance of tumor cells. However, the effect of NO on CSCs was shown to be reversible indicating that the induction of the

CSC phenotype is dependent upon the availability of NO and the surrounding microenvironment. Furthermore, NO promoted the CSC phenotype in glioma stem cells that were found to be mechanistically distinct from their less tumorigenic counterparts with respect to NO production via NOS2. Similarly, NOS2 inhibition lessened tumor growth in a xenograft glioma model, suggesting that the effect of NO on aggressive CSC behaviors is reversible [67]. Additionally, a recent study demonstrated that serum depletion in oncogenic H-Ras-transformed cells induced NO production that promoted CSC formation as indicated by CSC markers expression and resistance to anticancer agents [68]. Taken together, these studies support the role of NO in CSC regulation and its effects on tumor growth and metastasis.

CONCLUSION

Accumulating evidence demonstrates the oncogenic activity of NO in various types of cancer, ranging from tumor initiation of cellular transformation to tumor progression through the metastatic cascade and resistance to chemotherapy. CSCs have garnered attention due to their enhanced ability to migrate, invade, form tumors, and resist anoikis, subsequently becoming critical determinants for metastasis, chemoresistance and relapse as well as a therapeutic target for cancer therapy. However, the precise relationship between NO and the aggressive cancer phenotype in the context of CSCs and the underlying mechanisms have not been established. Evidence to date points to the role of CSCs in mediating the effects of NO on tumor growth and invasiveness via apoptosis and EMT signaling pathways. Furthermore, NO regulates tumor metastasis by affecting CSC regulatory proteins such as CAV-1 that was found to be critical in NO-mediated aggressiveness of CSCs. Based on recent studies, there is evidence to suggest that the ability of CSCs to metastasize is enhanced by NO and that NO could be one of the key signaling molecules that regulates CSC-mediated metastasis. These new findings may provide further insights into the molecular mechanisms of CSC regulation and may aid in the design of novel therapeutic strategies aimed at overcoming cancer chemoresistance and relapse.

Acknowledgements

This work was supported by grants from the National Institutes of Health (R01-ES022968 and R01-EB018857) and National Science Foundation (CBET-1434503).

Declaration of Interest: The authors reported no declarations of interest.

Disclaimers: Research findings and conclusions are those of the authors and do not necessarily represent the views of the National Institute for Occupational Safety and Health.

References

[1] Hermann PC, Huber SL, Herrler T, Aicher A, Ellwart JW, Guba M, et al. Distinct populations of cancer stem cells determine tumor growth and metastatic activity in human pancreatic cancer. Cell Stem Cell 2007;1:313–23.

[2] Marie-Egyptienne DT, Lohse I, Hill RP. Cancer stem cells, the epithelial to mesenchymal transition (EMT) and radioresistance: potential role of hypoxia. Cancer Lett 2012;341:63–72.

[3] Perona R, López-Ayllón BD, Castro Carpeño J, Belda-Iniesta C. A role for cancer stem cells in drug resistance and metastasis in non-small cell lung cancer. Clin Transl Oncol 2011;13:289–93.

[4] O'Brien CA, Pollett A, Gallinger S, Dick JE. A human colon cancer cell capable of initiating tumour growth in immunodeficient mice. Nature 2007;445:106–10.
[5] Clevers H. The cancer stem cell: premises, promises and challenges. Nat Med 2011;17:313–9.
[6] Bao B, Ahmad A, Azmi AS, Ali S, Sarkar FH. Overview of cancer stem cells (CSCs) and mechanisms of their regulation: implications for cancer therapy. Curr Protoc Pharmacol 2013;C14:U14.25.
[7] Levina V, Marrangoni AM, DeMarco R, Gorelik E, Lokshin AE. Drug-selected human lung cancer stem cells: cytokine network, tumorigenic and metastatic properties. PLoS One 2008;3.
[8] Baccelli I, Trumpp A. The evolving concept of cancer and metastasis stem cells. J Cell Biol 2012;198(3):281–93. 6.
[9] Liu X, Fan D. The epithelial-mesenchymal transition and cancer stem cells: functional and mechanistic links. Curr Pharm Des 2015;21(10):1279–91.
[10] Borovski T, Melo FE, Vermeulen L, Medema JP. Cancer stem cell niche: the place to be. Cancer Res 2011;71-3:634–9.
[11] Cabarcas SM, Mathews LA, Farrar WL. The cancer stem cell niche—there goes the neighborhood? Int J Cancer 2011;129(10):2315–27.
[12] Keibel A, Singh V, Sharma MC. Inflammation, microenvironment, and the immune system in cancer progression. Curr Pharm Des 2009;15:1949–55.
[13] Wink DA, Vodovotz Y, Laval J, Laval F, Dewhirst MW, Mitchell JB. The multifaceted roles of nitric oxide in cancer. Carcinogenesis 1998;19(5):711–21.
[14] Williams EL, Djamgoz MB. Nitric oxide and metastatic cell behaviour. Bioessays 2005;27(12):1228–38.
[15] Vannini F, Kashfi K, Nath N. The dual role of iNOS in cancer. Redox Biol 2015;6:334–43.
[16] Forstermann U, Schmidt HH, Pollock JS, Sheng H, Mitchell JA, Warner TD, et al. Isoforms of nitric oxide synthase. Characterization and purification from different cell types. Biochem Pharmacol 1991;42(10):1849–57.
[17] Cheng H, Wang L, Mollica M, Re AT, Wu S, Zuo L. Nitric oxide in cancer metastasis. Cancer Lett 2014;353(1):1–7.
[18] Fukumura D, Kashiwagi S, Jain RK. The role of nitric oxide in tumour progression. Nat Rev Cancer 2006;6:521–34.
[19] Lala PK, Orucevic A. Role of nitric oxide in tumor progression: lessons from experimental tumors. Cancer Metastasis Rev 1998;17:91–106.
[20] Chanvorachote P, Nimmannit U, Lu Y, Talbott S, Jiang BH, Rojanasakul Y. Nitric oxide regulates lung carcinoma cell anoikis through inhibition of ubiquitin-proteasomal degradation of caveolin-1. J Biol Chem 2009;284:28476–84.
[21] Chanvorachote P, Nimmannit U, Stehlik C, Wang L, Jiang BH, Ongpipatanakul B, et al. Nitric oxide regulates cell sensitivity to cisplatin-induced apoptosis through S-nitrosylation and inhibition of Bcl-2 ubiquitination. Cancer Res 2006;66:6353–60.
[22] Vahora H, Khan MA, Alalami U, Hussain A. The potential role of nitric oxide in halting cancer progression through chemoprevention. J Cancer Prev 2016;21(1):1–12.
[23] Mocellin S, Bronte V, Nitti D. Nitric oxide a double-edged sword in cancer biology: searching for therapeutic opportunities. Med Res Rev 2007;27:317–52.
[24] Ridnour LA, Thomas DD, Donzell D, Espey MG, Roberts DD, Wink DA, et al. The biphasic nature of nitric oxide in carcinogenesis and tumour progression. Lancet Oncol 2001;2:149–56.
[25] Cinel L, Polat A, Aydin O, Düşmez D, Eğilmez R. Bcl-2, iNOS, p53 and PCNA expression in normal, disordered proliferative, hyperplastic and malignant endometrium. Pathol Int 2002;52:384–9.
[26] Wei L, Gravitt PE, Song H, Maldonado AM, Ozbun MA. Nitric oxide induces early viral transcription coincident with increased DNA damage and mutation rates in human papillomavirus-infected cells. Cancer Res 2009;69:4878–84.
[27] Gallo O, Masini E, Morbidelli L, Franchi A, Fini-Storchi I, Vergari WA, et al. Role of nitric oxide in angiogenesis and tumor progression in head and neck cancer. J Natl Cancer Inst 1998;90(8):587–96.
[28] Colakogullari M, Ulukaya E, Yilmaztepe A, Ocakoglu G, Yilmaz M, Karadag M, et al. Higher serum nitrate levels are associated with poor survival in lung cancer patients. Clin Biochem 2006;39:898–903.
[29] Esme H, Cemek M, Sezer M, Saglam H, Demir A, Melek H, et al. High levels of oxidative stress in patients with advanced lung cancer. Respirology 2008;13:112–6.
[30] Masri FA, Comhair SAA, Koeck T, Xu W, Janocha A, Ghosh S, et al. Abnormalities in nitric oxide and its derivatives in lung cancer. Am J Respir Crit Care Med 2005;172:597–605.
[31] Heinecke JL, Ridnour LA, Cheng RY, Switzer CH, Lizardo MM, Khanna C, et al. Tumor microenvironment-based feed-forward regulation of NOS2 in breast cancer progression. PNAS 2014;111:6323–8.
[32] Steeg PS, Theodorescu D. Metastasis: a therapeutic target for cancer. Nat Clin Pract Oncol 2008;5:206–19.

[33] Fidler I. The pathogenesis of cancer metastasis: the 'seed and soil' hypothesis revisited. Nat Rev Cancer 2003;3:453–8.
[34] Talmadge JE, Fidler IJ. AACR centennial Series: the biology of cancer metastasis: historical perspective. Cancer Res 2010;70:5649–69.
[35] Luanpitpong S, Chanvorachote P. Nitric oxide and aggressive behavior of lung cancer cells. Anticancer Res 2015;35(9):4585–92.
[36] O'Connor ML, Xiang D, Shigdar S, Macdonald J, Li Y, Wang T, et al. Cancer stem cells: a contentious hypothesis now moving forward. Cancer Lett 2014;344:180–7.
[37] Muntané J, Mata MD. Nitric oxide and cancer. World J Hepatol 2010;2(9):337–44.
[38] Prueitt RL, Boersma BJ, Howe TM, Goodman JE, Thomas DD, Ying L. Inflammation and IGF-I activate the Akt pathway in breast cancer. Int J Cancer 2007;120:796–805.
[39] Lo HW, Hsu SC, Ali-Seyed M, Gunduz M, Xia W, Wei Y. Nuclear interaction of EGFR and STAT3 in the activation of the iNOS/NO pathway. Cancer Cell 2005;7:575–89.
[40] Biswas N, Mahato SK, Chowdhury AA, Chaudhuri J, Manna A, Vinayagam J. ICB3E induces iNOS expression by ROS-dependent JNK and ERK activation for apoptosis of leukemic cells. Apoptosis 2012;17:612–26.
[41] Siegert A, Rosenberg C, Schmitt WD, Denkert C, Hauptmann S. Nitric oxide of human colorectal adenocarcinoma cell lines promotes tumour cell invasion. Br J Cancer 2002;86:1310–5.
[42] Bravo-Cordero JJ, Hodgson L, Condeelis J. Directed cell invasion and migration during metastasis. Curr Opin Cell Biol 2012;24:277–83.
[43] Sanuphan A, Chunhacha P, Pongrakhananon V, Chanvorachote P. Long-term nitric oxide exposure enhances lung cancer cell migration. Biomed Res Int 2013;2013:186972.
[44] Shiomi T, Okada Y. MT1-MMP and MMP-7 in invasion and metastasis of human cancers. Cancer Metastasis Rev 2003;22(2–3):145–52.
[45] Ishii Y, Ogura T, Tatemichi M, Fujisawa H, Otsuka F, Esumi H. Induction of matrix metalloproteinase gene transcription by nitric oxide and mechanisms of MMP-1 gene induction in human melanoma cell lines. Int J Cancer 2003;103(2):161–8.
[46] Jespersen C, Doller A, Akool el S, Bachmann M, Muller R, Gutwein P, et al. Molecular mechanisms of nitric oxide-dependent inhibition of TPA-induced matrix metalloproteinase-9 (MMP-9) in MCF-7 cells. J Cell Physiol 2009;219(2):276–87.
[47] Gupta SK, Vlahakis NE. Integrin alpha9beta1 mediates enhanced cell migration through nitric oxide synthase activity regulated by Src tyrosine kinase. J Cell Sci 2009;122:2043–54.
[48] Wang S, Jiang J, Liang X, Tang Y. Links between cancer stem cells and epithelial–mesenchymal transition. J Onco Targets Ther 2015;8:2973–80.
[49] Frisch SM, Screaton RA. Anoikis mechanisms. Curr Opin Cell Biol 2002;14:563–8.
[50] Moreno JG, Miller MC, Gross S, Allard WJ, Gomella LG, Terstappen LW. Circulating tumor cells predict survival in patients with metastatic prostate cancer. Urology 2005;65:713–8.
[51] Numajiri N, Takasawa K, Nishiya T, Tanaka H, Ohno K, Hayakawa W, et al. On-off system for PI3-kinase-AKT signaling through S-nitrosylation of phosphatase with sequence. PNAS 2011;108:10349–54.
[52] Mawji IA, Simpson CD, Hurren R, Gronda M, Williams MA, Filmus J, et al. Critical role of Fas associated death domain-like interleukin-1-converting enzyme like inhibitory protein in anoikis resistance and distant tumor formation. J Natl Cancer Inst 2007;99:811–22.
[53] Pongrakhananon V, Nimmannit U, Luanpitpong S, Rojanasakul Y, Chanvorachote P. Curcumin sensitizes non-small cell lung cancer cell anoikis through reactive oxygen species-mediated BCL2 downregulation. Apoptosis 2010;15:574–85.
[54] Ho CC, Huang PH, Huang HY, Chen YH, Yang PC, Hsu SM. Up-regulated caveolin-1 accentuates the metastasis capability of lung adenocarcinoma by inducing filopodia formation. Am J Pathol 2002;161:1647–56.
[55] Folkman J. Role of angiogenesis in tumor growth and metastasis. Semin Oncol 2002;29:15–8.
[56] Sun MH, Han XC, Jia MK, Jiang WD, Wang M, Zhang H, et al. Expressions of inducible nitric oxide synthase and matrix metalloproteinase-9 and their effects on angiogenesis and progression of hepatocellular carcinoma. World J Gastroenterol 2005;11(38):5931–7.
[57] Yasuoka H, Tsujimoto M, Yoshidome K, Nakahara M, Kodama R, Sanke T, et al. Cytoplasmic CXCR4 expression in breast cancer: induction by nitric oxide and correlation with lymph node metastasis and poor prognosis. BMC Cancer 2008;8:340.

[58] Nakamura Y, Yasuoka H, Tsujimoto M, Yoshidome K, Nakahara M, Nakao K, et al. Nitric oxide in breast cancer: induction of vascular endothelial growth factor-C and correlation with metastasis and poor prognosis. Clin Cancer Res 2006;12(4):1201–7.

[59] Li S, Li Q. Cancer stem cells and tumor metastasis. Int J Oncol 2014;44:1806–12.

[60] Bao S, Wu Q, Sathornsumetee S, Hao Y, Li Z, Hjelmeland AB, et al. Stem cell–like glioma cells promote tumor angiogenesis through vascular endothelial growth factor. Cancer Res 2006;66:7843–8.

[61] Kalluri R, Weinberg RA. The basics of epithelial mesenchymal transition. J Clin Investig 2009;119:1420–8.

[62] Brabletz T. EMT and MET in metastasis: where are the cancer stem cells? Cancer Cell 2012;22:699–701.

[63] Chanvorachote P, Pongrakhananon V, Chunhacha P. Prolonged nitric oxide exposure enhances enhanced anoikis resistance and migration through epithelial-mesenchymal transition and caveolin-1 up-regulation. Biomed Res Int 2014;2014. 941359.

[64] Ridnour LA, Windhausen AN, Isenberg JS, Yeung N, Thomas DD, Vitek MP, et al. Nitric oxide regulates matrix metalloproteinase-9 activity by guanylyl-cyclase-dependent and independent pathways. PNAS 2007;104:16898–903.

[65] Switzer CH, Glynn SA, Cheng RY, Ridnour LA, Green JE, Ambs S, et al. S-nitrosylation of EGFR and Src activates an oncogenic signaling network in human basal-like breast cancer. Mol Cancer Res 2012;10(9):1203–15.

[66] Yongsanguanchai N, Pongrakhananon V, Mutirangura A, Rojanasakul Y, Chanvorachote P. Nitric oxide induces cancer stem cell-like phenotypes in human lung cancer cells. Am J Physiol Cell Physiol 2015;308(2):C89–C100.

[67] Eyler CE, Wu Q, Yan K, MacSwords JM, Chandler-Militello D, Misuraca KL, et al. Glioma stem cell proliferation and tumor growth are promoted by nitric oxide synthase-2. Cell 2011;146:53–66.

[68] Monji K, Uchiumi T, Hoshizawa S, Yagi M, Matsumoto T, Setoyama D, et al. Serum depletion induced cancer stem cell-like phenotype due to nitric oxide synthesis in oncogenic H-Ras transformed cells. Oncotarget 2016. http://dx.doi.org/10.18632/oncotarget.12117 [Epub ahead of print].

CHAPTER

12

Nitric Oxide Interacting with Glutathione Transferases

*Mario Lo Bello**, *Lorien J. Parker*[†,‡], *Michael W. Parker*[†,‡], *Giorgio Ricci**

[*]University of Rome Tor Vergata, Rome, Italy
[†]St. Vincent's Institute of Medical Research, Fitzroy, VIC, Australia
[‡]The University of Melbourne, Parkville, VIC, Australia

Abstract

This short review deals with nitric oxide interacting with proteins (e.g., glutathione transferases) in the presence of GSH and iron. GST P1-1 may act as a NO carrier protein when GSH depletion occurs in the cell. Conversely, in the presence of GSH and traces of ferrous ions, the new formed dinitrosyl-diglutathionyl-iron complexes (DNDGICs) may act as activity modulators with GST P1-1 through competitive inhibition and as strong oxidants with glutathione reductase through irreversible inhibition. DNICs formation inside the cell is followed by a strong interaction with a number of proteins including GSTs and has been object of a number of papers where the most recently evolved GSTs bind DNICs at the active site with the extraordinary affinity. Crystallographic studies of GST P1-1 in complex with DNICs have shown that the active site Tyr-7 coordinates to the iron atom through its phenolate group by displacing one of the GSH ligands. Despite the high binding affinity for this complex and the presence of GST P1-1 in bacteria in large amounts, the disappearance of the complex is clearly dependent on iron availability. In other cells, the same complex present at a cytosolic concentration of 0.19 mM, as strongly bound to alpha-class GSTs, exerts full protection of glutathione reductase activity from NO where it might act as irreversible inhibitor. Thus, we suggest that specific sequestering of DNDGIC is a physiological protective mechanism when cells are exposed to excessive levels of nitric oxide. In this case, other studies have assessed the role of GST P1-1 (as long-term storage of DNICs) and the role of MRP1 in the transport of DNICs outside the cell. It appears that both proteins exert through an opposite but combined effect a great protection from the cytotoxic effects of NO.

Abbreviations

CDNDGIC	di-GSH DNIC complex
DNDGIC	dinitrosyl-diglutathionyl-iron complex
DNICs	dinitrosyl-iron complexes
GSH	glutathione

Nitric Oxide (Donor/Induced) in Chemosensitization, Volume 1
http://dx.doi.org/10.1016/B978-0-12-811020-1.00012-0

GSNO	nitroso-glutathione
GST P1-1	glutathione transferase P1-1
NO^-	nitroxyl anion
NO^+	nitrosonium ion

INTRODUCTION

NO is a radical that can be readily reduced to a nitroxyl anion (NO^-) or oxidized to a nitrosonium ion (NO^+) depending on the redox status of the cell. For example, NO^- can be obtained in the presence of reduced glutathione (GSH) leading eventually to nitroso-glutathione (GSNO) in activated neutrophils [1]. This last species can be produced also in the presence of glutathione transferases [2] and may modulate the activity of a variety of different enzymes [3]. NO can bind also to iron sulfur clusters present in cytoplasmic proteins or to thiolate groups of proteins in the presence of free iron.

CELL BIOCHEMISTRY OF DNICs

The presence of free cellular thiols (e.g., cysteine or GSH) gives rise to low-molecular-weight dinitrosyl-iron complexes (DNICs). A number of studies in vitro have outlined the importance of the interaction between these DNICs and proteins. It has been reported that DNICs bind albumin forming paramagnetic NO-Fe protein complexes that can be detected by ESR spectroscopy [4]. Keese et al. [3] found that DNICs ternary complexes may act as irreversible inhibitors of glutathione reductase (K_i of 20 nM) and are reversible inhibitors of glutathione transferase P1-1(GST P1-1).They further investigated by X-ray crystallography the structures of the glutathione reductase in complex with GSNO and DNICs, respectively [5], and showed that the irreversible inactivation of this enzyme by DNICs is due to oxidation of an active site cysteine to cysteine-sulfinic acid. In the case of human GST P1-1, we have found a different scenario depending on the absence or presence of GSH. In the absence of this tripeptide, S-nitroso-GSH causes rapid and stable S-nitrosylation of both the Cys47 and Cys101 residues ruling out the possibility of S-glutathionylation and confirms the occurrence of a poly-S-nitrosylation. S-Nitrosylation of Cys47 lowers the affinity 10-fold for GSH, but this negative effect is minimized by a half-site reactivity mechanism that protects one Cys47/dimer from nitrosylation. GST P1-1, retaining most of its original activity, may act as a NO carrier protein when GSH depletion occurs in the cell. Conversely, in the presence of GSH and traces of ferrous ions, the new formed dinitrosyl-diglutathionyl-iron complex (DNDGIC) binds with extraordinary affinity to one active site subunit of GST P1-1 ($K_i < 10^{-12}$ M) and triggers negative cooperativity in the vacant subunit ($K_i < 10^{-9}$ M) [6]. Thus, it appears that DNICs may act as activity modulators with GST P1-1 through competitive inhibition and as strong oxidants with glutathione reductase through irreversible inhibition. DNICs, when bound to GST P1-1, is stable for hours, whereas in the free form and at low concentrations is stable only for few minutes.

DNICs FORMATION

In vitro synthesis is promptly accomplished in solution by adding GSNO to GSH (molar ratio 1:10) in the presence of substoichiometric amounts of iron. It is remarkable that the formation of DNICs, as observed by UV and ESR spectroscopy, is completed within 20 min and is unaffected by GST P1-1 [6] and this particular protocol adds to a complex that is stable up to 180 min in contrast to published methods where the lifetime is much shorter. The same species observed in cells have a very short half-life (less than 1 min) [4]. The fact that DNICs formation inside the cell is also followed by a strong interaction with a number of proteins including GSTs poses the question as to whether its stabilization is mediated by this protein interaction.

GSTs AND DNICs

Studies in our lab by means of presteady and steady-state kinetics, fluorometry, electron paramagnetic resonance, and radiometric experiments, using representative members of this GST superfamily, demonstrated an extraordinary affinity of this complex to the active site of GSTs with the GSTA1-1 isoform exhibiting the strongest interaction ($K_D < 10^{-10}$ M). From an evolutionary point of view, the most ancient GST classes (such as theta- and beta-classes) showed less or no affinity for DNICs suggesting that the capacity to utilize NO as a signal molecule was acquired in the most recently evolved GSTs [7]. In heterogenous systems, such as homogenates, DNICs are stabilized for hours only, in the presence of GSTs, where they are bound in stoichiometric amounts suggesting the possibility that these enzymes may act as storage proteins for NO [8]. Crystallographic studies of GST P1-1 in complex with DNICs revealed the molecular basis of this stabilization and represent the only example of a crystallographically characterized protein-bound DNICs [9]. In this complex, the active site Tyr-7 coordinates to the iron atom through its phenolate group by displacing one of the GSH ligands (Fig. 1). The crucial importance of this catalytic residue in binding the NO donor was demonstrated by site-directed mutagenesis of this residue to His, Cys, or Phe residues. The relative binding affinity for the complex was strongly reduced in all three mutants by about three orders of magnitude with respect to the wild type [9]. In vivo studies, at least in bacteria, suggest that, while formation of DNICs is fast and independent of the presence of GST P1-1, its disappearance is faster than that found in the in vitro studies, despite the presence of GST P1-1 in large amounts and is clearly dependent on iron availability [9]. Therefore, we have hypothesized that this complex may be stabilized in vivo through binding to GST P1-1. The strong interaction of this complex with GST was also studied in vitro in hepatocytes where the di-GSH DNIC complex (CDNDGIC) is bound to alpha-class GSTs with very high affinity (KD 10^{-10} M) [10]. In these cells, the complex, formed by NO-mediated extraction of iron from ferritin and transferrin in the presence of GSH, may reach a cytosolic concentration of 0.19 mM. However, about 20% of this alpha-class GST complex is associated with subcellular components including the nucleus. This high concentration of the complex, strongly bound to alpha-GSTs, exerts full protection of glutathione reductase activity from NO where it might act as irreversible inhibitor. Thus, we suggest that specific sequestering of DNDGIC is a physiological protective mechanism when cells are exposed to excessive levels of nitric oxide.

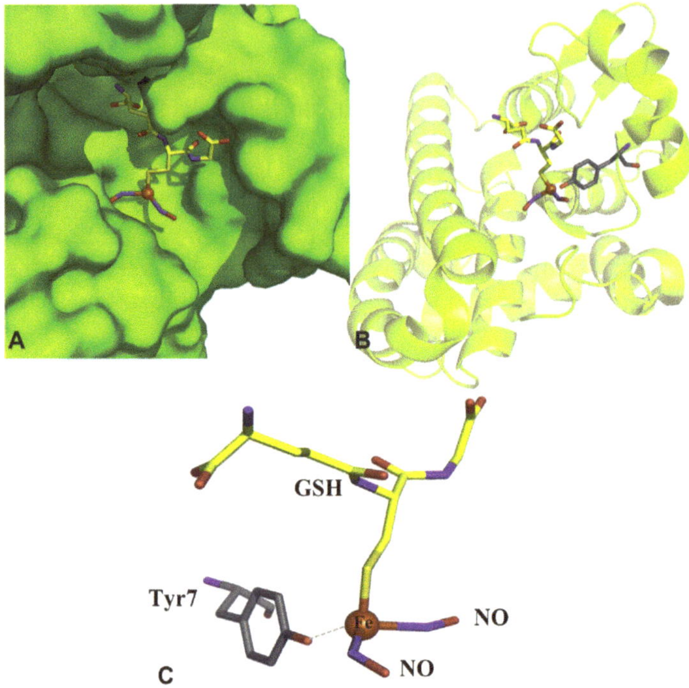

FIG. 1 Crystal structure of the human GSTP1-1 DNDGIC complex. (A) Close-up view of the active site cleft with the protein shown as a solvent-accessible surface (green) and the DNDGIC ligand in stick. (B) Ribbon representation of a GST P1-1 monomer with bound DNDGIC and Tyr 7 shown as stick. (C) The structure of DNDGIC, shown in atomic coloring, when bound to GST P1-1. The GST ligand, Tyr 7, is shown in grey bonds.

TRANSPORT OF DNICs

The close interaction between NO, iron, and GSH in the cytotoxic mechanisms of activated macrophages has been demonstrated by Watts et al. [11] who suggested that the iron release from cells is mediated by MRP1 under the presence of DNIC complex [11]. Further studies have assessed the role of GST P1-1 (as long-term storage of DNICs) and the role of MRP1 in the transport of DNICs outside the cell. Cells transfected with *GSTP1*, but not with *GSTA1* or *GSTM1*, showed a significant decrease NO-mediated ^{59}Fe from cells indicating an active role of this protein in preventing transport of DNICs outside of the cells by MRP1. Thus, it appears that both proteins exert through an opposite but combined effect a great protection from the cytotoxic effects of NO [12].

CONCLUDING REMARKS

Understanding the mechanisms involved in the metabolism of NO when it is present at higher concentrations into the cells is of fundamental importance in cellular signaling and for the development of new therapeutic targets [13]. The interaction between GSTs and NO has been considered also from the evolutionary point of view. It has been shown that the catalytic residues Cys or Ser present in the active site of the most ancient GSTs have been replaced by tyrosine in the more recently evolved GSTs (alpha-, mu-, and pi-classes) for a better protection from nitric oxide. Indeed, using a large panel of GSTs, only those containing tyrosine as catalytic residue displays a great affinity for DNDGIC complex, and in this way, they protect the cells against the excess of NO [14].

Acknowledgments

We acknowledge the support of the Victorian Government Operational Infrastructure Support Scheme to St. Vincent's Institute. L.J.P. and M.W.P. are National Health and Medical Research Council of Australia C.J. Martin and research fellows, respectively.

Conflict of Interest: No potential conflicts of interest were disclosed.

References

[1] Clancy RM, Levartovsky D, Leszczynska-Piziak J, Yegudin J, Abramson SB. Nitric oxide reacts with intracellular glutathione and activates the hexose monophosphate shunt in human neutrophils: evidence for S-nitrosoglutathione as a bioactive intermediary. Proc Natl Acad Sci U S A 1994;91(9):3680–4.

[2] Meyer DJ, Kramer H, Ketterer B. Human glutathione transferase catalysis of the formation of S-nitrosoglutathione from organic nitrites plus glutathione. FEBS Lett 1994;351(3):427–8.

[3] Keese MA, Böse M, Mülsch A, Schirmer RH, Becker K. Dinitrosyl-dithiol-iron complexes, nitric oxide (NO) carriers in vivo, as potent inhibitors of human glutathione reductase and glutathione-S-transferase. Biochem Pharmacol 1997;54(12):1307–13.

[4] Boese M, Mordvintcev PI, Busse R, Vanin AF, Mülsch A. S-nitrosation of serum albumin by dinitrosyl-iron complex. J Biol Chem 1995;270(49):29244–9.

[5] Becker K, Savvides SN, Keese M, Schirmer RH, Karplus PA. Enzyme inactivation through sulfhydryl oxidation by physiologic NO-carriers. Nat Struct Biol 1998;5(4):267–71.

[6] Lo Bello M, Nuccetelli M, Caccuri AM, Stella L, Parker MW, Rossjohn J, et al. Human glutathione transferase P1-1 and nitric oxide carriers; a new role for an old enzyme. J Biol Chem 2001;276(45):42138–45.

[7] De Maria F, Pedersen JZ, Caccuri AM, Antonini G, Turella P, Stella L, et al. The specific interaction of Dinitrosyl-Diglutathionyl-Iron complex, a natural NO carrier, with the Glutathione Transferase superfamily. J Biol Chem 2003;278(43):42283–93.

[8] Turella P, Pedersen JZ, Caccuri AM, De Maria F, Mastroberardino P, Lo Bello M, et al. Glutathione transferase superfamily behaves like storage proteins for dinitrosyl-diglutathionyl-iron complex in heterogeneous systems. J Biol Chem 2003;278(43):42294–9.

[9] Cesareo E, Parker LJ, Pedersen JZ, Nuccetelli M, Mazzetti AP, Pastore A, et al. Nitrosylation of human glutathione transferase P1-1 with dinitrosyl-diglutathionyl iron complex in vitro and in vivo. J Biol Chem 2005;280 (51):42172–80.

[10] Pedersen JZ, De Maria F, Turella P, Federici G, Mattei M, Fabrini R, et al. Glutathione transferases sequester toxic dinitrosyl-iron complexes in cells. J Biol Chem 2007;282(9):6364–71.

[11] Watts RN, Richardson DR. Nitrogen monoxide (no) and glucose: unexpected links between energy metabolism and no-mediated iron mobilization from cells. J Biol Chem 2001;276(7):4724–32.

[12] Lok HC, Suryo Rahmanto Y, Hawkins CL, Kalinowski DS, Morrow CS, Townsend AJ, et al. Nitric oxide storage and transport in cells are mediated by glutathione S-transferase P1-1 and multidrug resistance protein 1 via dinitrosyl iron complexes. J Biol Chem 2012;287(1):607–18.

[13] Suryo Rahmanto Y, Kalinowski DS, Lane DJ, Lok HC, Richardson V, Richardson DR. Nitrogen monoxide (NO) storage and transport by dinitrosyl-dithiol-iron complexes: long-lived NO that is trafficked by interacting proteins. J Biol Chem 2012;287(10):6960–8.

[14] Bocedi A, Fabrini R, Farrotti A, Stella L, Ketterman AJ, Pedersen JZ, et al. The impact of nitric oxide toxicity on the evolution of the glutathione transferase superfamily: a proposal for an evolutionary driving force. J Biol Chem 2013;288(34):24936–47.

CHAPTER

13

Application of MRI to Study the Role of Nitric Oxide in Cancer Therapy

Claudia Weidensteiner

Department of Radiology/Medical Physics, Medical Center - University of Freiburg, Faculty of Medicine, University of Freiburg, Freiburg, Germany

Abstract

Noninvasive imaging methods are increasingly used to study the function and the drug effects of nitric oxide (NO) and NO donors and to evaluate their potential as an anticancer agent in vivo. Magnetic resonance imaging (MRI) is a very versatile imaging modality with excellent soft-tissue contrast and high spatial resolution. In this article, we will review the applications of MRI for the detection of NO-mediated effects mainly in animal models of solid tumors: first, morphological MRI; second, functional MRI methods to assess vascular function; and third, first attempts of imaging biodistribution with molecular MRI. Technical aspects and pros and cons also in comparison with other methods are discussed.

Abbreviations

BLI	bioluminescence imaging
BOLD	blood-oxygen-level-dependent
CA	contrast agent
CT	X-ray computed tomography
DCE-MRI	dynamic contrast-enhanced MRI
DSC-MRI	dynamic susceptibility contrast MRI
EPR	electron paramagnetic resonance imaging
fBV	fractional blood volume
F-MISO	fluoromisonidazol
Gd-DOTA	gadolinium complex of tetraazacyclododecanetetraacetic acid
Gd-DTPA	gadolinium complex of diethylenetriaminepentaacetate
iAUC	initial area under the concentration curve
iNOS	inducible nitric oxide synthase
MION	monocrystalline iron-oxide-based nanoparticle
MRI	magnetic resonance imaging
NO	nitric oxide
NOS	nitric oxide synthase
PET	positron emission tomography

Nitric Oxide (Donor/Induced) in Chemosensitization, Volume 1
http://dx.doi.org/10.1016/B978-0-12-811020-1.00013-2

phMRI	pharmacological MRI
RANO	response assessment in neuro-oncology
RARE	rapid acquisition with relaxation enhancement
RECIST	response evaluation criteria in solid tumors
ROI	region of interest
SSFP	steady-state free precision
USPIO	ultrasmall superparamagnetic iron oxide

INTRODUCTION

Magnetic resonance imaging (MRI) provides soft-tissue contrast, relatively high spatial resolution, is noninvasive and free of ionizing radiation. It can provide structural, biochemical, and functional information. Therefore, it is widely used in clinical routine and in clinical studies for diagnosis, staging of disease, therapy monitoring, and follow-up.

In preclinical studies, MRI is used for the characterization of animal models of disease and for in vivo profiling of potential drug candidates. These studies are often performed in dedicated animal MR scanners. These scanners have a smaller bore and smaller coils adjusted to the dimension of rodents, and they are operated at a higher magnetic field strength compared with clinical MR scanners. The higher sensitivity of animal scanners allows a higher spatial resolution. Due to the noninvasiveness, MRI scans can be repeated in the same animal at multiple time points to monitor therapeutic effects in a longitudinal study. The number of animals can be reduced since animals can act as their own controls and culling of animals for invasive analysis can be avoided.

MRI displays a unique soft-tissue contrast. Tissue-specific parameters affect the MR signal: the proton density (in water and fat) and the MR relaxation times (T_1, T_2, T_2*) that describe evolution and decay of the MR signal. These parameters vary from tissue to tissue and give rise to the contrast between tissue types and even between normal and abnormal tissues in an MR image. The MRI signal intensity and therefore the image contrast can be manipulated by the choice of MR sequence and its operator-selectable acquisition parameters; for example, in T_1-, T_2-, or T_2*-weighted sequences, the signal intensity is determined by the corresponding relaxation time. The contrast can be further enhanced by the use of a contrast agent (CA). These manipulations provide different morphological and functional information and make MRI a very versatile imaging modality.

The role of nitric oxide (NO) in cancer is complex; it has both pro- and antineoplastic activity. In anticancer drug development, various strategies are investigated. The in vivo production of NO is manipulated with substances acting on the nitric oxide synthase (NOS) that exists in three isoforms, namely, neuronal, endothelial, and inducible NOS. Exogenous delivery is performed via NO donors. NO has certain cytotoxic activity. More importantly, it has chemo- and radiosensitizing effects that are of interest for the development of combination therapies [1–3]. NO is a vasodilator and increases the permeability of vessel walls. The increase in tumor blood flow and oxygenation improves the efficacy of radio- and chemotherapies. NO donors may also be useful for targeted drug delivery. Several NO prodrugs are activated by enzymes particularly overexpressed in cancer tissue (in particular glutathione

S-transferase) [4]. In this way, the cytotoxic or radio-/chemosensitizing effects are restricted to the tumor.

Several in vitro and in vivo studies showed encouraging results of the anticancer activity of NO. The majority of these in vivo studies are still restricted to the preclinical phase. Noninvasive imaging methods are increasingly used to study the function and the drug effects of NO and NO donors and to evaluate their potential as anticancer agents in vivo [5]. MRI serves as the premier anatomical imaging tool in clinical studies of cancer. This applies also to studies focusing on NO. The primary outcome measures and endpoints are tumor size and burden. Additional—especially functional—MRI readouts come increasingly into focus of researchers in the field of NO in oncology. In this overview article, the focus is on already established and new promising applications of MRI to study the role of NO in cancer therapy mainly in preclinical studies (Table 1).

TABLE 1 Overview of the MRI Methods Presented in This Article

MRI Method	Tumor Property Measured	CA	Already Performed in NO Cancer Studies In Vivo?	Suited for Human Studies?	References
Morphological MRI	Tumor size, necrosis extent	No	Yes	Yes, routinely used	[6–8]
Morphological MRI with CA	Tumor size, necrosis extent, enhancing region	Gd-based	Yes	Yes, routinely used	[9]
Susceptibility contrast MRI	Blood volume	Iron-oxide-based	Yes	In principle yes, but only limited approval of iron-oxide-based CAs. Rather, DSC-MRI is used in clinical studies	[10–12]
DCE-MRI	Permeability, perfusion	Gd-based	Yes	Yes	[7,9,13]
BOLD MRI	Perfusion, oxygenation	No	Yes	Yes	[13]
BOLD MRI with breathing challenge	Tumor vessel functionality and maturity	No	Yes	Yes	[10]
Molecular MRI of NO	Biodistribution of NO	Spin traps	No, only in a sepsis model	No	[14]
Molecular MRI of NOS	Expression of NOS	Antibody coupled to Gd-based CA	Once	Not yet	[8]

MORPHOLOGICAL MRI FOR TUMOR SIZE MEASUREMENTS

Tumor size is the most commonly used marker for assessing tumor response in solid tumors, both in preclinical and clinical studies [15]. This applies also to in vivo studies with NO and NO donors: their efficacy to inhibit tumor growth in vivo due to their cytotoxic or radio-/chemosensitizing effects is first and foremost judged by tumor size measurements. In clinical studies, tumor size is assessed via the RECIST criteria [16] or Macdonald/RANO criteria for brain tumors [17] using MRI, X-ray computed tomography (CT), or positron emission tomography in combination with CT (PET/CT). MRI scans are acquired with standard protocols from clinical routine.

In preclinical studies, tumor size is routinely assessed with caliper measurements in subcutaneous tumor models or with bioluminescence imaging (BLI) in models based on transfected luciferase-expressing tumor cell lines. Both methods are fast and easy to use. However, caliper measurements cannot be applied in orthotopic tumor models and are only applicable to palpable tumors. BLI can be used to track tumor growth and also metastasis both in subcutaneous and orthotopic (e.g., intracranial) tumor models. It is sensitive enough to detect microscopic tumors. However, tumor size is only measured indirectly as light intensity emitted at the surface of the body and therefore prone to errors.

MRI allows 3D visualization of tumors; the volume can be measured accurately. It is suited for all types of tumor models (orthotopic or subcutaneous, luciferase-expressing or not, intracranial, etc.) except for microscopic ones that cannot be clearly identified. The size measurement of infiltrative lesions is also problematic since they cannot be clearly delineated on the images. Compared with the other two methods, MRI measurements are time-consuming especially in studies with a high number of animals.

High-resolution MRI datasets consist of multiple consecutive slices covering the whole tumor. Even though (semi)automatic algorithms have been demonstrated mainly in research environments [18], the tumor borders are usually still traced manually on all slices of the MRI dataset using standard software, and the tumor volume is calculated. By repeating these MR measurements over time, the tumor progression or its shrinkage under therapy can be monitored.

Soft-tissue contrast between tumor and surrounding healthy tissue can be achieved by T_1- and T_2-weighting of the MR sequence. We and others routinely use T_2-weighted RARE (also called turbo spin echo or fast spin echo) sequences with an in-plane resolution around 100 μm in mice and rats [6,7] (Fig. 1C). An alternative for morphological imaging are balanced SSFP sequences (also called TrueFISP), for example, in [19]. In the article by Edrei et al., the metastatic burden in the liver in a mouse colorectal cancer model was assessed as total metastatic volume [6]. Several hepatic nodules were detected on the T_2-weighted datasets. An antiangiogenic drug combination with the NOS inhibitor L-NAME proved to be very effective in eliminating metastasis. MR CAs based on gadolinium are used for an improved visualization of the tumor. Examples for commonly used, clinically approved CAs are Gd-DTPA (Magnevist, Bayer HealthCare) and Gd-DOTA (Dotarem, Guerbet). They are applied intravenously during the examination, in rodents via the tail vein, and are excreted by the kidneys. After the CA application, the vascularized, vital part of the tumor shows up bright in T_1-weighted images. This is especially useful in brain tumors

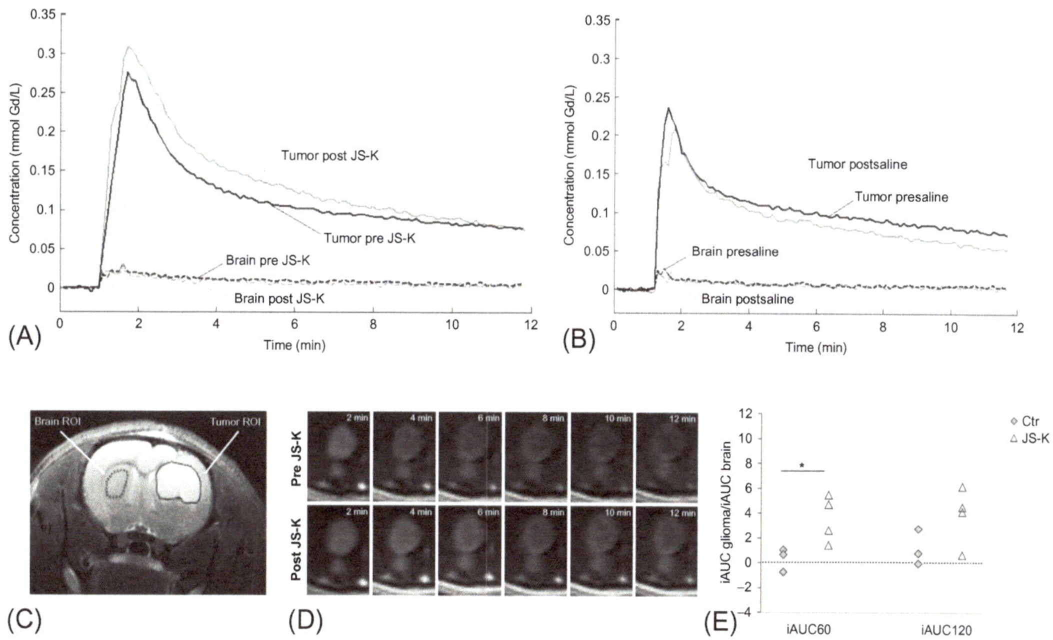

FIG. 1 DCE-MRI study in an orthotopic U87 glioma rat model to investigate the effect of the NO donor JS-K on tumor vessel permeability. (A) CA concentration curves of the first (black curves, pre-JS-K, and baseline) and second (gray curves and post-JS-K) DCE-MRI scan in tumor (solid upper curves) and contralateral brain (dashed lower curves). The second DCE-MRI scan was started 18 min after the end of the first DCE-MRI scan. JS-K injection was 3 min before the start of the second DCE-MRI scan. The gray area visualizes the $iAUC_{60}$ (=iAUC up to 60 s post-CA injection) for the pre-JS-K time course. (B) CA concentration curves in a control rat pre- and post-saline injection. (C) ROIs for DCE-MRI in tumor (shown right, solid) and contralateral brain (shown left, dashed) depicted on an anatomical T_2-weighted image in the rat shown in (A). (D) Time series of DCE-MRI scans showing the CA-induced change in image contrast in the tumor pre-(upper row) and post(lower row)-JS-K injection (other rat than in (A)–(C)). CA was administered at time point 1 min. JS-K was injected 3 min before the second DCE-MRI time series. (E) Change in normalized $iAUC_{60}$ and $iAUC_{120}$ (=iAUC up to 120 s post-CA injection) in the control group ($n=3$) and the JS-K group ($n=4$) after injection of saline (Ctr; gray diamonds) or JS-K (white triangles). A selective increase in blood-tumor barrier permeability upon i.v. JS-K delivery leads to a higher CA uptake in gliomas. *$P \leq 0.05$ t-test compared with control. *Reproduced with permission from Weidensteiner C, Reichardt W, Shami PJ, Saavedra JE, Keefer LK, Baumer B, Werres A, Jasinski R, Osterberg N, Weyerbrock A. Effects of the nitric oxide donor JS-K on the blood-tumor barrier and on orthotopic U87 rat gliomas assessed by MRI. Nitric Oxide 2013;30:17–25.*

where the tumor tissue shows enhancement due to CA extravasation, whereas the healthy tissue with an intact blood-brain barrier does not. This is the routine procedure for tumor patients in clinics, and it can also be applied in animal models, for example, in an orthotopic glioma model in [9].

Apart from tumor dimensions, also the necrosis extent within the tumor can be detected in MR images. For example, in Towner et al., the necrotic lesion volume in several orthotopic rat glioma models was measured [8]. The lesions appeared dark in both T_1- and T_2-weighted images.

MONITORING CHANGES IN THE TUMOR VASCULAR FUNCTION USING CONTRAST AGENTS

In addition to tumor size, its position and other anatomical information, functional readouts can be collected during the same examination with specialized MR imaging techniques. Functional MRI methods that permit indirect measurement of hemodynamics in tissues and tumors are powerful tools to study the vasoactive effects of NO in vivo. The assessment of tumor perfusion is important for the assessment of drug delivery and radiosensitivity. Tumor perfusion can also be studied using ultrasound imaging with microbubbles as CA, for example, in a study with an NO donor in a mouse tumor model [3].

Susceptibility contrast MRI is a method to estimate blood volume and is primarily applied in brain and brain tumors. It is based on the change of the transverse relaxation rate $R_2^* = 1/T_2^*$ after intravenous application of iron-oxide-based CAs and is mainly performed in animal studies [20]. These superparamagnetic particles (called USPIO, i.e., ultrasmall superparamagnetic iron oxide, or MION, i.e., monocrystalline iron-oxide-based nanoparticle) remain intravascular and change the contrast in T_2^*-weighted MR images that are acquired with multigradient echo sequences [21]. From the change in image signal intensity pre- versus post-USPIO injection, the fractional blood volume can be estimated [22]. R_2^* maps are calculated on a voxel-by-voxel basis from the multigradient echo datasets. The change in R_2^*, $\Delta R_2^* = R_2^*{}_{\text{post-USPIO}} - R_2^*{}_{\text{pre-USPIO}}$, is proportional to the fractional blood volume (fBV). For correct assessment of the blood volume with this method, it is essential that the CA remains intravascular and does not extravasate. Susceptibility contrast MRI provides an in vivo measurement of the hemodynamic effect of drugs and is, therefore, referred to as pharmacological MRI (phMRI) [23]. In principle, ΔR_2^* is measured before and at several time points after injection of the drug.

Papaevangelou et al. measured the fractional blood volume in an s.c. glioma model in mice (injection of 5.2 mgFe/kg of USPIO) [10]. They manipulated endogenous NO production by a switchable system for iNOS expression. The ΔR_2^* maps were used to distinguish vascularized tumor tissue from necrotic/poorly perfused regions (Fig. 2A third row, C and D). In vascularized regions, ΔR_2^* values and therefore the fractional blood volume exceeded a certain threshold. In tumors where iNOS was downregulated, the tumor growth rate and also necrosis were reduced compared with iNOS-expressing tumors. Mistry et al. applied the method in an orthotopic rat glioma model after i.v. nitrite injection (multigradient echo sequence with multiple slices, in-plane resolution 313 μm, and injection of 6 mg/kg MION) [11]. They observed a steady increase in mean ΔR_2^* in the tumor that reached a plateau 1.5 h after nitrite injection. They concluded that nitrite is converted into NO under the hypoxic conditions within the tumor that led to an increase in vascular permeability (but not blood volume) and subsequent CA extravasation. Kostourou et al. studied blood volume in an s.c. mouse glioma model injecting 2.5 mgFe/kg of USPIO [12]. ΔR_2^* maps revealed a necrotic core with decreased blood volume in tumors where inducible NOS (iNOS) was knocked down.

Iron-oxide-based CAs are only approved for a limited number of indications in humans, for example, in liver tumors [20]. In clinical studies of brain tumors, a slightly different method, namely, dynamic susceptibility contrast MRI (DSC-MRI), is used [24]. Here, the rapid change in contrast in a time series of T_2^*-weighted images caused by an intravascular

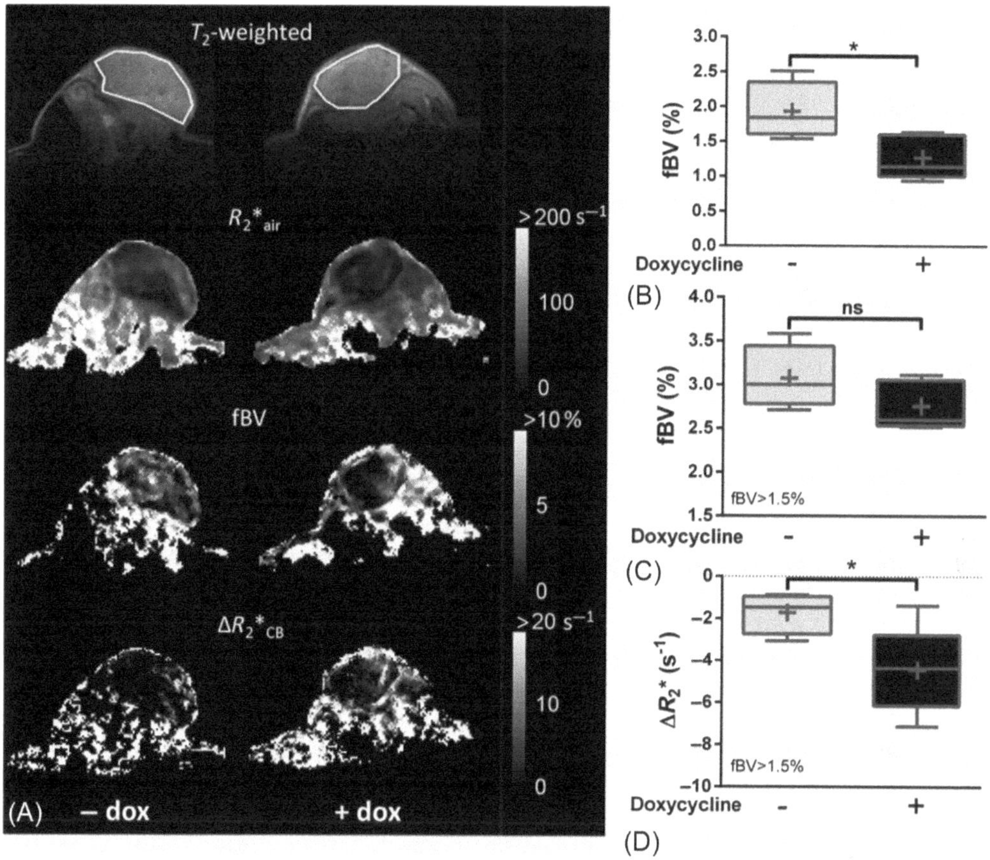

FIG. 2 Effects of NO downregulation on tumor vascular development as determined by MRI in an s.c. glioma mouse model (C6 pTet-off asiNOS xenografts). iNOS expression is induced by administration of doxycycline (dox +). (A) Representative T_2-weighted images, parametric maps of baseline $R_2{}^*{}_{air}$, fractional blood volume (fBV) measured with susceptibility contrast MRI using USPIO, and $\Delta R_2{}^*{}_{CB}$ in response to carbogen measured with functional MRI based on the BOLD effect. The ROIs encompassing the whole tumor volume in the central slice are demarcated on the T_2-weighted images. (B–D) Box plot distributions of the median fBV values measured over the whole tumor volume (B) and the median fBV values (C) and the median $\Delta R_2{}^*{}_{CB}$ values (D) in vascularized regions, of $n=4$ tumors with reduced iNOS expression (dox –) and $n=5$ with induced iNOS expression (dox +). Vascularized tumor tissue is defined as regions where fBV is higher than the threshold (threshold = median fBV value of all the data). Vasculature of iNOS-expressing tumors shows a higher reactivity to carbogen depicted as a bigger drop in R_2^*, that is, it is more mature. $^*P<0.05$. *Reproduced with permission from Papaevangelou E, Whitley GS, Johnstone AP, Robinson SP, Howe FA. Investigating the role of tumour cell derived iNOS on tumour growth and vasculature in vivo using a tetracycline regulated expression system. Int J Cancer 2016;138:2678–2687.*

bolus of clinically approved Gd-based CA is measured, and blood volume and blood flow can be calculated. However, leakage of this type of CA in tumors with disrupted blood-tumor barrier leads to errors in blood volume calculation [25].

Dynamic contrast-enhanced MRI (DCE-MRI) is based on a time series of MR images acquired before, during, and after administration of a low-molecular-weight CA such as Gd-DTPA, which readily diffuses from the blood to the extravascular-extracellular space.

CA leads to a change in the MR relaxation time T_1 and consequently to a change in signal intensity in T_1-weighted images especially in tissue with high vascular leakage. By acquiring a series of rapid MR images, the time course of the CA-induced change in T_1 can be followed, and the CA concentration can be calculated in a region of interest (ROI). More or less complex data processing strategies from simply calculating the initial area under the concentration curve (iAUC) up to pharmacokinetic modeling [26] can be applied to these CA concentration curves. The calculated (semi)quantitative biomarkers are sensitive to changes in a number of hemodynamic parameters, including blood flow, blood volume, and vessel permeability [27]. The extravasation depends on the size (molecular weight) of CA; thus, the permeability is specific for the size of the CA molecule. DCE-MRI has been used in many preclinical and clinical studies for monitoring of antiangiogenic therapies of tumors, including vascular endothelial growth factor (VEGF) inhibition in animal tumor models and in cancer patients [28,29].

DCE-MRI is the in vivo method of choice for investigating and quantifying blood-brain barrier impairment in tumors and other diseases [30,31]. So far, therapeutic blood-brain barrier disruption was predominantly studied with invasive techniques such as quantitative autoradiography [32] or postmortem methods like Evans blue staining.

DCE-MRI was applied in just a few preclinical studies involving NO donors and NOS inhibitors. Côté et al. studied opening of the blood-brain barrier after treatment with a B1 receptor agonist in combination with an NOS inhibitor [9]. They measured the CA concentration time courses in a glioma rat model using a T_1-weighted multislice sequence with 51 s temporal resolution and 270 μm in-plane resolution. They injected two CAs of different molecular weights Gd-DTPA (0.5 kDa) and Gadomer (17 kDa, Bayer HealthCare) as probes for the permeability of substances of different sizes. The enhancing regions were segmented on the images to yield a CA distribution volume as a measure for vascular permeability. Upregulated permeability resulted in an increased CA distribution volume for both types of CA. Jordan et al. reported a higher increase in signal intensity after injection of the NO donor isosorbide dinitrate versus control in an s.c. mouse liver tumor model [13]. They used a T_1-weighted spin echo sequence (single slice, 55 s temporal resolution, 270 μm in-plane resolution, and injection of 0.1 mmol/kg Gd-DTPA) and calculated the mean signal intensity increase, but did not quantify CA concentrations. They attributed the enhancement to an increase in tumor blood flow (in conjunction with results from BOLD effect measurements, see next chapter). Our group studied the acute effect of the NO donor in an orthotopic glioma rat model [7]. JS-K is activated by glutathione S-transferase selectively in the tumor [4]. We acquired DCE-MRI scans with an inversion recovery TrueFISP sequence pre-JS-K and 3 min post-JS-K infusion (single slice, temporal resolution 6 s, in-plane resolution 230 × 310 μm, and injection of 0.1 mmol/kg Gd-DTPA) (Fig. 1d). CA concentration curves were calculated from ROIs containing the whole tumor in the imaging slice (Fig. 1A and B). The significant increase in the DCE-MRI readout iAUC indicated an acute increase in permeability selectively in the glioma after JS-K treatment (Fig. 1E).

MONITORING CHANGES IN THE TUMOR VASCULAR FUNCTION USING THE BOLD EFFECT

The MR blood-oxygen-level-dependent (BOLD) contrast method is a functional MRI method that detects change in the oxygenation level in the blood [33]. The relaxation rate

$R_2{}^* = 1/T_2{}^*$ increases with the concentration of deoxygenated hemoglobin in the blood. A change in the fraction of deoxygenated to oxygenated blood (e.g., due to increased blood flow caused by the hemodynamic response in activated brain areas) leads to a change in mean $T_2{}^*$ and, therefore, to a change in image intensity in $T_2{}^*$-weighted MR images. The BOLD method is mainly used to study brain activation but also to assess tumor oxygenation and perfusion. Therefore, it can also serve as a phMRI technique to study drugs affecting hemodynamics.

Jordan et al. used a $T_2{}^*$-weighted gradient echo sequence (2 min temporal resolution and 270 μm in-plane resolution) and reported an increase in the mean signal intensity in an s.c. mouse liver tumor model a few minutes after injection of the NO donor isosorbide dinitrate [13]. This result is in accordance with an NO-mediated increase in tumor blood flow.

Functional MRI using the BOLD effect can also be combined with hyperoxic/hypercapnic challenge [34–36]. It assesses the maturation and functionality of tumor blood vessels by measuring the reactivity of the vessels to the breathing challenges. Papaevangelou et al. checked the tumor vessel reactivity with functional MRI under carbogen breathing challenge in their model with inducible iNOS expression that was already mentioned in the previous report [10]. Perfused tissue in tumors with reduced iNOS expression had a reduced $\Delta R_2{}^*$ response to carbogen, for example, the vessels were less mature (Fig. 2A bottom row and 2D). They conclude that NO derived from iNOS promotes tumor growth and vascular maturation that makes iNOS a promising target for antivascular therapy.

Advantage of the BOLD method is that it does not require the administration of a contrast agent. However, it cannot quantify tumor oxygenation since the BOLD signal depends also on the blood volume/flow. Tumor oxygenation can be measured indirectly with other imaging modalities: with electron paramagnetic resonance imaging EPR after injection of oxygen-sensitive probes and with PET using fluoromisonidazole (19F-MISO) as tracer [37]. For invasive measurements, polarographic electrodes and luminescence-based optical sensors are used [38].

An alternative CA-free approach to assess perfusion mainly in the brain and brain tumors is arterial spin labeling [39]. It was applied in both animal models and brain tumor patients [40,41], but not yet to study NO-mediated effects specifically. In arterial spin labeling, the inflowing blood is magnetically labeled using a special MR pulse sequence. The labeled blood changes the signal intensity in the imaging slice depending on the blood flow.

MOLECULAR IMAGING OF NO AND NOS EXPRESSION

The term molecular imaging can be defined as the in vivo characterization and measurement of biologic processes at the cellular and molecular levels [42]. Imaging techniques offering as much sensitivity and spatial resolution as possible are applied in conjunction with molecular probes. Molecular imaging can be used to either image an endogenous molecule or an administered drug directly (i.e., in the case of NO, following its distribution and target binding of NO or NO donors) or the target itself (e.g., expression of NO receptors or NOS) [43]. Molecular imaging has to cope with low target concentrations. So the central challenge is the development of specific reporter probes and amplification strategies to differentiate target information from the nonspecific background noise [42,43]. The basic principle is to couple the targeted agent (e.g., an antibody) to a reporter moiety that can be detected by a given imaging modality. The probe has to be designed and administered in such a way that it is capable to reach the target, that is, it is capable to overcome biological delivery barriers as, for example, it is able to extravasate from the vasculature after i.v. injection [42]. Otherwise, specific binding is not possible, and the method does not work correctly.

The most severe drawback of MR methods to study molecular processes is the inherent low sensitivity compared with optical and nuclear imaging methods. Therefore, the molecular markers are not measured directly. In fact in MRI, the targeted agent is bound to a CA, and the signal intensity change induced by the CA part of the molecular probe is detected. In this way, one takes advantage of the high spatial resolution of MRI.

NO is highly reactive, and it can be converted very quickly; thus, it is very difficult to detect and monitor the level of NO in vivo. Generally, imaging of NO relies on trapping the molecule with an agent [5]. Imaging of NO is performed with optical imaging (fluorescence), EPR imaging, and MRI [5]. NOS expression has been investigated using optical and PET imaging [5] and MRI [8]. All these approaches are not yet applicable for clinical studies. In the case of MRI, there are just a few reports of in vivo applications in animals. There is just one article about MRI of NO distribution in animals. The study was performed in a model of septic shock [14]—that is, a condition where relatively large amounts of NO are released [37]. NO was complexed with the Fe(II)-chelate spin trap. This complex resulted in a decrease of both T_1 and T_2 relaxation times. Enhancement of the liver was visible in T_1-weighted images after i.p. injection of the spin trap in rats with septic shock. The distribution of NO donors in vivo was not studied yet with MRI. Instead, PET imaging can be used after radiolabeling the NO donor molecule [44].

There is one study by Towner et al. about MR imaging of iNOS expression [8]. For this, an MRI probe for detecting iNOS expression was developed and applied in orthotopic rat glioma models. The probe consists of an anti-iNOS antibody coupled to albumin-Gd-DTPA-biotin. The rat brains were imaged at 20–120 min intervals post-i.v. injection of the probe (1 mg antibody/kg; 0.05 mmol Gd/kg) with a T_1-weighted RARE sequence (Fig. 3a–c). The probe crossed the blood-tumor barrier (but not the intact blood-brain barrier in healthy brain tissue) and led to an increase in signal intensity in the gliomas. The specific binding was validated by ex vivo fluorescence detection of the probe via tagging the biotin moiety of the probe (Fig. 3I).

DISCUSSION AND CONCLUSION

The main focus of MRI in the drug development process is the novel application of validated MRI techniques to gain additional and valuable information in preclinical and clinical studies. Routine MRI scanning procedures are widely available in radiological centers and are ready to be used in clinical studies. More advanced MRI methods need input from MRI experts both in study planning/protocol setup and in data analysis. Experimental MRI methods that were demonstrated only in a limited number of cases and often need further optimization may not serve as a robust tool in drug trials. However, they are of scientific interest for developers of imaging methods and imaging biomarkers. Preclinical MRI studies are performed by dedicated research groups in academia and large pharmaceutical companies who have

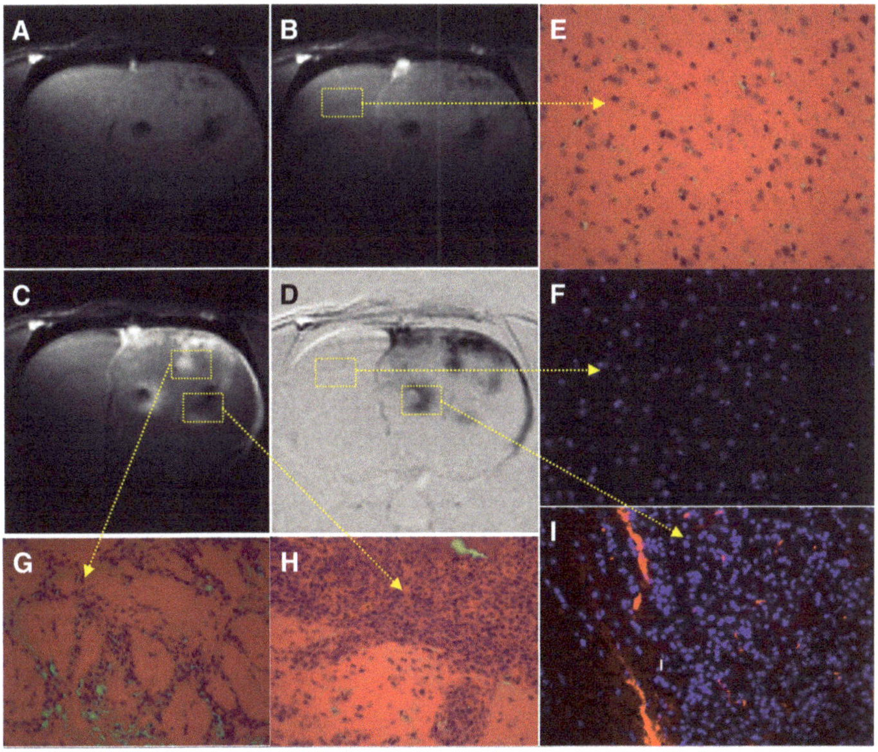

FIG. 3 Molecular MRI using an anti-iNOS probe in an orthotopic C6 glioma rat model. T_1-weighted MR images (A) preadministration and (B) 20 min and (C) 120 min postadministration of the targeting agent (anti-iNOS antibodies tagged with a biotin-albumin-Gd-DTPA-based CA). Dark voids depict necrotic regions, enhanced regions depict CA uptake. The tumor margins cannot be identified on T_1-weighted images, as opposed to T_2-weighted images (not shown). (D) Difference image between (A) and (C) depicting dark areas of anti-iNOS probe uptake. Corresponding histological images (H and E) (E) in contralateral brain tissue, (G) within the glioma tissue, and (H) at the tumor boundary were obtained in the regions outlined in the MRIs (B and C, respectively). After MRI scans, tissues from (F) the contralateral brain and (I) glioma were obtained for fluorescence imaging (original magnification 20 ×) of the anti-iNOS probe via tagging the biotin moiety of the probe with streptavidin-Cy3. Only the glioma tissue had specific binding of the anti-iNOS probe, indicating overexpression of iNOS. *Reproduced with permission from Towner RA, Smith N, Doblas S, Garteiser P, Watanabe Y, He T, Saunders D, Herlea O, Silasi-Mansat R, Lupu F. In vivo detection of inducible nitric oxide synthase in rodent gliomas. Free Radic Biol Med 2010;48:691–703.*

both the equipment and the expertise. Bottleneck of the application of MRI in large studies is its limited throughput capability, especially in animal studies.

MRI offers several advantages. First, it is noninvasive. Measurements can be performed in vivo and in situ. Second, MRI is very versatile. Several anatomical, morphological, functional, and molecular readouts can be assessed in one experimental setup. Third, MRI is a translational tool. Biomarkers and imaging methods applied in preclinical studies can in principle be translated to clinical studies and vice versa. An imaging biomarker is defined as a validated characteristic that is objectively measured by an imaging method as an indicator of normal/pathogenic processes or response to therapy [45]. Biomarkers help to identify potentially successful drugs in the development pipeline. In this respect, the measurement of tumor size is a very successful (even translational) MRI biomarker in oncological drug development. Also DCE-MRI readouts are considered useful biomarkers [45]. Other MRI methods, like BOLD, susceptibility-weighted MRI, or arterial spin labeling are less frequently used for decision-making in the drug development process.

Regarding NO research, MRI is an emerging modality and apart from morphological imaging not yet frequently used, as can be seen in the limited number of published articles. Molecular imaging with MRI in the field of NO is still very experimental. Currently, other molecular imaging modalities are more useful to study the biodistribution of NO and NO donors. However, MRI is capable of monitoring the different strategies of NO therapy in tumors [32], namely,

- cytotoxic effect with morphological MRI,
- attempts to improve the drug delivery into tumors by increasing blood flow and vessel permeability with functional MRI methods (BOLD MRI, susceptibility contrast MRI, and DCE-MRI),
- approaches to decrease tumor blood supply and inhibit angiogenesis by modulation of NO signaling with functional MRI (BOLD MRI, susceptibility contrast MRI, and DCE-MRI),
- attempts to induce chemo-/radiosensitivity by increasing the blood supply and therefore the oxygenation with BOLD and susceptibility contrast MRI,
- efficacy of combination therapy with morphological MRI.

Therefore, MRI is capable of providing valuable information in NO research and therapy development, especially functional methods like susceptibility contrast and BOLD MRI.

Current cancer research focuses on the roles of the tumor microenvironment, vasculature, and immune system in tumor development, growth, and metastasis. In vitro studies are not sufficient to study these complex systemic processes. Therefore, noninvasive, functional imaging methods like MRI that are applicable in both animal and patient studies and which may even qualify as a biomarker will gain more and more importance in the drug development process.

Acknowledgments

I want to thank my colleagues Martin Büchert and Wilfried Reichardt for carefully reading the manuscript and my coworkers Astrid Weyerbrock and Nadja Osterberg for introducing me to the field of NO and for doing the JS-K-MRI studies with Wilfried and me.

Conflict of Interest: No potential conflicts of interest were disclosed.

References

[1] Frederiksen LJ, Sullivan R, Maxwell LR, Macdonald-Goodfellow SK, Adams MA, Bennett BM, et al. Chemosensitization of cancer in vitro and in vivo by nitric oxide signaling. Clin Cancer Res 2007;13:2199–206.

[2] Weyerbrock A, Baumer B, Papazoglou A. Growth inhibition and chemosensitization of exogenous nitric oxide released from NONOates in glioma cells in vitro. J Neurosurg 2008;110:128–36.

[3] Scicinski J, Oronsky B, Ning S, Knox S, Peehl D, Kim MM, et al. NO to cancer: the complex and multifaceted role of nitric oxide and the epigenetic nitric oxide donor, RRx-001. Redox Biol 2015;6:1–8.

[4] Chakrapani H, Kalathur RC, Maciag AE, Citro ML, Ji X, Keefer LK, et al. Synthesis, mechanistic studies, and anti-proliferative activity of glutathione/glutathione S-transferase-activated nitric oxide prodrugs. Bioorg Med Chem 2008;16:9764–71.
[5] Hong H, Sun J, Cai W. Multimodality imaging of nitric oxide and nitric oxide synthases. Free Radic Biol Med 2009;47:684–98.
[6] Edrei Y, Gross E, Corchia N, Abramovitch R. Improved efficacy of a novel anti-angiogenic drug combination (TL-118) against colorectal-cancer liver metastases; MRI monitoring in mice. Br J Cancer 2012;107:658–66.
[7] Weidensteiner C, Reichardt W, Shami PJ, Saavedra JE, Keefer LK, Baumer B, et al. Effects of the nitric oxide donor JS-K on the blood-tumor barrier and on orthotopic U87 rat gliomas assessed by MRI. Nitric Oxide 2013;30:17–25.
[8] Towner RA, Smith N, Doblas S, Garteiser P, Watanabe Y, He T, et al. In vivo detection of inducible nitric oxide synthase in rodent gliomas. Free Radic Biol Med 2010;48:691–703.
[9] Côté J, Bovenzi V, Savard M, Dubuc C, Fortier A, Neugebauer W, et al. Induction of selective blood-tumor barrier permeability and macromolecular transport by a biostable kinin B1 receptor agonist in a glioma rat model. PLoS One 2012;7:e37485.
[10] Papaevangelou E, Whitley GS, Johnstone AP, Robinson SP, Howe FA. Investigating the role of tumour cell derived iNOS on tumour growth and vasculature in vivo using a tetracycline regulated expression system. Int J Cancer 2016;138:2678–87.
[11] Mistry N, Stokes AM, Gambrell JV, Quarles CC. Nitrite induces the extravasation of iron oxide nanoparticles in hypoxic tumor tissue. NMR Biomed 2014;27:425–30.
[12] Kostourou V, Cartwright JE, Johnstone AP, Boult JKR, Cullis ER, Whitley G, et al. The role of tumour-derived iNOS in tumour progression and angiogenesis. Br J Cancer 2011;104:83–90.
[13] Jordan BF, Misson P, Demeure R, Baudelet C, Beghein N, Gallez B. Changes in tumor oxygenation/perfusion induced by the no donor, isosorbide dinitrate, in comparison with carbogen: monitoring by EPR and MRI. Int J Radiat Oncol Biol Phys 2000;48:565–70.
[14] Fujii H, Wan X, Zhong J, Berliner LJ, Yoshikawa K. In vivo imaging of spin-trapped nitric oxide in rats with septic shock: MRI spin trapping. Magn Reson Med 1999;42:235–9.
[15] Parker GJM. Seeing within: magnetic resonance imaging in drug development. Drug Discov World 2015;Winter 2015/2016:19–22.
[16] Eisenhauer EA, Therasse P, Bogaerts J, Schwartz LH, Sargent D, Ford R, et al. New response evaluation criteria in solid tumours: revised RECIST guideline (version 1.1). Eur J Cancer Oxf 2009;45:228–47. 1990.
[17] Wen PY, Macdonald DR, Reardon DA, Cloughesy TF, Sorensen AG, Galanis E, et al. Updated response assessment criteria for high-grade gliomas: response assessment in neuro-oncology working group. J Clin Oncol 2010;28:1963–72.
[18] Gordillo N, Montseny E, Sobrevilla P. State of the art survey on MRI brain tumor segmentation. Magn Reson Imaging 2013;31:1426–38.
[19] Laufer S, Mazuz A, Nachmansson N, Fellig Y, Corn BW, Bokstein F, et al. Monitoring brain tumor vascular heamodynamic following anti-angiogenic therapy with advanced magnetic resonance imaging in mice. PLoS One 2014;9:e115093.
[20] Weinstein JS, Varallyay CG, Dosa E, Gahramanov S, Hamilton B, Rooney WD, et al. Superparamagnetic iron oxide nanoparticles: diagnostic magnetic resonance imaging and potential therapeutic applications in neurooncology and central nervous system inflammatory pathologies, a review. J Cereb Blood Flow Metab 2010;30:15–35.
[21] Corot C, Robert P, Idée J-M, Port M. Recent advances in iron oxide nanocrystal technology for medical imaging. Adv Drug Deliv Rev 2006;58:1471–504.
[22] Chen Y-CI, Mandeville JB, Nguyen TV, Talele A, Cavagna F, Jenkins BG. Improved mapping of pharmacologically induced neuronal activation using the IRON technique with superparamagnetic blood pool agents. J Magn Reson Imaging 2001;14:517–24.
[23] Perles-Barbacaru T-A, Procissi D, Demyanenko AV, Jacobs RE. Quantitative pharmacologic MRI in mice. NMR Biomed 2012;25:498–505.
[24] Barajas RF, Cha S. Benefits of dynamic susceptibility-weighted contrast-enhanced perfusion MRI for glioma diagnosis and therapy. CNS Oncol 2014;3:407–19.
[25] Shiroishi MS, Castellazzi G, Boxerman JL, D'Amore F, Essig M, Nguyen TB, et al. Principles of T2*-weighted dynamic susceptibility contrast MRI technique in brain tumor imaging. J Magn Reson Imaging 2015;41:296–313.
[26] Tofts PS. Modeling tracer kinetics in dynamic Gd-DTPA MR imaging. J Magn Reson Imaging 1997;7:91–101.

[27] Padhani AR, Khan AA. Diffusion-weighted (DW) and dynamic contrast-enhanced (DCE) magnetic resonance imaging (MRI) for monitoring anticancer therapy. Target Oncol 2010;5:39–52.
[28] Checkley D, Tessier JJL, Wedge SR, Dukes M, Kendrew J, Curry B, et al. Dynamic contrast-enhanced MRI of vascular changes induced by the VEGF-signalling inhibitor ZD4190 in human tumour xenografts. Magn Reson Imaging 2003;21:475–82.
[29] Morgan B, Thomas AL, Drevs J, Hennig J, Buchert M, Jivan A, et al. Dynamic contrast-enhanced magnetic resonance imaging as a biomarker for the pharmacological response of PTK787/ZK 222584, an inhibitor of the vascular endothelial growth factor receptor tyrosine kinases, in patients with advanced colorectal cancer and liver metastases: results from two phase I studies. J Clin Oncol 2003;21:3955–64.
[30] Larsson HB, Stubgaard M, Frederiksen JL, Jensen M, Henriksen O, Paulson OB. Quantitation of blood-brain barrier defect by magnetic resonance imaging and gadolinium-DTPA in patients with multiple sclerosis and brain tumors. Magn Reson Med 1990;16:117–31.
[31] Sourbron S, Ingrisch M, Siefert A, Reiser M, Herrmann K. Quantification of cerebral blood flow, cerebral blood volume, and blood-brain-barrier leakage with DCE-MRI. Magn Reson Med 2009;62:205–17.
[32] Weyerbrock A, Walbridge S, Saavedra JE, Keefer LK, Oldfield EH. Differential effects of nitric oxide on blood–brain barrier integrity and cerebral blood flow in intracerebral C6 gliomas. Neuro Oncol 2011;13:203–11.
[33] Ogawa S, Lee TM, Kay AR, Tank DW. Brain magnetic resonance imaging with contrast dependent on blood oxygenation. Proc Natl Acad Sci U S A 1990;87:9868–72.
[34] Abramovitch R, Dafni H, Smouha E, Benjamin LE, Neeman M. In Vivo Prediction of Vascular Susceptibility to vascular endothelial growth factor withdrawal. Cancer Res 1999;59:5012–6.
[35] Baudelet C, Cron GO, Gallez B. Determination of the maturity and functionality of tumor vasculature by MRI: correlation between BOLD-MRI and DCE-MRI using P792 in experimental fibrosarcoma tumors. Magn Reson Med 2006;56:1041–9.
[36] Howe FA, Robinson SP, McIntyre DJ, Stubbs M, Griffiths JR. Issues in flow and oxygenation dependent contrast (FLOOD) imaging of tumours. NMR Biomed 2001;14:497–506.
[37] Sonveaux P, Jordan BF, Gallez B, Feron O. Nitric oxide delivery to cancer: why and how? Eur J Cancer 2009;45:1352–69.
[38] Jordan BF, Sonveaux P. Targeting tumor perfusion and oxygenation to improve the outcome of anticancer therapy. Front Pharmacol 2012;3:94.
[39] Detre JA, Leigh JS, Williams DS, Koretsky AP. Perfusion imaging. Magn Reson Med 1992;23:37–45.
[40] Silva AC, Kim S-G, Garwood M. Imaging blood flow in brain tumors using arterial spin labeling. Magn Reson Med 2000;44:169–73.
[41] Warmuth C, Günther M, Zimmer C. Quantification of blood flow in brain tumors: comparison of arterial spin labeling and dynamic susceptibility-weighted contrast-enhanced MR imaging. Radiology 2003;228:523–32.
[42] Weissleder R, Mahmood U. Molecular imaging. Radiology 2001;219:316–33.
[43] Rudin M, Weissleder R. Molecular imaging in drug discovery and development. Nat Rev Drug Discov 2003;2:123–31.
[44] Pippin A, Voll R, Liotta D, Smith M, Goodman M. In vivo biodistribution of the nitric oxide donor, PRG150, using PET imaging in rats. J Nucl Med 2016;57:1078.
[45] O'Connor JPB, Aboagye EO, Adams JE, Aerts HJWL, Barrington SF, Beer AJ, et al. Imaging biomarker roadmap for cancer studies. Nat Rev Clin Oncol 2016;14:169–86.

CHAPTER

14

S-Nitrosothiol Metabolism in Cancer and Therapeutic Implications

Jon Peñarando, Laura M. López-Sánchez*, Amanda Cañas*, Antonio Rodríguez-Ariza*,†*

*Maimónides Institute of Biomedical Research of Córdoba (IMIBIC), Córdoba, Spain

†Oncology Department, Córdoba University Hospital, Córdoba, Spain

Abstract

The complex chemistry and biochemistry of nitric oxide (NO) and their relation to both physiological and pathological processes is one of the main reasons of the considerable difficulties in understanding the precise roles of this small molecule in cancer biology. One of the modifications mediated by NO is the coupling of a nitroso group moiety to a reactive thiol group in specific cysteine residues in peptides or proteins resulting in S-nitrosothiol (SNO) formation, a process commonly known as S-nitrosylation. In many cases, an altered production of NO and/or impaired SNO homeostasis promotes an aberrant S-nitrosylation of proteins. Importantly, this alteration in SNO homeostasis and S-nitrosylation of proteins plays an important role in a broad range of diseases, including cancer. Recent research in cell metabolism of SNOs has led to the identification and characterization of enzyme-mediated processes (nitrosylases and denitrosylases) that are implicated in targeted S-nitrosylation or denitrosylation of proteins in physiological settings. In this chapter, we will review these studies that may help to unravel the complex interrelationships between NO and the molecular pathways in the initiation and progression of tumors and their response to therapy.

Abbreviations

ADH5	alcohol dehydrogenase 5
Biotin-HPDP	*N*-[6-(biotinamido)hexyl]-3′-(2′-pyridyldithio)propionamide
BST	Biotin-switch technique
CCL2	C-C chemokine ligand 2
CCR2	C-C chemokine receptor type 2
cGMP	cyclic guanosine monophosphate
C-PTIO	2-(4-carboxyphenyl)-4,4,5,5-tetramethylimidazoline-1-oxyl-3 oxide
CTL	cytotoxic T lymphocyte
cysTMT	cysteine-reactive tandem mass tag
eNOS	endothelial nitric oxide synthase

Nitric Oxide (Donor/Induced) in Chemosensitization, Volume 1
http://dx.doi.org/10.1016/B978-0-12-811020-1.00014-4

ER	estrogen receptor
GAPDH	glyceraldehyde-3-phosphate dehydrogenase
GR	GSSG reductase
GSH	glutathione
GSNHOH	glutathione S-hydroxysulfenamide
GSNO	S-nitrosoglutathione
GSNOR	S-nitrosoglutathione reductase
GSSG	oxidized glutathione
HER2	human epidermal growth factor receptor 2
HNO	nitroxyl
HSF1	heat shock factor protein 1
iNOS	inducible nitric oxide synthase
iodoTMT	iodoacetyl tandem mass tag
LC-MS/MS	liquid chromatography tandem mass spectrometry
MDSC	Myeloid-derived suppressor cells
MMTS	methyl-methanethiosulfonate
MS	mass spectrometry
NADH	nicotinamide adenine dinucleotide
NADPH	nicotinamide adenine dinucleotide phosphate
nNOS	neuronal nitric oxide synthase
NO	nitric oxide
NOS	nitric oxide synthase
PD-L1	programmed death-ligand 1
SH	sulfhydryl
SIRT1	sirtuin 1
SNO	S-nitrosothiol
STAT3	signal transducer and activator of transcription 3
TCR	T-cell receptor
Tregs	regulatory T cells
Trk	tyrosine kinase receptor
Trx	thioredoxin
TrxR	thioredoxin reductase

INTRODUCTION

Nitric oxide (NO) is a small short-lived molecule implicated in different signaling pathways in living organisms [1]. In addition to all its physiological functions, NO has been related with the development of several pathologies, including cancer [2,3]. This small molecule is synthesized by the metabolism of L-arginine to L-citrulline through a reaction catalyzed by the enzyme nitric oxide synthase (NOS), of which there are three isoforms in mammals: neuronal NOS (nNOS), endothelial NOS (eNOS), and inducible NOS (iNOS) [4]. Depending on the enzyme the output of the synthesized NO varies from picomolar to nanomolar range for short periods to a micromolar range for long periods [5]. Thus, nNOS and eNOS produce low and transient levels of NO [4], whereas iNOS is able to synthesize higher amounts of NO for many hours or even days [6]. The overexpression of the different NOS isoforms in human solid tumors strongly suggests a key role of NO in tumor processes [7]. However, there is still a long way to go to completely understand how this molecule contributes in tumorigenesis as studies have demonstrated the dual role of NO in the pathogenesis of cancer. Thus, high levels of NO generate reactive nitrogen species that can react with reactive oxygen species causing a nitrosative and oxidative stress and resulting in DNA damage, mitochondrial dysfunction and increasing apoptosis [8], whereas low levels of NO may have antiapoptotic effects and stimulate angiogenesis [9]. It is clear that the considerable difficulties in understanding the precise roles of NO in cancer biology derive from the complex chemistry and biochemistry of this small molecule and their relation to both physiological and pathological processes.

Many of NO actions are mediated by the specific interaction of NO with the heme group of the hemoprotein soluble guanylyl cyclase resulting in the production of the second messenger cGMP, which acts on cGMP-dependent protein kinases, cGMP-gated cation channels, and phosphodiesterases [10]. However, NO can activate or modulate cell signaling by modifying other proteins through the coupling of a nitroso moiety to a reactive thiol group in specific cysteine residues leading to the formation of all S-nitrosothiol (SNO), a process usually known as S-nitrosylation [11]. Significantly, the deregulation of protein S-nitrosylation has been related with a number of human diseases, including cardiovascular and pulmonary disease, neurological disorders, and cancer [12,13]. Recent research in the cell metabolism of S-nitrosothiols has led to the identification and characterization of enzyme-mediated processes in the nitrosylation and denitrosylation of proteins. These studies may definitely help to unravel the complex interrelationships between NO and the molecular pathways in the initiation and progression of tumors and their response to therapy.

THE S-NITROSOPROTEOME

It is estimated that over 1000 mammalian proteins are susceptible to modification by S-nitrosylation [14,15]. However, detection of SNO proteins is complicated by the labile nature of the S-NO bond and the low levels present in the cells and tissues. Thus, detection of the modification is a challenging task, and in recent years, a variety of new strategies and improved methods for detection and identification of SNO protein have been established. The biotin-switch technique (BST) developed by Jaffrey and coworkers [16] is the most used methodological tool to detect S-nitrosylated proteins. In this approach, the S-nitrosylated cysteines of the proteins are selectively labeled with biotin through three steps. First, free thiols in proteins are specifically blocked with the thiol-specific methylating agent methylmethanethiosulfonate (MMTS). Next, the protein SNO groups are selectively reduced to thiols with ascorbate; then, these newly formed thiols are reacted with the sulfhydryl-specific biotinylating reagent *N*-[6-(biotinamido)hexyl]-3′-(2′-pyridyldithio)propionamide (biotin-HPDP). The biotinylated proteins can be readily detected by immunodetection with antibiotin antibodies, recognition via streptavidin, or they can be enriched and purified on avidin-agarose for mass spectrometry (MS) analysis. Different strategies based on modifications of the BST method were soon developed to improve the sensitivity, specificity, and throughput of the identification of SNO proteins and also to identify those Cys residues that are S-nitrosylated in these proteins. Thus, improvements for the quantitative proteomic analysis of S-nitrosylation were developed through the introduction of a tryptic digestion step after biotinylation and before streptavidin purification, thus allowing the purification of the biotinylated-Cys residues that can be directly analyzed by MS to identify the SNO-sites [17]. Other variations of the BST method are based on using other labels instead of biotin. Thus, in another BST-based approach, isobaric labeling methods have been used, where a set of cysteine-reactive tandem mass tag (cysTMT) reagents are used instead biotin, conferring the advantage of multiple isobaric tags with reporter ions and allowing the quantification and identification of S-nitrosylated proteins [18]. However, the cysTMT tag is reversible and not compatible with reduction before trypsin digestion to MS analysis, and therefore, these reagents have been recently replaced by thiol-reactive iodoacetyl tandem mass tag (iodoTMT) sixplex reagents [19]. The iodoTMT reagents irreversibly bind to Cys residues resulting in a stable tagging modification that can be detected with an anti-TMT antibody or analyzed by LC-MS/MS analysis.

Despite of all these advances and developments, nitrosoproteomic analyses have generated large discrepancies among identified sets of SNO proteins. In this regard, when Chung et al. [20] recently used the two labeling reagents (cysTMT and iodoTMT) to compare whether both were able to identify the same SNO sites, they observed that only 30% of the SNO proteins were identified by both tags. Thus, there could be differences in the chemistry between the reagents and the SNO residues. Therefore, when they used a method combining parallel cysTMT and iodoTMT reagents a considerable increase in the coverage of SNO sites identified was observed, they concluded that for studies of S-nitrosoproteomes identification, the use of a parallel combination of two tags for dual labeling maximizes the identification of SNO proteins. It is clear that new approaches to detect specific SNOs, along with a careful validation of methods with respect to different SNO subtypes, are required to fully reveal the essential role of S-nitrosylation in tumor biology [21].

NITROSYLASES AND DENITROSYLASES

Although the chemical reactions involved in S-nitrosothiols formation are relatively well-known, the mechanisms governing S-nitrosylation of proteins remain obscure. A number of factors, such as the thiol microenvironment that can render a pK of the sulfhydryl group anomalously low or regions of local hydrophobicity within proteins that may arise from tertiary protein structure and protein-protein interactions, are known contributors to the specificity of S-nitrosylation. Other factors are metal ion availability and the redox status of the immediate environment that is related not only to subcellular localization of the protein but also to the proximity to nitric oxide and reactive oxygen sources or to hydrophobic environments [22,23]. In addition to these specificity determinants, recent studies have shown that cellular S-nitrosylation levels and, consequently, the S-nitrosoproteome are substantially regulated by the presence of several enzymes that promote S-nitrosylation (nitrosylases) or denitrosylation reactions (denitrosylases).

Nitrosylases

Nitrosylases are proteins involved in transnitrosylation (Fig. 1), which is the transfer of a nitroso moiety from one thiol to another, and contribute to target specificity in the S-nitrosylation reactions [13,24]. As discussed below, several transnitrosylases that have been shown to function in physiological setting might play a significant role in the S-nitrosothiol homeostasis of cancer cells.

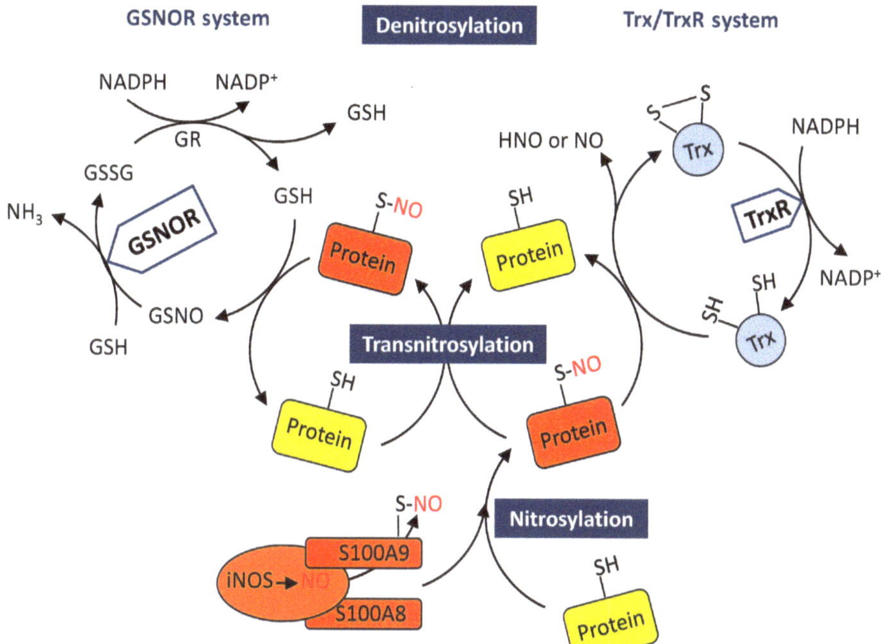

FIG. 1 Enzyme-mediated processes governing targeted S-nitrosylation or denitrosylation of proteins. Protein thiols can undergo S-nitrosylation and generate S-nitrosothiols (SNO) by distinct reactions involving NO groups and different thiols substrates. The heterotrimeric S-nitrosylase complex consisting of inducible nitric oxide synthase (iNOS) and S100A8 and S100A9 proteins directs specific S-nitrosylation on multiple target proteins. The NO moiety can also be transferred directly from one thiol to another in a process termed transnitrosylation. SNO homeostasis depends on enzymatic denitrosylating systems, including S-nitrosoglutathione reductase (GSNOR) and the thioredoxin and thioredoxin reductase (Trx/TrxR) systems. GSNO, S-nitrosoglutathione; GSSG, oxidized glutathione; GR, GSSG reductase. See text for further details.

S-Nitrosylated-glyceraldehyde-3-phosphate dehydrogenase (SNO-GAPDH) functions as a physiological nitrosylase for a subset of nuclear proteins, thereby implementing apoptotic and transcriptional programs [25]. In addition to its role as a glycolytic enzyme, upon apoptotic stimulation, GAPDH is nitrosylated at its active site Cys150 promoting its binding to the E3 ubiquitin ligase Siah1 and cotranslocation to the nucleus [26]. Kornberg et al. subsequently showed that nuclear SNO-GAPDH binds to and transnitrosylates several nuclear proteins including SIRT1, which is a lysine deacetylase, inhibiting its downstream effects on transcription [27]. Interestingly, SIRT1 has been reported to stimulate proliferation and glycolysis gene expression in pancreatic cancer [28]. However, the role of S-nitrosylated GAPDH in cancer cells remains unknown. Recently, Jia et al. [29] have described a stimulus-inducible, heterotrimeric S-nitrosylase complex consisting of iNOS and S100A8 and S100A9 proteins that directs specific S-nitrosylation on multiple target proteins, including GAPDH (Fig. 1). In this nitrosylase complex, S100A8 and S100A9 proteins coordinately contributed to site selectivity, whereas S100A9 exhibited transnitrosylase activity, shuttling NO from iNOS to the target protein [29]. Notably, S100A8 and S100A9 levels are elevated in cancer compared with normal tissues, and their increased expression is associated with tumor aggressiveness and metastasis [30]. Therefore, since iNOS is overexpressed in most tumors, the iNOS-S100A8/A9 S-nitrosylase complex may play an important role in governing the S-nitrosothiol homeostasis in cancer cells.

The thioredoxin system, which consists of thioredoxin (Trx) and thioredoxin reductase (TrxR), plays a key role in cellular redox homeostasis and signaling [31]. A number of recent studies have stressed the importance of the cross talk between NO and Trx system, particularly in the context of protein nitrosylation/denitrosylation [32]. In addition to the active site, mammalian Trx has two additional cysteines (Cys69 and Cys73) that have emerged as physiological targets of S-nitrosylation and that participate in transnitrosylation reactions with other target proteins [32,33]. Thus, SNO-Trx transnitrosylates caspase-3 to form SNO-caspase-3, thereby preventing apoptosis [33,34]. Additionally, the overexpression of a mutant disulfide reductase-inactive Trx in HeLa cells promoted the nitrosylation of specific target proteins and allowed the identification of 47 Trx transnitrosylation target proteins [35]. Importantly, the Trx/TrxR system also mediates denitrosylation of many SNO proteins, as discussed below.

Denitrosylases

Denitrosylases are enzymes that catalyze the elimination of NO groups from the Cys thiol side chain of low-molecular-weight thiols or proteins. S-Nitrosoglutathione reductase (GSNOR), a highly evolutionarily conserved enzyme, was one of the first denitrosylating enzymatic systems described [36]. GSNOR utilizes NADH to convert GSNO to glutathione S-hydroxysulfenamide (GSNHOH), which in turn is converted to oxidized glutathione (GSSG) [37]. The denitrosylation cycle is completed through reduction of GSSG by glutathione reductase; thereby, GSSG reductase activity is required for physiological denitrosylation of GSNO [38]. Thus, GSNOR governs protein S-nitrosylation by influencing the cellular equilibrium between SNO-proteins and GSNO (Fig. 1). For this reason, GSNOR plays an important role in the maintenance of nitrosothiols levels and functions to ameliorate nitrosative stress [39]. Thus, we have demonstrated that an increase in GSNOR expression and activity is critical to return the SNO protein content to basal levels in L-nitrosocysteine-challenged human hepatocytes [40]. Moreover, inhibition of NO production in cholestatic rats ameliorated hepatocellular injury by reverting the reduction in GSNOR expression and activity, improving S-nitrosothiol homeostasis, and thereby reducing the SNO-protein levels during cholestasis [41]. On the other hand, GSNOR deficiency has been shown to impair DNA repair through S-nitrosylation and proteasomal degradation of the key DNA repair protein O(6)-alkylguanine-DNA alkyltransferase and to promote hepatocarcinogenesis [42], which can be prevented by iNOS inhibition [43].

Recent studies suggest that N6022, a pyrrole derivative class with potent GSNOR inhibiting activity, has beneficial effects in animal models of asthma, obstructive pulmonary disease, and chronic inflammatory bowel disease, since such disorders have been associated with low levels of GSNO and other S-nitrosothiols [44–46]. Recently, we have shown that the inhibition of GSNOR activity with N6022 in HER2-overexpressing breast cancer cells augments protein S-nitrosylation and reduces the antitumoral efficacy of trastuzumab, a therapeutic monoclonal antibody against HER2/neu, which is overexpressed in 15%–20% of breast cancers [47]. Significantly, GSNOR inhibition restored the activation of survival signaling pathways involved in the resistance to anti-HER2 therapies, including the AKT survival pathway, Src and c-Abl kinases, and tyrosine kinase receptors (TrkA/NRTK1, TrkB/NRTK2, EphA1, and EphA3) and reduced the apoptotic effect of trastuzumab. Accordingly, GSNOR inhibition augmented the S-nitrosylation of proteins with important roles in the regulation and modulation of various aspects of the cellular apoptosis machinery, including Apaf-1, pSer73/63 c-Jun, calcineurin subunit α and HSF1. Moreover, we showed that HER2 expression in breast tumors was associated with lower expression of GSNOR protein and that high ADH5/GSNOR gene expression was associated with high patient survival rates in HER2 tumors [47]. These data stress that an altered SNO homeostasis contributes to the progression of HER2 + breast tumors cancer and could facilitate the development of therapeutic options to counteract the resistance to anti-HER2 therapies.

As mentioned above, the Trx/TrxR system has also emerged as another physiologically relevant denitrosylase [48] (Fig. 1). Early work by Holmgren and colleagues showed that the Trx/TrxR enzymatic system catalyzes the cleavage of GSNO into GSH and NO [49]. Subsequent studies showed evidence that Trx/TrxR not only reduces various low-molecular-weight SNOs but also directly catalyzes the denitrosylation of a wide range of cellular proteins [50–52]. Of note, Trx/TrxR-dependent denitrosylation plays a role in the physiological regulation of cellular functions since recent studies have shown that Trx-/TrxR-mediated denitrosylation regulates a range of NO-dependent processes, including the heme protein maturation [53,54], inflammation [55], exocytosis [56], and apoptosis [51]. However, so far, there are few studies that have addressed the importance of Trx-/TrxR-dependent denitrosylation and S-nitrosothiol homeostasis in tumor biology. We recently demonstrated that the impairment of S-nitrosothiol homeostasis with auranofin, a highly specific TrxR inhibitor, conferred survival advantages to estrogen receptor (ER)-positive breast cancer cells [57]. Furthermore, positive ER status in breast tumors was found to be associated with significantly lower levels of TrxR expression compared with ER-negative tumors. Therefore, our study suggests that the ER status in breast cancer may dictate the tumor response to impaired SNO homeostasis and that the molecular mechanisms involved may also play a significant role in the development of resistance against hormonal therapies. In this study, we also demonstrated that TrxR-specific inhibition sensitized breast cancer cells to the growth inhibitory effects of high levels of S-nitrosothiols, with a marked increase in protein S-nitrosylation [57].

The importance of Trx/TrxR in protecting cancer cells from nitrosative stress has been highlighted by other studies that have also shown that targeting the Trx/TrxR system sensitizes cancer cells to nitrosylating agents [58,59]. In this regard, a recent report revealed that TrxR is itself susceptible to nitrosylation-mediated inhibition, with both thiol and selenol modifications at the two redox active centers of the enzyme [60]. Moreover, depletion of cellular GSH augmented S-nitrosylation and inactivation of TrxR in HeLa tumor cells, leading to a marked Trx1 oxidation and extensive cell death. This suggests that GSH-dependent pathways play an important role in limiting the nitrosylation of TrxR and maintaining its enzymatic activity. These observations led the authors to suggest the possibility of exploiting the nitrosylation susceptibility of TrxR for killing tumor cells [60].

However, other studies illustrate the complex relationships between protein S-nitrosylation and tumor behavior, implying that caution should be exercised with the possibility of using nitrosylating drugs as an antitumor therapy. For instance, Shen et al. [61] recently reported an S-nitrosylation-dependent pathway by which glioma cells recruit and exploit microglia for their proliferation and invasion abilities. In this microglia-glioma cell to cell communication signaling pathway, the NO produced by NOS2 in glioma cells led to S-nitrosylation-dependent inhibition of mitochondrial Trx2 activity in microglia, which in turn resulted in increased S-nitrosylation and inhibition of caspase-3, an event that promoted the tumor-supportive phenotype of microglia [61]. Microglia are brain-resident macrophages that are part of the innate arm of the immune system in the central nervous system and can promote cell growth and invasion in the case of glioma [62]. Therefore, this study also highlights the relevance of S-nitrosothiol metabolism in the complexity of the cancer-immune system interplay and in the promise of immunotherapy for the treatment of cancer, as we discussed below.

NITROSOTHIOLS AND CANCER IMMUNOTHERAPY

Nitrosothiols have been shown to participate in signaling pathways involved both in innate and adaptive immunity, such as macrophage activation and self-protection by denitrosylases and T-cell differentiation and function [63]. Evidence is rapidly accumulating that the immune system plays an important role in host defenses against tumors. Actually, many current immunotherapeutics, which can unleash the power of immune system to fight cancer, are yielding remarkable and durable responses for some cancer patients [64]. Our scientific knowledge of the immune system and how it interacts with cancer cells is rapidly increasing [3], and certainly, the role of NO in the immune system also impacts the immune responses to tumors [65]. S-Nitrosothiol metabolism and S-nitrosylation influence the activity of key signaling and adaptor proteins and participate in the immunosuppressive mechanisms deployed by tumors. For instance, tumors induce an aberrant expansion of iNOS-expressing myeloid-derived suppressor cells (MDSCs) that cause reactive nitrosylation of proteins such as the T-cell receptor (TCR), CD3, and CCR2, thus hampering T-cell function [66]. Besides suppressing T-cell activation, MDSCs have been recently reported to promote breast tumor formation by enhancing cancer cell stemlike properties through the NO-mediated activation of NOTCH and prolonged STAT3 activation [67]. Also, NO exerts additional immunosuppressive activities within the tumor by inducing S-nitrosylation of the chemoattractant chemokine CCL2 and promoting its capacity to recruit MDSCs but abolishing its ability to attract CD8+ T cells to the inner core of tumor tissues [68]. On the other hand, a tumor-expressing iNOS has been reported to play an active role in the recruitment and activation of MDSCs, and iNOS inhibition reversed tumor-mediated immunosuppression in mouse melanoma [69]. Another main cell type contributing to tumor immunosuppression are the regulatory T cells (Tregs), which are present at a high frequency in tumors. It is known that a subpopulation of Tregs is derived from T cells induced by NO, and these NO-Tregs suppress CD4+ CD25-effector T cells [70]. Therefore, NO-derived therapies could be used in combination with current strategies aimed to control Tregs for the activation of antitumor immunity [71].

The energy metabolism of most cancer cells differs markedly from that of normal cells. This metabolic reprogramming in tumor cells is necessary to satisfy their bioenergetic and biosynthetic demands for rapid cell proliferation and to adapt to the tumor microenvironment. Recent evidence suggests that tumor infiltrating immune cells also undergo metabolic reprogramming that contributes to their pro- or antitumor functions [72]. Notably, NO is a key intermediate in the metabolic alterations of activated immune cells, and S-nitrosylation of cellular targets is an important mechanism by which NO exerts its effects

[72,73]. It remains to be investigated whether alterations of nitrosylation/denitrosylation pathways are involved in the metabolic switch of tumor-associated immune cells and their role in cancer progression.

The inhibition of NO-mediated pathways may constitute an appropriate strategy for abating the immunosuppressive tumor microenvironment and could potentially be combined with other immunomodulatory approaches, such as antitumor vaccination or adoptive immunotherapy with cytotoxic T lymphocytes (CTLs). Remarkably, a recent preclinical study has reported that targeting NO as mediator of the immunosuppressive activity of MDSCs is an appropriate strategy to restore impaired CTL function and improve the efficacy of immunotherapy. The antitumor activity of CTLs was augmented in mice receiving the NO radical scavenger C-PTIO the day after CTL injection, with the aim of eliminating NO produced by CTL-induced MDSCs [74].

In summary, cancer immunotherapy holds an extraordinary promise for the future, with new agents and treatment strategies on the horizon and new developments occurring at a relentless pace. However, one main concern is the absence of biomarkers to better select patients, given that less than half of patients respond to these immunotherapies. This is an important issue as it will avoid treatment-related toxicity and cost in those patients that are unlikely to benefit from these treatments [75]. Conceivably, NO-related mechanisms, including nitrosylation/denitrosylation pathways, hold a predictive value which, in combination with other predictive biomarkers, such as mutational load, lymphocyte infiltrates, or PD-L1 expression [64], may improve their capacity to predict the immunotherapy outcome.

CONCLUSIONS

The complexity of the role of the signaling NO network in cancer biology confounds the simple translation of preclinical data from bench to bedside. However, recent studies have greatly contributed to unravel the intricate web of interrelationships between NO and critical molecular pathways involved in the initiation and progression of tumors. In this regard, the identification and characterization of enzyme-mediated processes governing targeted S-nitrosylation or denitrosylation of proteins will have obvious implications in the recognition of potential targets for antitumor therapy. Moreover, these mechanisms participating in the control of SNO homeostasis may also lead to identify important prognostic factors and also predictive of response to therapy. Certainly, future studies will not only expand on our current knowledge of the role of SNO metabolism in cancer but also contribute to overcome the challenges of translating NO-based anticancer therapies to the clinic. For instance, available evidence suggests that the combination of inhibition of NO-/SNO-mediated pathways with other immunomodulatory approaches, such as antitumor vaccination or adoptive immunotherapy, may constitute an appropriate strategy for abating the immunosuppressive tumor microenvironment and must be tested in future clinical cancer trials.

Acknowledgments

This work was supported by grants from the Plan Estatal de I+D+I, ISCIII-Subdirección General de Evaluación y Fomento de la Investigación, Spanish government, and European Fund for Regional Development (FEDER) (grants PI10-00428 and PI13-00553).

Conflict of Interest: No potential conflicts of interest were disclosed.

References

[1] Heinrich TA, da Silva RS, Miranda KM, Switzer CH, Wink DA, Fukuto JM. Biological nitric oxide signalling: chemistry and terminology. Br J Pharmacol 2013;169:1417–29.

[2] Choudhari SK, Chaudhary M, Bagde S, Gadbail AR, Joshi V. Nitric oxide and cancer: a review. World J Surg Oncol 2013;11:118.

[3] Aranda E, Lopez-Pedrera C, de la Haba-Rodriguez JR, Rodriguez-Ariza A. Nitric oxide and cancer: the emerging role of S-nitrosylation. Curr Mol Med 2012;12:50–67.

[4] Knowles RG, Moncada S. Nitric oxide synthases in mammals. Biochem J 1994;298(Pt 2):249–58.

[5] Alderton WK, Cooper CE, Knowles RG. Nitric oxide synthases: structure, function and inhibition. Biochem J 2001;357:593–615.

[6] Ambs S, Merriam WG, Ogunfusika MO, Bennett WP, Ishibe N, Hussain SP, et al. p53 and vascular endothelial growth factor regulate tumor growth of NOS2-expressing human carcinoma cells. Nat Med 1998;4:1371–6.

[7] Hirst D, Robson T. Nitric oxide in cancer therapeutics: interaction with cytotoxic chemotherapy. Curr Pharm Des 2010;16:411–20.

[8] Wink DA, Vodovotz Y, Laval J, Laval F, Dewhirst MW, Mitchell JB. The multifaceted roles of nitric oxide in cancer. Carcinogenesis 1998;19:711–21.

[9] Carpenter AW, Schoenfisch MH. Nitric oxide release: part II. Therapeutic applications. Chem Soc Rev 2012;41:3742–52.

[10] Francis SH, Busch JL, Corbin JD, Sibley D. cGMP-dependent protein kinases and cGMP phosphodiesterases in nitric oxide and cGMP action. Pharmacol Rev 2010;62:525–63.

[11] Hess DT, Stamler JS. Regulation by S-nitrosylation of protein post-translational modification. J Biol Chem 2012;287:4411–8.

[12] Foster MW, Hess DT, Stamler JS. Protein S-nitrosylation in health and disease: a current perspective. Trends Mol Med 2009;15:391–404.

[13] Anand P, Stamler JS. Enzymatic mechanisms regulating protein S-nitrosylation: implications in health and disease. J Mol Med (Berl) 2012;90:233–44.

[14] Seth D, Stamler JS. The SNO-proteome: causation and classifications. Curr Opin Chem Biol 2011;15:129–36.

[15] Chen YJ, Lu CT, Su MG, Huang KY, Ching WC, Yang HH, et al. dbSNO 2.0: a resource for exploring structural environment, functional and disease association and regulatory network of protein S-nitrosylation. Nucleic Acids Res 2015;43:D503–11.

[16] Jaffrey SR, Erdjument-Bromage H, Ferris CD, Tempst P, Snyder SH. Protein S-nitrosylation: a physiological signal for neuronal nitric oxide. Nat Cell Biol 2001;3:193–7.

[17] Hao G, Derakhshan B, Shi L, Campagne F, Gross SS. SNOSID, a proteomic method for identification of cysteine S-nitrosylation sites in complex protein mixtures. Proc Natl Acad Sci U S A 2006;103:1012–7.

[18] Murray CI, Uhrigshardt H, O'Meally RN, Cole RN, Van Eyk JE. Identification and quantification of S-nitrosylation by cysteine reactive tandem mass tag switch assay. Mol Cell Proteomics 2012;11:M111.

[19] Qu Z, Meng F, Bomgarden RD, Viner RI, Li J, Rogers JC, et al. Proteomic quantification and site-mapping of S-nitrosylated proteins using isobaric iodoTMT reagents. J Proteome Res 2014;13:3200–11.

[20] Chung HS, Murray CI, Venkatraman V, Crowgey EL, Rainer PP, Cole RN, et al. Dual labeling biotin switch assay to reduce bias derived from different cysteine subpopulations: a method to maximize S-nitrosylation detection. Circ Res 2015;117:846–57.

[21] Seth D, Stamler JS. SNOs differ: methodological and biological implications. Circ Res 2015;117:826–9.

[22] Lopez-Sanchez LM, Lopez-Pedrera C, Rodriguez-Ariza A. Proteomics insights into deregulated protein S-nitrosylation and disease. Expert Rev Proteomics 2012;9:59–69.

[23] Lopez-Sanchez LM, Lopez-Pedrera C, Rodriguez-Ariza A. Proteomic approaches to evaluate protein S-nitrosylation in disease. Mass Spectrom Rev 2014;33:7–20.

[24] Smith BC, Marletta MA. Mechanisms of S-nitrosothiol formation and selectivity in nitric oxide signaling. Curr Opin Chem Biol 2012;16:498–506.

[25] Tristan C, Shahani N, Sedlak TW, Sawa A. The diverse functions of GAPDH: views from different subcellular compartments. Cell Signal 2011;23:317–23.

[26] Hara MR, Agrawal N, Kim SF, Cascio MB, Fujimuro M, Ozeki Y, et al. S-nitrosylated GAPDH initiates apoptotic cell death by nuclear translocation following Siah1 binding. Nat Cell Biol 2005;7:665–74.

[27] Kornberg MD, Sen N, Hara MR, Juluri KR, Nguyen JV, Snowman AM, et al. GAPDH mediates nitrosylation of nuclear proteins. Nat Cell Biol 2010;12:1094–100.

[28] Pinho AV, Mawson A, Gill A, Arshi M, Warmerdam M, Giry-Laterriere M, et al. Sirtuin1 stimulates the proliferation and the expression of glycolysis genes in pancreatic neoplastic lesions. Oncotarget 2016. http://dx.doi.org/10.18632/oncotarget.11013. Aug 2.

[29] Jia J, Arif A, Terenzi F, Willard B, Plow EF, Hazen SL, et al. Target-selective protein S-nitrosylation by sequence motif recognition. Cell 2014;159:623–34.

[30] Lim SY, Yuzhalin AE, Gordon-Weeks AN, Muschel RJ. Tumor-infiltrating monocytes/macrophages promote tumor invasion and migration by upregulating S100A8 and S100A9 expression in cancer cells. Oncogene 2016. http://dx.doi.org/10.1038/onc.2016.107. Apr 18.

[31] Lu J, Holmgren A. The thioredoxin antioxidant system. Free Radic Biol Med 2014;66:75–87.

[32] Sengupta R, Holmgren A. Thioredoxin and thioredoxin reductase in relation to reversible S-nitrosylation. Antioxid Redox Signal 2013;18:259–69.

[33] Mitchell DA, Morton SU, Fernhoff NB, Marletta MA. Thioredoxin is required for S-nitrosation of procaspase-3 and the inhibition of apoptosis in Jurkat cells. Proc Natl Acad Sci U S A 2007;104:11609–14.

[34] Mitchell DA, Marletta MA. Thioredoxin catalyzes the S-nitrosation of the caspase-3 active site cysteine. Nat Chem Biol 2005;1:154–8.

[35] Wu C, Liu T, Chen W, Oka S, Fu C, Jain MR, et al. Redox regulatory mechanism of transnitrosylation by thioredoxin. Mol Cell Proteomics 2010;9:2262–75.

[36] Liu L, Hausladen A, Zeng M, Que L, Heitman J, Stamler JS. A metabolic enzyme for S-nitrosothiol conserved from bacteria to humans. Nature 2001;410:490–4.

[37] Staab CA, Alander J, Brandt M, Lengqvist J, Morgenstern R, Grafstrom RC, et al. Reduction of S-nitrosoglutathione by alcohol dehydrogenase 3 is facilitated by substrate alcohols via direct cofactor recycling and leads to GSH-controlled formation of glutathione transferase inhibitors. Biochem J 2008;413:493–504.

[38] Forrester MT, Foster MW, Stamler JS. Assessment and application of the biotin switch technique for examining protein S-nitrosylation under conditions of pharmacologically induced oxidative stress. J Biol Chem 2007; 282:13977–83.

[39] Foster MW, Liu L, Zeng M, Hess DT, Stamler JS. A genetic analysis of nitrosative stress. Biochemistry 2009;48:792–9.

[40] Lopez-Sanchez LM, Corrales FJ, Gonzalez R, Ferrin G, Munoz-Castaneda JR, Ranchal I, et al. Alteration of S-nitrosothiol homeostasis and targets for protein S-nitrosation in human hepatocytes. Proteomics 2008;8:4709–20.

[41] Lopez-Sanchez LM, Corrales FJ, Barcos M, Espejo I, Munoz-Castaneda JR, Rodriguez-Ariza A. Inhibition of nitric oxide synthesis during induced cholestasis ameliorates hepatocellular injury by facilitating S-nitrosothiol homeostasis. Lab Invest 2010;90:116–27.

[42] Wei W, Li B, Hanes MA, Kakar S, Chen X, Liu L. S-nitrosylation from GSNOR deficiency impairs DNA repair and promotes hepatocarcinogenesis. Sci Transl Med 2010;2:19ra13.

[43] Tang CH, Wei W, Hanes MA, Liu L. Hepatocarcinogenesis driven by GSNOR deficiency is prevented by iNOS inhibition. Cancer Res 2013;73:2897–904.

[44] Colagiovanni DB, Drolet DW, Langlois-Forget E, Piche MP, Looker D, Rosenthal GJ. A nonclinical safety and pharmacokinetic evaluation of N6022: a first-in-class S-nitrosoglutathione reductase inhibitor for the treatment of asthma. Regul Toxicol Pharmacol 2012;62:115–24.

[45] Green LS, Chun LE, Patton AK, Sun X, Rosenthal GJ, Richards JP. Mechanism of inhibition for N6022, a first-in-class drug targeting S-nitrosoglutathione reductase. Biochemistry 2012;51:2157–68.

[46] Sun X, Qiu J, Strong SA, Green LS, Wasley JW, Blonder JP, et al. Structure-activity relationship of pyrrole based S-nitrosoglutathione reductase inhibitors: carboxamide modification. Bioorg Med Chem Lett 2012; 22:2338–42.

[47] Canas A, Lopez-Sanchez LM, Penarando J, Valverde A, Conde F, Hernandez V, et al. Altered S-nitrosothiol homeostasis provides a survival advantage to breast cancer cells in HER2 tumors and reduces their sensitivity to trastuzumab. Biochim Biophys Acta 1862;2016:601–10.

[48] Benhar M. Nitric oxide and the thioredoxin system: a complex interplay in redox regulation. Biochim Biophys Acta 1850;2015:2476–84.

[49] Nikitovic D, Holmgren A. S-nitrosoglutathione is cleaved by the thioredoxin system with liberation of glutathione and redox regulating nitric oxide. J Biol Chem 1996;271:19180–5.

[50] Stoyanovsky DA, Tyurina YY, Tyurin VA, Anand D, Mandavia DN, Gius D, et al. Thioredoxin and lipoic acid catalyze the denitrosation of low molecular weight and protein S-nitrosothiols. J Am Chem Soc 2005;127:15815–23.

[51] Benhar M, Forrester MT, Hess DT, Stamler JS. Regulated protein denitrosylation by cytosolic and mitochondrial thioredoxins. Science 2008;320:1050–4.

[52] Benhar M, Thompson JW, Moseley MA, Stamler JS. Identification of S-nitrosylated targets of thioredoxin using a quantitative proteomic approach. Biochemistry 2010;49:6963–9.

[53] Chakravarti R, Stuehr DJ. Thioredoxin-1 regulates cellular heme insertion by controlling S-nitrosation of glyceraldehyde-3-phosphate dehydrogenase. J Biol Chem 2012;287:16179–86.

[54] Chakravarti R, Gupta K, Majors A, Ruple L, Aronica M, Stuehr DJ. Novel insights in mammalian catalase heme maturation: effect of NO and thioredoxin-1. Free Radic Biol Med 2015;82:105–13.

[55] Kelleher ZT, Sha Y, Foster MW, Foster WM, Forrester MT, Marshall HE. Thioredoxin-mediated denitrosylation regulates cytokine-induced nuclear factor kappaB (NF-kappaB) activation. J Biol Chem 2014;289:3066–72.

[56] Ito T, Yamakuchi M, Lowenstein CJ. Thioredoxin increases exocytosis by denitrosylating N-ethylmaleimide-sensitive factor. J Biol Chem 2011;286:11179–84.

[57] Canas A, Lopez-Sanchez LM, Valverde-Estepa A, Hernandez V, Fuentes E, Munoz-Castaneda JR, et al. Maintenance of S-nitrosothiol homeostasis plays an important role in growth suppression of estrogen receptor-positive breast tumors. Breast Cancer Res 2012;14:R153.

[58] Arai RJ, Ogata FT, Batista WL, Masutani H, Yodoi J, Debbas V, et al. Thioredoxin-1 promotes survival in cells exposed to S-nitrosoglutathione: Correlation with reduction of intracellular levels of nitrosothiols and up-regulation of the ERK1/2 MAP Kinases. Toxicol Appl Pharmacol 2008;233:227–37.

[59] Edes K, Cassidy P, Shami PJ, Moos PJ. JS-K, a nitric oxide prodrug, has enhanced cytotoxicity in colon cancer cells with knockdown of thioredoxin reductase 1. PLoS One 2010;5.

[60] Engelman R, Ziv T, Arner ES, Benhar M. Inhibitory nitrosylation of mammalian thioredoxin reductase 1: Molecular characterization and evidence for its functional role in cellular nitroso-redox imbalance. Free Radic Biol Med 2016;97:375–85.

[61] Shen X, Burguillos MA, Osman AM, Frijhoff J, Carrillo-Jimenez A, Kanatani S, et al. Glioma-induced inhibition of caspase-3 in microglia promotes a tumor-supportive phenotype. Nat Immunol 2016;17:1282–90.

[62] Hambardzumyan D, Gutmann DH, Kettenmann H. The role of microglia and macrophages in glioma maintenance and progression. Nat Neurosci 2016;19:20–7.

[63] Hernansanz-Agustin P, Izquierdo-Alvarez A, Garcia-Ortiz A, Ibiza S, Serrador JM, Martinez-Ruiz A. Nitrosothiols in the immune system: signaling and protection. Antioxid Redox Signal 2013;18:288–308.

[64] Farkona S, Diamandis EP, Blasutig IM. Cancer immunotherapy: the beginning of the end of cancer? BMC Med 2016;14:73.

[65] Bogdan C. Nitric oxide synthase in innate and adaptive immunity: an update. Trends Immunol 2015;36:161–78.

[66] Creelan BC. Update on immune checkpoint inhibitors in lung cancer. Cancer Control 2014;21:80–9.

[67] Peng D, Tanikawa T, Li W, Zhao L, Vatan L, Szeliga W, et al. Myeloid-derived suppressor cells endow stem-like qualities to breast cancer cells through IL6/STAT3 and NO/NOTCH cross-talk signaling. Cancer Res 2016;76:3156–65.

[68] Molon B, Ugel S, Del PF, Soldani C, Zilio S, Avella D, et al. Chemokine nitration prevents intratumoral infiltration of antigen-specific T cells. J Exp Med 2011;208:1949–62.

[69] Jayaraman P, Parikh F, Lopez-Rivera E, Hailemichael Y, Clark A, Ma G, et al. Tumor-expressed inducible nitric oxide synthase controls induction of functional myeloid-derived suppressor cells through modulation of vascular endothelial growth factor release. J Immunol 2012;188:5365–76.

[70] Niedbala W, Cai B, Liu H, Pitman N, Chang L, Liew FY. Nitric oxide induces CD4+CD25+ Foxp3 regulatory T cells from CD4+CD25 T cells via p53, IL-2, and OX40. Proc Natl Acad Sci U S A 2007;104:15478–83.

[71] Takeuchi Y, Nishikawa H. Roles of regulatory T cells in cancer immunity. Int Immunol 2016;28:401–9.

[72] Biswas SK. Metabolic reprogramming of immune cells in cancer progression. Immunity 2015;43:435–49.

[73] Kelly B, O'Neill LA. Metabolic reprogramming in macrophages and dendritic cells in innate immunity. Cell Res 2015;25:771–84.

[74] Hirano K, Hosoi A, Matsushita H, Iino T, Ueha S, Matsushima K, et al. The nitric oxide radical scavenger carboxy-PTIO reduces the immunosuppressive activity of myeloid-derived suppressor cells and potentiates the antitumor activity of adoptive cytotoxic T lymphocyte immunotherapy. Oncoimmunology 2015;4.

[75] Schumacher TN, Kesmir C, van Buuren MM. Biomarkers in cancer immunotherapy. Cancer Cell 2015;27:12–4.

Index

Note: Page numbers followed by *f* indicate figures, and *t* indicate tables.

D

E

Kolibri 360 GmbH
Ferdinand-Jühlke-Straße 7
99095 Erfurt, Deutschland
produktsicherheit@kolibri360.de